J. Wright Clarke

Plumbing Practice

J. Wright Clarke

Plumbing Practice

ISBN/EAN: 9783337106423

Printed in Europe, USA, Canada, Australia, Japan

Cover: Foto ©ninafisch / pixelio.de

More available books at **www.hansebooks.com**

WILLIAM BLEWS & SONS,

(ESTD. A.D. 1782.)

CHANDELIER & ELECTROLIER MAKERS.

CHURCH BELLS,

SHIP BELLS,

MUSICAL HAND BELLS,

SCHOOL BELLS, &c.

Weights and Measures of all Countries.

GAS FITTINGS, COCKS,
WATER COCKS, &c.

New Bartholomew Street, BIRMINGHAM, ENGLAND.

RUFFORD & CO.,

ROYAL PORCELAIN BATH AND GLAZED BRICK WORKS,

STOURBRIDGE,

ENGLAND.

LONDON SHOW ROOMS:—

331, *Farringdon Street, E.C.* (Nr. Holborn Viaduct.)

MANUFACTURERS AND ORIGINAL PATENTEES OF

THE ROYAL PORCELAIN BATH,

Moulded and Glazed in one piece, for which

GOLD ISIS MEDAL OF THE SOCIETY OF ARTS

WAS AWARDED.

HOUSEHOLD AND OTHER GLAZED SINKS,

ALSO

Glazed Bricks, White and Coloured, &c., &c.

LLOYD & LLOYD,

Albion Tube Works, BIRMINGHAM.

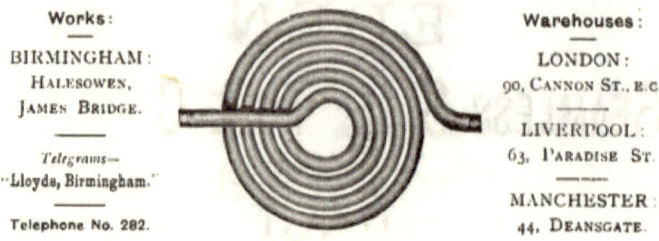

Works:
BIRMINGHAM:
HALESOWEN,
JAMES BRIDGE.

Telegrams—
"Lloyds, Birmingham."

Telephone No. 282.

Warehouses:
LONDON:
90, CANNON ST., E.C.

LIVERPOOL:
63, PARADISE ST.

MANCHESTER:
44, DEANSGATE.

"WROUGHT-IRON TUBES & FITTINGS."

Fittings and all Sockets stamped with

TRADE **L & L** MARK.

"Perkins High-pressure Hot-water Tube."

Manufacturers of Every Description of

WROUGHT-IRON AND STEEL TUBES AND FITTINGS.

Coils for Hot-water and Other Purposes

To any Size or Shape.

GAS, WATER, AND STEAM TUBES,

GALVANIZED TUBES,

BOILER TUBES AND COILS.

TELEGRAMS—"CREDENDA, BIRMINGHAM."

THE
CREDENDA
SEAMLESS STEEL TUBE COMPANY,
(LIMITED),

BIRMINGHAM.

Patentees and Manufacturers
OF
COLD-DRAWN SEAMLESS STEEL TUBES
FOR

LOCOMOTIVE, MARINE, & OTHER BOILERS; FERRULES & STAYS, HYDRAULIC PRESSES, AIR RESERVOIRS, DIAMOND BORING RODS, SPINDLES, BUSHES, COUPLINGS, HOLLOW SHAFTING, COLLARS, SOCKETS, COPS, ROLLERS,

AND

BICYCLE BACK-BONES, FORKS & RIMS, TRICYCLE FRAMES,

AND OTHER USES.

TRADE CREDENDA MARK.

PRIZE MEDAL.	DIPLOMA OF HONOUR AND COMMEMORATIVE GOLD MEDAL.	PRIZE MEDAL.
London, 1885.	Anglo-Danish Exhibition, LONDON, 1888.	Newcastle-on-Tyne, 1887.

ESTABLISHED 1790.

JAMES WOODWARD & ROWLEY,

SWADLINCOTE, near Burton-on-Trent,

SANITARY POTTERS,

(Telegrams to "ROWLEY," Swadlincote.)   SOLE MANUFACTURERS OF

THE "WASH-OUT" CLOSET (PATENT).

THREE AWARDS AT THE INTERNATIONAL MEDICAL SANITARY EXHIBITION, SOUTH KENSINGTON, 1881.

 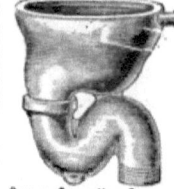

"Improvement on Sanitary Condition of Houses," from a Paper by J. CORBETT, Esq., read before the Social Science Congress, Manchester, 1879:—"We replace defective closet appliances by the simple 'Wash-out' Closet, which is of white earthenware without any valve, and so perfectly self-cleansing as to require very little attention."

REFUSE IMITATIONS—See that the registered Trade Mark "WASH-OUT" is printed inside the Basin. *NONE are GENUINE without this.*   SPECIAL CHEAP NEW PATTERN.

NOTICE.—This title was registered on 11th March, 1878.

Specialities in Closets, Traps, Urinals, Lavatories, Sinks, and other Sanitary Fittings.

PRICE LISTS AND FULL PARTICULARS ON APPLICATION.

SPECIAL OFFER.

NO matter how successful, every paper likes to increase the number of its subscribers. Many of the readers of this book might like to possess some of the bound volumes of

The Engineering and Building Record

AND

THE SANITARY ENGINEER.

The subscription price of THE ENGINEERING AND BUILDING RECORD is 20s.

The price of the Bound Volumes from Nos. 5 to 17 is 15s. each. Special Terms when a number are taken.

The following special offers are made:

For every new subscriber, one bound volume will be sent half-price.

For two new subscribers and 40s., one bound volume will be sent free.

For four new subscribers and 80s., two bound volumes will be sent free, and two at half-price.

Address,

THE ENGINEERING AND BUILDING RECORD,

92 & 93, Fleet Street,

LONDON, E.C.

By Her Majesty's Royal Letters Patent.

THOUSANDS IN USE. THE "ASPHYXIATOR." REGISTERED TRADE MARK No. 36872.

Under the Patronage of
H.R.H.
The Prince of Wales,

The War Department,
and the
Lords of the Admir..lty.

FOR APPLYING THE SMOKE TEST TO DRAINS. FOR DISINFECTING PURPOSES.
FOR THE DESTRUCTION OF VERMIN IN HOLES.

The "ASPHYXIATOR" is universally acknowledged to be the only reliable machine by which the smoke-test can be applied to drains. It is used by Sanitary Authorities, Sanitary Associations, Unions, Medical Officers of Health, Architects, Plumbers, and Builders throughout the United Kingdom and abroad. The "ASPHYXIATOR" is also applicable for disinfecting purposes.

CERTIFICATE OF MERIT (ONLY AWARD), SANITARY EXHIBITION, DUBLIN, 1884.

Descriptive Circular, with Testimonials, Price List of Machines, Fumigating Materials, and full instructions for use, post free on application to the Manufacturers.

JOHN WATTS & CO., BROAD WEIR WORKS, BRISTOL.

Telegraphic Address—"ASPHYXIATOR," BRISTOL.

CAUTION.—In consequence of spurious imitations by unprincipled firms, Buyers should observe that every genuine Machine bears our *Registered Trade Mark* "ASPHYXIATOR."

WINSER & CO.,
Sanitary Engineers & Manufacturing Plumbers.
52, BUCKINGHAM PALACE ROAD, LONDON, S.W.

Sole Licensees for the United Kingdom for the
"DECECO" CLOSET.
PRICE, complete, in White Ware, with Cistern,
£6 5s. 0d.
PIPES AND WOODWORK EXTRA.

The "DECECO."   The "WASH-OUT."

The above cuts show the several differences between the "DECECO" and "WASH-OUT" Closets: a depth of water of 7 in. as against 1½ in.; water seal 4 in. as against 2 in.; and, in the case of the "DECECO," the trap in full view with the walls of the outlet under water and odourless.

WINSER & CO. are Patentees and Manufacturers of several approved Appliances, such as Channel Bends, Traps, Manhole Covers, &c., &c., of which particulars may be had on application.

They are prepared to undertake and carry out work of any description, either in large or small jobs, at the lowest remunerative price. They are, however, neither able nor willing to compete with "skin plumbers," as they only put in good work and material.

Sole Address:—

WINSER & CO.,
52, Buckingham Palace Road,
LONDON, S.W.

Works:—
BELGRAVE BUILDINGS.
S.W.

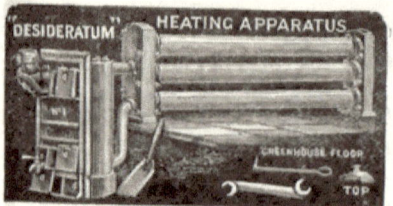

VENTILATION

For Public Halls, Churches, Theatres,
Billiard Rooms, Yachts, Ships, Sewers, Drains, Soil Pipes, &c.

BANNER'S

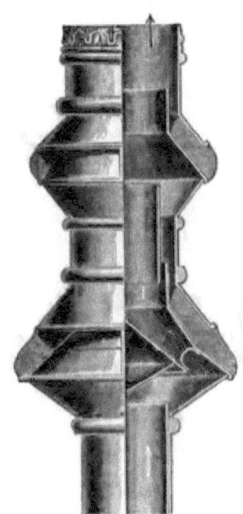

Atmospheric Ventilators.

(Proof of over 10,000 in use in the United Kingdom.)

Pneumatic Ventilators.

IMPROVED

Hydraulic Ventilators.

(WATERSPRAY (IMPROVED) VENTILATORS.)

Mechanical Ventilators.

*Exhaust, "Inlet" and
"Outlet," Ventilators.*

HIGHEST PRIZES at all the most important Exhibitions since 1880.

ONE GOLD MEDAL, THREE SILVER, ONE BRONZE,
At Health Exhibition, 1884.

ALL VENTILATORS OF THE ABOVE DESCRIPTIONS AT REDUCED—AND NOW THE CHEAPEST—PRICES, FOR THE MOST EFFICIENT AND ORNAMENTAL FORMS.

For further particulars, &c., apply to

Banner Sanitation Company,

WESSEX HOUSE, NORTHUMBERLAND AVENUE,
CHARING CROSS, LONDON.

BY

J. WRIGHT CLARKE,

PLUMBER,

First National Honors Medalist for Plumbers' Work, 1881.
Lecturer on Technical and Instructor of Practical Plumbing at the Polytechnic, London.

ILLUSTRATED.

LONDON:
THE ENGINEERING AND BUILDING RECORD,
92 AND 93, FLEET STREET, E.C.

1888.

NEW YORK:
22-24, FULTON STREET.

Spongy Iron Filter.

EXTRACTS *from* OFFICIAL REPORTS *on* PREVENTION *of* TYPHOID, &c., *by* SPONGY IRON FILTERS,

THE ONLY REPORTS OF THE KIND IN EXISTENCE

Free on Application.

Special Ball-cock Pattern. Fitted with Asbestos Cloth.

From COL. SIR JOHN C. COWELL, K.C.B., R.E., Master of Her Majesty's Household.

"BUCKINGHAM PALACE, Sept. 25th, 1878.

"You are at liberty to state that your Company has supplied "the Royal Residences of the Queen with Filters, and so high "is my opinion of the advantages which they possess, that I "have obtained none of any other kind since I became aware "of their purifying action and great merits."

XIX. "Army Medical Report," *pp.* 170-171.

SPONCY IRON.—"This is a very powerful filtering substance. The water filtered shows no tendency to favour the growth of low forms of life, and may be stored with impunity."

VI. "Report on Rivers' Pollution," *p.* 220.

ANIMAL CHARCOAL.—"The property, which Animal Charcoal possesses in a high degree, of *favouring* the growth of the low forms of organic life, is a serious drawback to its use as a filtering medium for potable waters." *(See also XIX. Army Med. Rep., p. 170.)*

XIX. "Army Medical Report," *p.* 170.

CHARCOAL IN POROUS BLOCKS.—"Water filtered through them and stored, shows signs of the formation of low forms of life."

? ? Is it reasonable to assume that Animal Charcoal, which, upon the above official evidence, FAVOURS the development of Bacteria, will DESTROY such Bacteria as the Bacillus of Cholera or Typhoid ? **? ?**

PRICE LISTS, &c., FREE.

SPONGY IRON FILTER COMPANY,
22, *New Oxford Street, London, W.C.*

SOLE PURVEYORS TO H.M. THE QUEEN BY APPOINTMENT.

PREFACE.

THIS Treatise on Plumbing Practice contains the subject-matter of a series of papers contributed to THE ENGINEERING & BUILDING RECORD, New York, in 1883-84-85-86-87. The papers are now re-arranged, and, to some extent, re-written and added to, with a view to presenting them in a more complete form. That plumbers shall command respect, and their advice be taken with implicit confidence, is most earnestly to be desired. To attain this end, the head, as well as the hand, must be trained. Workshop practice is of the utmost importance, but if the workman has no mental training he is only one degree removed from a piece of machinery. A man crammed full of technical knowledge but lacking the ability to execute any desired work may be classed with the non-producers; but the plumber who has the skill to execute, and the knowledge to plan or design, must command both the respect and confidence of his patrons.

In the hope that this work may in a measure supply a want to some who have either practical or technical training, but lack the combination, it is offered to the author's fellow-workers.

LONDON, 1888.

JAMES KEITH.

Gas, Hydraulic, Heating & Ventilating Engineer, LONDON, EDINBURGH & ARBROATH.

Awarded
GOLD, SILVER, and BRONZE MEDALS,
at LONDON, EDINBURGH, GLASGOW and ABERDEEN.
Works—ARBROATH.
SHOWROOMS: 130 George Street, EDINBURGH.
and 57 Holborn Viaduct, LONDON, E.C.
Established 1871.

EDINBURGH 1886.
HIGHEST AWARD FOR HOT WATER BOILERS.

EDINBURGH 1886.
HIGHEST AWARD FOR HYDRAULIC RAMS.

KEITH'S
PATENTED SPECIALTIES
STAND AT THE HEAD.

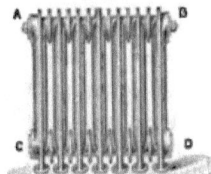

KEITH'S PATENT BOILERS
For Hot-Water Heating,
Require no Building Work.

KEITH'S PATENT UNIVERSAL COILS
For Hot-Water Heating,
Made in One Piece, Many Sizes.

KEITH'S PATENT HYDRAULIC RAMS,
And Hydraulic Ram-Pumps,
Self-Acting for Raising Water.

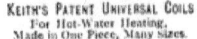

KEITH'S PATENT SYSTEM OF WARMING WATER FOR SWIMMING and OTHER BATHS without the use of Steam—as exemplified at the Putney, Hampstead, and Battersea Public Baths, London—stands Unrivalled for Efficiency, Rapidity, Safety, and Economy.

KEITH'S PATENT MINERAL OIL GAS WORKS,
For Lighting Mansions, &c., in the Country. Adopted by the Commissioners of Northern Lights and the Board of Trade.

KEITH'S PATENT SYSTEM OF HEATING AND VENTILATING TURKISH BATHS BY STEAM.

KEITH'S PATENT ORNAMENTAL OPEN FIRE HOT WATER APPARATUS,
Combine Warmth, Cheerfulness and Ventilation.

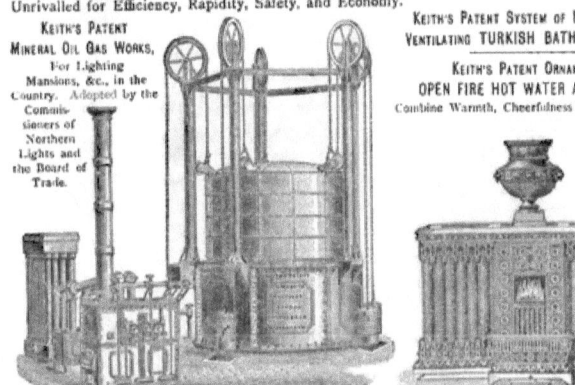

CONTRACTOR TO HER MAJESTY'S GOVERNMENT.

TABLE OF CONTENTS.

	PAGE.
CHAPTER I.—METALS. Lead ores—Smelting of ores—Separation of other metals from lead—Extraction of silver—Physical and chemical properties of lead—Lead alloys—Market forms of lead—Sheet-lead casting—Sheet-lead on external ornament—Uses of sheet-lead	17–26
CHAPTER II.—LEAD PIPES. Methods of making, by hand and press—Soldered seams—Burnt seams	27–33
CHAPTER III.—HAND-MADE PIPES. Methods of soldering	34–39
CHAPTER IV.—PIPE BENDING AND ELBOWS. "Bobbins" and "followers"—Dummies—Depressions in the bends	40–50
CHAPTER V.—PIPE BENDING AND ELBOWS—*continued*. Bending for S-traps—Flasks for use in bending—Use of sand—Water bends—Soldered bends—Soldered elbows—Templates—Setting out	51–59
CHAPTER VI.—PIPE BENDING AND ELBOWS—*continued*. Use of geometry—Elbows from sheet-lead—Square pipe double elbows—Elbows to fit angles—Bends in square pipe	60–68
CHAPTER VII.—JOINT MAKING. Preparation of the pipe—"Soil" and "touch"—Catching waste solder—Strength and length of joints—Table of lengths for various sized pipes	69–75
CHAPTER VIII.—JOINT MAKING—*continued*. Wiping cloths—Overcasting joints—Cast-lead joints—Block joints—Tafting—Special form of compasses recommended for use where lead of light substance is used—How to prepare branch joints	76–86
CHAPTER IX.—JOINT MAKING—*continued*. Fixing joints for wiping—Rolled joints—Copper-bit joints—Welted joints	87–94

TABLE OF CONTENTS.

CHAPTER X.—PIPE FIXING.
Wall-hooks—Fillets for carrying horizontal pipe—Tacks—Crowding pipes in chase 95–102

CHAPTER XI.—RAIN-WATER PIPES.
Leaden v. Iron—Fixing—Astragals—How to make a mould for casting astragals—Bends in length of iron pipe—How to cast lead ears 103–107

CHAPTER XII.—LINING SINKS AND CISTERNS.
How to cut out and prepare the lead—Weight of suitable lead—Angles—Shave-hooks—Wiping the angles—Nailing the angles . . . 108–116

CHAPTER XIII.—LINING SINKS.
Lead should be thicker than for cisterns—Sloping sided sinks—Expansion of lead with hot water—Capping for sinks—Waste-hole—Gratings and plugs—Overflows to sinks . 117–121

CHAPTER XIV.—SEWERAGE AND SEWERS.
Ventilation of sewers—Drs. Shirley Murphy, and Lyon Playfair on the results to human life of better sanitation—Sir William Jenner's opinion of the sewers of the Metropolis—Sewer-flaps and tide-valves—Rats in sewers—Asphyxiator 122–127

CHAPTER XV.—HOUSE DRAINS.
Brick drains—Drain pipes—Connection with sewers—Precautions to be taken—The trench—Joints—Fall . 128–134

CHAPTER XVI.—DRAINS AND TRAPS.
Junction pipes—Branches—Channel pipes and inspection chambers—Brick and syphon traps—Taper pipes . . 135–143

CHAPTER XVII.—IRON DRAINS.
Rusting—Thickness of pipe—How to test the pipes—Joints—Table of weights of pipes—Concrete beds—Pipes passing through walls . 144–151

CHAPTER XVIII.—IRON DRAINS—continued.
Testing for leakage—How to make a pressure gauge—Removal of obstructions—Junction pipes—Duck's foot bends—Reducing diameter—Calking—The tools used for setting up joints 152–160

CHAPTER XIX.—IRON DRAINS—continued.
Defective connections of soil pipe and drains—Joints for connecting lead and iron pipe 161–165

CHAPTER XX.—DRAIN-VENTILATION.
Syphon principle—Water-spray principle—Induced current of air through drains—Brick piers and gratings—Manholes 166–174

TABLE OF CONTENTS.

CHAPTER XXI.—DRAIN-VENTILATION—*continued*.
Air Inlets—Mica valves—Flush tanks . 175-183

CHAPTER XXII.—DRAIN-TRAPS FOR SURFACE-WATER & WASTE PIPES
Cesspool traps for sinks—Bell-traps—Grease-traps—Scullery sinks—Draining-board 184-192

CHAPTER XXIII.—TRAPS AND WASTE PIPES.
Lip-traps—Arrangement of waste pipe from sink—Gulley traps—Open channels—Wired ends to vent pipes—Evils of enclosed sinks 193-199

CHAPTER XXIV.—SLOP SINKS.
Use and abuse—Expansion joints—How to connect branch waste pipes 200-207

CHAPTER XXV.—BATHS.
Various materials used—Various ways of arranging—Waste water used for flushing drains—Supply cocks—Overflow and waste fittings 208-213

CHAPTER XXVI.—BATHS—*continued*.
Waste and overflow arrangements—Relative position to hot-water cistern—Ventilation of bath-room—Shower, needle, and sitz-baths . . . 214-223

CHAPTER XXVII.—WASH-HAND BASINS.
Rims—Waste, supply, and overflow arrangements—Trapping 224-331

CHAPTER XXVIII.—WASH-HAND BASINS—*continued*.
Plug wastes—Patent wastes—Cabinet enclosures—Traps 232-239

CHAPTER XXIX.—WASH-HAND BASINS—*continued*.
Mechanical traps—Ventilation—Ranges for public places—Shower and spray fittings 240-245

CHAPTER XXX.—URINALS.
Iron stall, trough, and basin urinals—Water supply 246-254

CHAPTER XXXI.—URINALS—*continued*.
Hotel and club—Cabinet enclosed—Automatic flush tanks for urinals 255-259

CHAPTER XXXII.—SOIL PIPES.
Drain pipes as soil and ventilating pipes—Iron rain-water pipes as soil pipes—Combined lead and iron pipes . 260-264

CHAPTER XXXIII.—SOIL PIPES—*continued*.
Position in the house—Position with regard to windows—Traps on branches—Air from soil pipes into bedrooms—Water syphons out of traps—Old arrangement of D-traps 265-273

TABLE OF CONTENTS.

PAGE

CHAPTER XXXIV.—SOIL PIPES—*continued*.
Red lead joint to trap—Branch soil pipes—Waste from sink into water-closet trap—Causes of smells in buildings 274–281

CHAPTER XXXV.—SOIL PIPES AND TRAPS.
Traps usually fixed under water-closets—Self-cleansing traps—Improperly fixed traps—How to set out a trap and soil pipe—Advantage of understanding architects' drawings 282–290

CHAPTER XXXVI.—WATER-CLOSETS.
Inventors of water-closets—Descriptions of good and bad kinds of water-closets—Overflow pipes to water-closets 291–296

CHAPTER XXXVII.—WATER-CLOSETS—*continued*.
Water-closets made in one piece of earthenware—Pan closets—"Washout" closets—"Washdown" closets—Closets without enclosures—Floors of closets—Water-closets for Schools and Institutions 297–304

CHAPTER XXXVIII.—HOT-WATER BOILERS OR WATER-BACKS.
Open kitchen ranges—Boilers and feed-cisterns fixed to ranges 305–310

CHAPTER XXXIX.—CIRCULATION BOILERS.
On fixing boilers—Improperly made pipe connections to boilers—Boiler manholes 311–315

CHAPTER XL.—HOT-WATER FITTINGS.
Fixing circulation pipes—Bending iron pipes—Pipes used for hot water—Safety valves—Fusible plugs fixed to boilers—Hot-water cylinders 316–321

CHAPTER XLI.—CYLINDERS AND HOT-WATER CIRCULATION
Enclosing cylinders to prevent loss of heat—Connections between boilers and cylinders—On lead pipes for hot-water—Coupling unions for lead pipes—Examples of hot-water arrangements in mansions—Air binding of hot-water pipes—Bends in pipes—Expansion joints—Hot-water pipes improperly arranged 322–337

LIST OF ILLUSTRATIONS.

FIGURE		PAGE
1.	Reverberatory furnace for smelting ores	18
2.	Casting-shop and fittings	22
3–5.	External ornaments in sheet-lead	24, 25
6.	Mould for casting lead pipes, A.D. 1639	27
7.	Section of hand-made lead-pipe (old style)	28
8.	Casting flange on lead pipes	28
9	Bramah's pipe press	30
10–12.	Pipe-making machines	31, 32
13, 14.	Sections of hand-made pipes, showing variety of seam	34
15.	Gauge-hook	35
16.	Cone-swab	36
17.	Section of soldered seam in hand-made pipe	37
18.	"Swabbed" seam	38
19–27.	Pipe-bends (bent cold)	42–46
28.	"Dummies" for working out depressions in the throat of bent pipes	47
29, 30.	Pipe-bending	49
31.	Pipe bent to form a syphon trap	50
32.	Bending-dresser	51
33.	Bend	51
34.	Lead "flask" for making small traps	52
35.	Bends made with sand cores	53
36.	Knot traps	54
37.	Method of making bends in halves	55
38.	Setting out a bend with chalk	57
39–43.	Soldered elbows	58, 59
44–50.	Elbow-making from sheet-lead	61–66
51.	Square section bends from round pipe	67
52–55.	Preparing joints on lead pipes	69–71
56.	Shaving hook	71
57, 58.	Collar for soldering, and its application	73
59–62.	Wiped joints	73, 74
63, 64.	Joints for connecting brass-work to lead pipe	75

LIST OF ILLUSTRATIONS.

FIGURE.		PAGE.
65, 66.	Overcast joints	77
67.	Lead joint	78
68, 69.	Block joints	78, 79
70.	Taft joint	80
71, 72.	"Compass" legs for shaving circular work, &c.	81
73.	Welted joint	81
74–80.	Prepared pipes for branch-wiped joints	82–84
81.	Welted branch joint	85
82, 83.	Shaving joints	85
84, 85.	Fixing large-sized branches for soldering	88, 89
86.	"Rolled" joint	89
87–92.	Copper-bit joints	91, 92
93.	Horizontal pipe on wall-hooks	95
94.	Section of horizontal pipe on wooden fillet	95
95.	Section of vertical pipe on wall-hook	96
96.	Section of horizontal pipe with "tack"	96
97, 98.	"Tacks" on perpendicular pipes	97
99, 100.	Sections of pipe, showing method of soldering on "tacks"	98
101, 102.	"Face-tacks" and nail	99
103.	Elevation and plan of leaden rain-water pipe, fixed with wall-hooks	104
104–106.	Making plaster moulds for casting lead astragals	105, 106
107.	Lead socket with ears and astragals for square pipe	107
108.	Lead lining for cistern	108
109, 110.	Sections showing soldered angles in cistern	109
111.	Lead lining for bottom and sides of cistern, in one piece	113
112.	Fitting end to cistern	113
113.	Bent shave-hook for cistern lining	115
114.	Spoon-hook for cistern lining	116
115.	Sink with sloping sides	119
116.	Section of lined rectangular sink, showing wooden fillet in angle	119
117, 118.	Section of front side of sink, lining buckled through improper finishing, and proper method of finishing	119
119.	Section of sink with hollowed plug, and sunk washer	120
120, 121.	Sections of sewer, showing side gullies and lamp holes as ventilators	122
122, 123.	Gulley traps	123
124–126.	Tide flaps	126
127.	Tide valve	126
128.	Scamped house-connection to sewer	129
129.	Effect of leaking drains laid in a sandy soil	130
130.	Section of crushed drain pipe, owing to hardness of trench	130
131.	Drain formed of bagged pipes	131
132.	Improperly laid pipe, with cement forced out of hub into pipe	132

LIST OF ILLUSTRATIONS.

FIGURE.		PAGE.
133.	Longitudinal section of drain, with a bricklayer's fall	133
134, 135.	Improper methods of changing direction of drain, without a bend	134
136–138.	Y & T junctions	135
139.	Section of small branch drain, laid to avoid eddies from main drain	135
140, 141.	Section and plan of main drain and branch, showing stoppage with paper, owing to faulty connection	136
142.	Manhole and faulty connection with channel pipe	136
143–146.	Sections and plan of channel pipes and branches	137
147, 148.	Channel pipe with branches and bend, in one piece	137
149, 150.	Method of introducing branches into channel pipes	138
151.	Dipstone trap, showing evils	139
152.	Sewer trap with stand pipe	140
153.	Improperly fixed sewer trap	141
154–156.	Self-cleansing sewer traps	141
157.	Drain syphon with lip	142
158–160.	Sewer traps with taper pipes	142, 143
161.	Taper pipe	143
162–165.	Iron drain pipes drawn to scale, showing relative thicknesses	145
166.	Spigot-end of pipe with bead	146
167.	Joint of iron pipe with yarn falling through	147
168, 169.	Iron pipe joints	147
170.	Iron pipe on brick pier in trench	150
171, 172.	Unsafe and safe methods of taking pipes through a wall	150
173.	Pressure-gauge for air testing of drains, &c.	152
174.	Inspection pipe and cap	155
175, 176.	Square junction and branch	156
177.	Faulty joint, owing to pipes falling the wrong way	156
178.	Section of collar for connecting two spigot ends of pipes	157
179.	"Duck's-foot" bend	157
180.	Section of small socket in a large pipe, with lead joint	158
181.	Yarning-iron	158
182–185.	Calking-irons	159
186–188.	Faulty joints between lead soil pipes and drains	161
189, 190, 191, 192, 194.	Connections between lead soil pipes and drains	162, 164
193.	Collar-joint for lead and earthenware pipe	163
195.	Sectional sketch, showing drain ventilation on the syphon system	167
196.	Section of an air-inlet with perforated water pipe	169
197.	Sectional sketch, showing by-pass ventilating the sewer through the house drains	170
198.	Sectional sketch of cesspool, showing drain trap and air-inlet	171
199.	Hollow brick pier for drain ventilation	172
200.	Section of manhole in a shrubbery	173
201.	Section of manhole with drain laid some distance for ingress of fresh air	173

LIST OF ILLUSTRATIONS.

FIGURE.		PAGE
202, 203, 205, 206, 207.—Plan and section showing air-inlets to drains in ordinary terrace house		175, 177, 178
204.—Mica flap valve in box		176
208–210.—Basement plans, showing drains and methods of ventilating		179–182
211.—Cesspool trap under sink		184
212–217.—Grease traps		186–190
218.—Range of scullery sinks with grease and sand trap		192
219.—Section of drainage boards in same		192
220.—Bell-trap		193
221.—Lip-trap		194
222, 223.—Gulley-trap, and sketch of same in position		195
224.—Gulley-trap with channel from waste pipes		196
225.—Mouth of ventilating pipe from sink-trap		198
226–230.—Slop sinks		200, 201
231.—Expansion joint for waste pipe from slop sink		202
232.—Mandrel for making same		202
233.—Expansion joint for bath-waste		203
234.—Elevation of pipe with iron bracket		204
235.—Vulcanized rubber cone for expansion joint for bath waste		204
236, 237.—Stacks of waste pipe with branches		205, 206
238.—Supply cocks to sinks		207
239.—Wooden bath-tub with sheet-lead lining		208
240.—Bath improperly fitted with common waste and supply cocks		209
241.—Standard cock with projecting nozzle		209
242.—Combination nozzle for hot and cold supply		210
243.—Bath with inside supply and waste		210
244.—Bath with side supply		210
245.—Bath with specially designed overflow		211
246–248.—Bath-waste valves		212–214
249, 250.—Plug and washer for bath-waste		214
251, 251A.—Section of porcelain bath with waste and overflow connections and gratings		214
252.—Flap valve for outer end of bath waste, not discharging into waste pipe		215
253.—Improper arrangement of bath waste and overflow		215
254.—Improved arrangement of bath waste and overflow		216
255.—Faulty connection of bath waste with slop sink		216
256.—Arrangement of bath and hot water cistern, whereby dirty bath water was supplied to the house		217
257–259.—Shower bath brackets		219, 220
260.—Needle-bath (elevation)		221
261, 262.—Arrangement of needle sprays in the hood		221
263, 264.—Face-plates, for names and arrangement of cocks		222
265.—Spray bath with glass sides for hospital		222

LIST OF ILLUSTRATIONS.

FIGURE.		PAGE.
266.—Sitz-bath		223
267-269.—Rims for wash-hand basins		224
270-273.—Cabinet-stand wash-hand basins		225
274, 275.—Basins with flushing-rims		226
276.—Basin with brass fan-rose		226
277.—Basin with supply through waste apparatus		226
278.—Supply valve for basin		227
279.—Cam-action supply valve		228
280.—Vent-pipe for overflow		229
281, 282.—Section of basin with overflow in the earthenware, and waste plug and washer for same		229
283.—Faulty arrangement of basin with overflow, waste and D-trap		230
284.—Arrangement of same as altered		231
285-287.—Ordinary and improved forms of plug-wastes		232
288.—Special form of brass plug and washer for waste		233
289.—"Trigger" connection to waste plug		234
290.—"Valve" waste		234
291.—"Cap" for waste, in lieu of grating		235
292.—Tip-up basin and receiver		236
293-295.—Cabinet enclosures for wash-hand basins		236, 237
296.—Two-and-a-half gallon trap, with "inlet" and outgo pipes, found under a 12-inch basin		238
297-299.—Forms of traps for wash basins		238
300-302.—Wash-basin traps; flap-valve, heavy, and floating ball		240
303.—Double trap on wash-basin waste		241
304-307.—Elevations and end views of ranges of wash basins and fittings, for use in public institutions, &c.		241, 242
308.—V-shaped slate trough, seen in Paris, for use in public institutions, minimising risk of infection from disease		243
309, 310.—Shower and spray fittings for lavatory use		244, 245
311-321.—Plans and elevations of ordinary forms of street urinals		247-251
322.—Public urinal with basins		251
323-325.—Sectional plans of basins		251
326.—Syphon urinal basin and trap in one piece of earthenware		252
327.—Method of arranging range of urinals, to retain water at a common level in all the basins		252
328.—Arrangement of service-pipes giving proportionate supply of water to each basin		253
329.—Treadle-apparatus		253
330, 331.—Plan of range of V-shaped urinals, in use in London, and section showing floor-flushing		254
332.—Drip-pan with waste		255
333.—Sectional elevation of range of urinals in a London club		255
334.—Urinal shown at the Health Exhibition, with bent glass back, porcelain foot bowl, and marble enclosure		256

LIST OF ILLUSTRATIONS.

FIGURE.		PAGE.
335, 336.—Mahogany enclosure containing urinal; closed and open		256
337.—Wash-basin and urinal in common enclosure		257
338.—Folding wall-urinal		257
339.—Urinettes for ladies' cloak-rooms		258
340-344.—Forms of automatic flush-tank, suitable for urinals		258, 259
345.—Sectional sketch of shaft used as a soil-conduit in an old building		260
346.—Earthen drain pipe in party wall, used as a ventilating shaft		261
347.—Common method of fixing soil pipes in small houses		261
348.—Effect of sun on joint in same		262
349-352.—Elevations showing results of combined iron rain-water and lead soil pipe, and sectional drawings of floors		262, 263
353.—Pan-closet and D-trap, with slip joint, found in the house of a medical officer of health		263
354, 355.—Elevation and sectional drawing, showing result of smoke test on light iron soil pipe, with rain-water leaders, fixed inside a house		265
356.—Elevation of house with light iron soil pipe within		266
357-359.—Section between two upper floors and roof of a West-End house, with open trough under flooring from gutter to soil pipe, and methods taken to improve same		267
360-362.—Rain-water leaders attached to soil pipes, ventilating same into the house		268
363, 364.—Elevation and plan of closet and safe discharging into outgo from closet trap, forming ventilation to drain, owing to syphonage		270
365.—Plan of range of closets with safes and traps, old style		271
366.—Large D-trap		271
367.—Useless weeping pipe		272
368.—Safe waste branched into cheek of closet D-trap		272
369.—Old water-closet trap with branch waste pipes		274
370.—Waste pipe air bound and red lead joint of water-closet trap to soil pipe		274
371, 372, 376, 377.—Good arrangement of branch soil pipes		275, 276
373, 374, 375, 378.—Bad arrangement of branch soil pipes		276, 277
379.—Waste water from sink passes through water-closet trap into the safe		278
380.—Improper ventilation of traps, the cause of intermittent smells		278
381, 382.—The usage of upper water-closets drives water out of lower traps		280
383-388.—Traps for fixing under water-closet apparati		282, 283
389, 390.—Machine made traps		284
391.—Result of improper fixing of trap		285
392-395.—How to set out the lines for trap and soil pipe		286-289
396, 397.—How to set out the lines for trap and soil pipe (a more difficult case)		289

LIST OF ILLUSTRATIONS.

FIGURE.		PAGE.
398.	Plan showing arrangement of pipes in a London Hotel	290
399.	Bramah's water-closet	292
400.	Modern valve water-closet	292
401.	Valve water-closet with trap above floor	296
402.	Plug water-closet	297
403.	Pan water-closet	298
404.	"Washout" water-closet	298
405–408.	"Washdown" water-closets	299, 300
409.	Common hopper water-closet	301
410–414.	Trough water-closets	301–303
415.	Open kitchen range	305
416, 418, 419, 420, 421, 422, 423.	Back boilers and feed cistern	306, 308, 309, 310
417.	Bad hot-water arrangement	307
424–427.	Boilers fixed unlevel	313
428, 429.	Boiler connections	314
430, 431.	Boiler manholes	315
432, 434.	Hot-water circulation pipes	316, 319
433.	Dead weight safety valve	318
435–437.	Hot-water cylinders	319–321
438, 439.	How to connect pipes to hot-water cylinders	323
440–442.	Screwed couplings for lead hot-water pipes	325, 326
443.	Bird's-eye view of hot-water pipes in a mansion	326
444, 459.	Examples of hot-water arrangements	327, 336
445, 446.	Example of "air-binding" in hot-water pipes	329
447.	Bends in hot-water pipes	331
448.	Elbow for hot-water pipes	331
449, 450.	Expansion joints and sockets	331
451.	Expansion of hot-water pipe	332
452.	Pipe bracket	332
453–455.	Improper arrangements of hot-water pipes	333
456.	Erratic hot-water circulation	334
457, 458.	Branches for hot-water pipes	335

THE PERFECTION OF CLEANLINESS, UTILITY, AND SIMPLICITY.

Twyford's "UNITAS,"

COMBINING

W.C. BASIN & TRAP,

URINAL, & SLOP SINK.

No Wood Fittings are required except a hinged seat, which, being raised, the Basin can be used as a Urinal or a Slop Sink, the "wetting," so objectionable in Closets having permanent seats, being avoided. Free access can thus be had to all parts of the Basin and Trap, so that everything about the Closet can be kept clean.

Made in Fine Earthenware. Plain or Decorated, and in Strong Fire Clay.

The flushing arrangements are so perfect, that with a flush of two gallons of water it is guaranteed that all the soil and paper will be completely removed from the Basin and through Trap, the whole of the inside being thoroughly washed, and with the aid of the Patent "After-flush" Chamber, the full quantity of water required to receive the soil is left in the bottom of the Basin.

For Prices and Catalogues apply to

TWYFORD, HANLEY, STAFFORDSHIRE.

Twyford's "National" Side Outlet Closet

WITH PATENT "AFTER-FLUSH CHAMBER."

Advantages:—SIMPLICITY, CLEANLINESS, EFFICIENCY, CHEAPNESS.

40,000 Now in Use.

Forming in itself a COMPLETE WATER-CLOSET BASIN and TRAP, it has no complicated metal working parts to FOUL or get OUT OF ORDER, hence its SIMPLICITY and CLEANLINESS are apparent. Its EFFICIENCY is proved by the fact that the DEMAND is continually INCREASING; and while thousands are being sold annually, wherever properly fixed they have given the most complete SATISFACTION. Prices are much lower than those of any first-class Closet claiming the same advantages.

Where a *Front Outlet is preferred to a Side Outlet, a new Closet, called the "ALLIANCE," on the same principle as the "National," can be substituted.*

Made either complete in one piece of earthenware or in two pieces with Basin and Trap separate. In two pieces it can be had with either S or P trap.

Sole Maker—THOMAS W. TWYFORD,

Manufacturer of all descriptions of Sanitary and Plumbers' Earthenware,

HANLEY, STAFFORDSHIRE.

CHAPTER I.

METALS.

It is of the utmost importance that plumbers should know something about the principal metal they use—namely, lead. The Latin name for lead is "plumbum," hence we get the title of "plumber"—a worker of lead. This title is only given to men who use lead when in a manufactured state, and not to those who extract it from the ores.

LEAD ORES.

There are several ores, but only two which contain lead in sufficient quantity to extract the metal from on a large scale: Galena—sulphuret of lead, lead, and sulphur; cerussite—carbonate of lead, lead, oxygen, and carbonic acid. The sulphur, oxygen, and carbonic acid are the chief impurities which have to be removed to get the pure metallic lead. Antimony, copper, and iron are also found in some specimens of galena, and nearly all contain silver.

Lead ores are found in several places in the British Isles, in the United States, Spain, Saxony, Australia, and other parts of the world.

SMELTING OF THE ORES.

The ores are first prepared by being sorted by hand, and then broken or crushed to small pieces. They are then washed to remove earth or any soluble matter. The sulphur has to be expelled by means of heat, and this process is called roasting. This operation is conducted in different places in different ways, which depend to a certain extent on the impurities which the ores contain. The principle is nearly the same in all cases, and consists in roasting the prepared ores in a kind of large oven called a reverberatory furnace.*
This is shown in section at Figure 1, where A is the hearth, B the

---

* Reverberate—to bound back.

fire, and C the chimney. The ores being spread over the hearth and the fire lighted, the flames rebound from the arch on to the ores, the sulphur in which enters into combination with the unconsumed oxygen brought in with the flames, and passes away up the chimney as sulphurous-acid gas. Part of the lead also combines with oxygen, forming oxide of lead. This being again roasted with fresh sulphide of lead, the sulphur and oxygen combine and pass away as sulphurous-acid gas, leaving the lead in a metallic state. Sometimes lime is thrown into the furnaces to act as a flux to form any earthy matter into a slag.

SEPARATION OF OTHER METALS.

Antimony, copper, and iron have the property of making lead very hard, and should always be extracted when soft lead is required. This extraction is done by means of the calcining or roasting furnace, which is a reverberatory furnace with a large, shallow

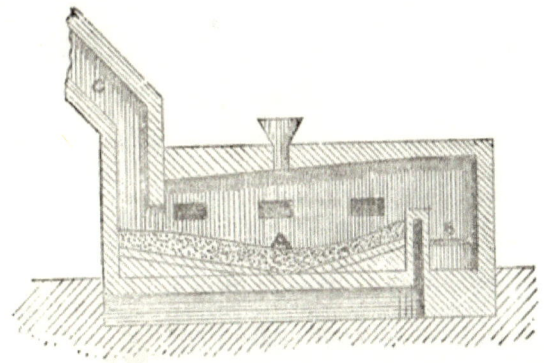

FIGURE 1.

cast-iron pan instead of a hearth. The melted metal is exposed to the action of the flames, when part of the lead, the copper, iron, and antimony are converted into oxides and float on the surface of the metal, and are skimmed off at intervals until the lead acquires the requisite degree of softness. Type-metal is an alloy of lead and antimony in proportion of about four of lead and one of antimony. Sometimes a small proportion of tin is added to improve it.

METALS. 19

EXTRACTION OF SILVER.

Silver is separated from lead by cupellation and by the Pattinson process. Mr. Pattinson's process is to melt the lead containing silver and to allow it to cool slowly, when part of the lead crystallizes. These crystals are fished out with a perforated ladle, the portion remaining being rich in silver. This is then placed in a cupel, when the remainder of the fluid lead is converted into an oxide of lead, sometimes called litharge, the air-blast used for this purpose driving the litharge over the further end of the cupel. The silver, which is not affected by oxygen, remains behind, cools, and sets in a large cake. Before the introduction of Mr. Pattinson's process the whole of the lead had to be converted into an oxide, but since this discovery it is only the liquid portion left after the lead crystals are fished out that has to be so converted. This saves a great deal of time, as the oxygen has to be expelled from a much less quantity of lead to reconvert it into its metallic condition. Some lead ores are richer in silver than others, but the value of silver is such that two parts in a thousand will pay for extraction.

PHYSICAL PROPERTIES.

The physical properties of lead make it one of the most useful of metals.

Its *tenacity* is so low that it is useless for any purposes where strength is required; neither can lead be rolled into very thin sheets, because of its want of tenacity. Lead, on account of its softness, is very *malleable*, and easily worked without the aid of heat, but because of its low tenacity care must be taken not to reduce its thickness too much, or it will break into holes. The application of heat, which should not exceed about 350° Fahr., makes lead softer and more malleable, but a higher degree of heat makes it brittle, while at about 620° Fahr. it is melted or converted into a fluid state.

The *ductility* of lead is so very low that it is impossible to draw it into a thin wire, and even if made into wire its want of tenacity prevents its application to any useful purpose where strength is required.

Lead is very heavy, *the specific gravity* being 11·36, or about

$11\frac{1}{3}$ times heavier than an equal volume of distilled water at a temperature of 60° Fahr.

Lead is not a good *conductor of either heat or electricity*. Compared with silver or copper, its conducting power is only about one-twelfth.

Fusibility.—The low temperature at which lead melts renders it of great utility for certain purposes, such as for making into pipes; for casting into useful forms without the aid of a blast furnace; for fastening iron into stone-work, although this is not so much practised now as formerly, on account of a galvanic action setting up between the metals when exposed to moisture, which results in the iron being eaten away.

The *softness* and *plasticity* of lead is such as to make it most easily worked, and applicable to various forms where it would be almost impossible to use any other metal. These properties, and the low price at which it can be produced, make lead one of the most useful of metals for especial purposes.

Lead has a very brilliant *lustre*, but it is so readily attacked by oxygen that in some cases its brightness is destroyed in a few minutes, and this takes place more readily when moisture is present.

CHEMICAL PROPERTIES.

The chemical properties of lead render it useful for some purposes. The following among several combinations with oxygen are all useful: "Litharge," PbO: "Massicot" (is a superior kind of litharge), and Minium Pb_3O_4, commonly called red lead. These are used as the basis of some paints, or else as driers to accelerate the drying or hardening of paint, the red lead facilitating the oxygenation. Litharge is also used in the operation of *lead*-glazing some kinds of earthenware.

Oxide of lead and carbonic acid, when in the proper combination, produce "ceruse," or white lead, $PbCO^3$, so much used, when ground and mixed with oil, for painting. Carbonate of lead, commonly called white lead, is found in a natural state in the form of crystals, but the white lead of commerce is produced by exposing thin sheets of lead to the action of air, vinegar, and carbonic acid in a chamber kept moderately warm by fermenting tan. It is reported that a way for producing white lead so as not to be injurious to the workers has been discovered, but up to now this is a trade secret.

Metallic lead enters into combination with other chemicals, forming plumbic sulphide (galena), plumbic iodide, plumbic nitrate, plumbic chloride, plumbic oxalate, plumbic phosphate, plumbic chromate, and other salts of more interest to the chemist than the plumber.

In addition to being used as paints, red and white lead mixed in certain proportions are used as cement for the joints of iron pipes, and for several other purposes.

LEAD ALLOYS.

Lead will alloy with several other metals. Mixed with tin in various proportions plumbers and other tradesmen use it as solder. Some cheap kinds of brass-work have lead as one of the constituents. A little lead added to brass-work makes it better for turning and filing.

Pewter is an alloy of lead and tin in the proportion of about four parts of tin to one of lead. Type-metal, as already mentioned, is composed principally of lead.

MARKET FORMS OF LEAD.

Lead is bought in the form of cast pigs, sheets, or pipes. The pigs are generally the pure lead as it leaves the furnaces, and weigh from one to one-and-a-half hundredweight each. Sheet-lead is either cast or milled. *Milled-lead* is manufactured by casting a cake of lead and then passing it to and fro between large rollers until it is reduced to the desired thickness. Milled sheets are made from 20 to 40 feet long, and from 6 feet 9 inches to 9 feet wide. Sheet-lead is described as being five, six, or seven-pound lead; this signifies that one square foot of the lead will weigh such a number of pounds. It is difficult to mill lead to a less thickness than three pounds (·051 of an inch), by reason of its want of tenacity. The weights of sheet-lead vary from three to fourteen pounds per foot superficial; above this weight it is usual to describe the milled-lead as plates.

Cast sheet-lead is generally made by plumbers in the workshop.

Old lead is sometimes used, but unless a certain proportion of new, or pig lead, is added the sheets crack as they cool. The operation of casting is as follows: At one end of the workshop, Figure 2, a large cast-iron pot, B, to hold about fifteen hundredweight, is set in brick-work, with a fire-place beneath, and flues all around to heat the lead. In the centre of the shop the casting-frame

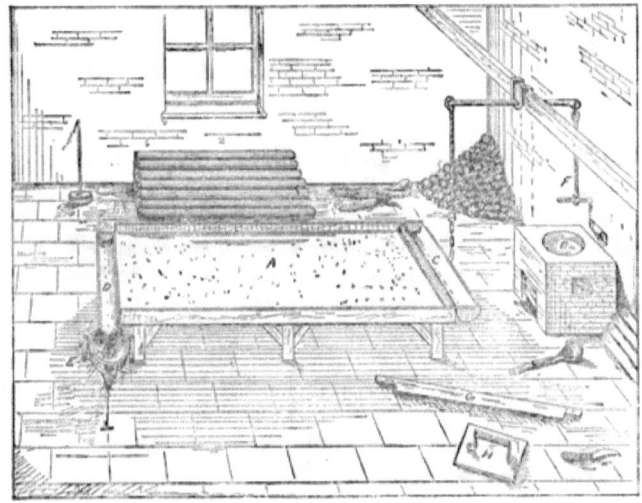

FIGURE 2.

should stand, with one end near the casting-pot. The bed of the frame, A, is covered with sand moistened to give it cohesion. This is made as true on the surface as possible, two men, one at each end, passing the strike, G, from end to end of the frame, and as they push the strike before them they remove all the superfluous sand and leave the remainder with a true surface. The next operation is to *plane* the sand so as to make it quite smooth. The plane, H, is made of sheet-copper, and is similar to a large plasterer's trowel. The edges of the copper are slightly curled upward.

A little bit of "touch" (or tallow) rubbed on the face of the plane makes it work better. A good lead-caster will so work up the surface of the sand that it will look quite bright when finished and ready for casting. When the frame is ready and the lead heated to the right degree, sufficient is ladled out of the pot into the cast-iron head-pan, C. The lead should be hot enough to allow for cooling by contact with the head-pan, or it would set into a solid mass. The lead in the head-pan should be kept in motion by stirring with a dry piece of wood; if this were not done part of the lead would chill and stick to the sides. As soon as the lead has cooled down to the proper degree of heat—usually judged by dipping in a piece of clean wood and noticing the extent of its charring—the third hand (man) pulls the cord, F, and upsets the head-pan so that the contents flow on to the frame. The first and second hands stand ready with the strike, and as soon as the lead is upset they push the strike before them, pressing the ends hard on the rim of the frame, and as quickly as possible remove all superfluous lead before it has time to congeal. At the bottom end of the frame is the tail-pan, D, placed to catch the spare lead removed by the strikers, and at the bottom end of the tail-pan is a large cast-iron bowl, E, on wheels, called the "wagon." All the spare lead runs into this wagon, which is then dragged to and the contents emptied into the pot. If this lead is found to be too much set for ladling out, an iron ring is held half immersed in it, so that when cold, hoisting tackle, with a hook on the end, can be used for bodily lifting the lead out of the wagon into the pot; or, if not too heavy, two men with handspikes can lift it.

The ends are now trimmed, and the sheet of lead rolled up and hoisted off the frame, after which the whole operation is repeated. If the lead is good, twelve or fourteen sheets of lead can be cast in one day, but sometimes, when the lead is hard, half of these sheets have to be cut up again and put into the pot; this being necessary from the number of "sand cracks" found in them, arising, probably, from unequal contraction of the metal as it cools, but this rarely occurs when all new lead is used.

Some lead-casters, but not all, will lay a strip of sheet-lead across the frame on the surface of the sand beneath the lip of the head-pan, to prevent the melted lead from washing a hole in the sand at that point. Cast sheet-lead should never be less than seven or eight pounds per foot superficial. In positions on house-roofs, exposed to great variations of temperature, arising from the action of the sun's rays followed by cold, cast-sheet lead is found to last much longer than milled sheet-lead. This probably arises from the fact that cast sheet-lead has all its particles in natural positions, but in milled-lead these particles are squeezed into unnatural contiguity when being passed through the mill.

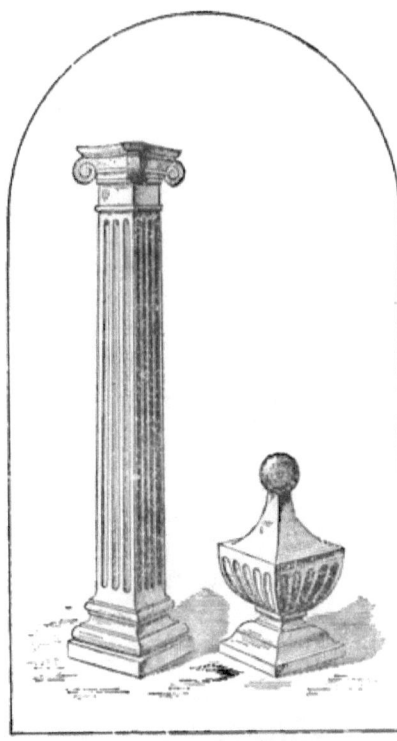

FIGURE 4.   FIGURE 3.

For lining tanks and cisterns the milled sheet-lead is the best, as it is more uniform in thickness. Milled sheet-lead also makes the neatest work, and for very fine chasing the members can be worked up sharper so as to show up distinct. Figure 3 is a specimen of a block of wood covered with one piece of six-pound lead. Figure 4 is another specimen. In this case the wooden column and base are covered with one piece. The capital was made separately, and is hollow, not having any wood inside, and it appears to have been a very difficult undertaking. Figure 5 is another specimen of first-class workmanship. The top portion was bossed up out of sheet-lead. The stem was made out of a piece of drawn-lead soil-pipe. It is intended as a ventilation end for a soil-pipe. These three specimens were made by two young plumbers (students at the Polytechnic in London), in their spare time at their own homes, and each have received prizes for their skill. It would have been very difficult to produce equal specimens

FIGURE 5.

of handicraft if cast sheet-lead had been used. Sheets of cast-lead are generally about 14 to 18 feet long by 6 feet wide.

Laminated lead is a very thin kind of sheet-lead; it is sometimes used for keeping back the dampness in walls. The weight of sheet-lead may be approximately arrived at by measuring the thickness and dividing that by ·017 of an inch, this being the thickness of one pound of lead spread over a surface one foot square; the result will be in pounds. Or, if the weight is known, the thickness can be found by multiplying the pounds contained in one square foot by ·017 of an inch, when the result will be in inches.

USES.

Sheet-lead is principally used for covering roofs of houses, or for the gutters, flashings, etc. It is also used for lining cisterns and sinks. Vitriol-chambers are lined with lead, as it resists the action of this acid to a very great extent.

Floors and passages are often covered with lead. If well fixed on stairs lead will last a long time, and it has the advantage that it does not wear smooth so that persons risk slipping and falling downstairs.

The weights of sheet-lead used in various positions are generally as follow: Lead flats, gutters, and valleys, six, seven, or eight pounds; hips and ridges, six or seven pounds; dormers, the tops, six, seven, or eight pounds; the sides, five, six, or seven pounds; aprons, step, and cover flashings, four, five, or six pounds.

CHAPTER II.

LEAD PIPES.

I HAVE tried to find out who first made lead pipe without seam, and have not succeeded in getting the inventor's name, but have seen a machine for casting small pipes. The date, 1639, was cast upon it. This was called a "staffing and burning machine," and consisted of an iron mould made in two halves, and a core or mandrel of the same material, slightly longer than the mould. The mould was about 21 inches long, and after being put together and fastened with screw-clamps, it was placed on its end, as shown at Figure 6, and melted lead poured into it. When the lead was set the mould was opened and the piece of pipe drawn upward on the mandrel, but not quite off it. The mould was then replaced and more molten lead poured in until it was full to overflowing, so that the end of the first piece that was cast became fused or melted and so joined to it. After this became cool enough it was drawn upward, and the whole operation repeated until a pipe was made of the length required. This machine was for making ¾-inch bore pipes.

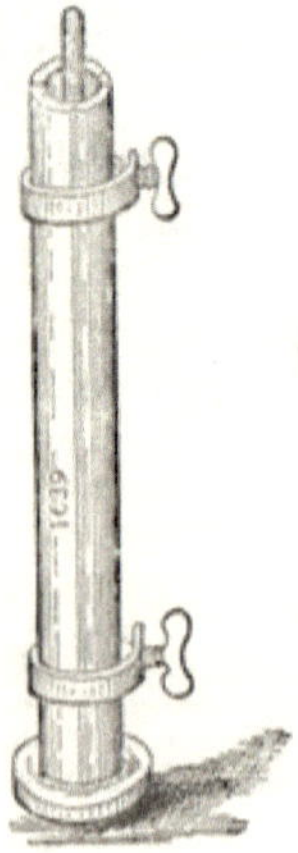

FIGURE 6.

For making larger pipes the lead was cast or rolled into sheets, and then cut to the size required and folded into the form of the pipe. This was then tightly filled with moistened sand. The edges of the lead to be joined were cleaned or shaved the required width, and pieces of iron, called "clams," placed on each side, so that the seam was left the proper size. Or, sometimes the pipe was buried in sand and a furrow left over the part to be joined, with weirs at intervals for excess of lead to flow away. Lead melted and heated to redness was then poured quickly on the part to be joined until it was fused, the superfluous molten metal flowing away over the weirs, or, as some plumbers call them, the "gates." After pouring lead on as described, and before it sets, it should be probed with a piece of wood to make sure that the edges of the lead pipe are melted, care being taken not to disturb the sand core, as if that were done the pipe would be rough inside. Where it is intended for use as a pump-barrel this would have an injurious effect on the pump-bucket, and also prevent the proper quantity of water being discharged at each stroke of the handle,

FIGURE 7.

by allowing part to escape past. The pipe should be laid on its side and carefully levelled before pouring the lead on the seam, or it would be thicker at one end than the other. After the lead is set the sand is all cleaned off and the raised seam trimmed up and left as shown in section, Figure 7. This old way of making lead pipes for pump-barrels is still followed in some parts of the country, and also in certain factories for special pipes for conveying certain kinds of acids, etc.

Bends of heavy lead are usually made in halves, and have a seam "burnt" (as it is commonly called) on each side, and when

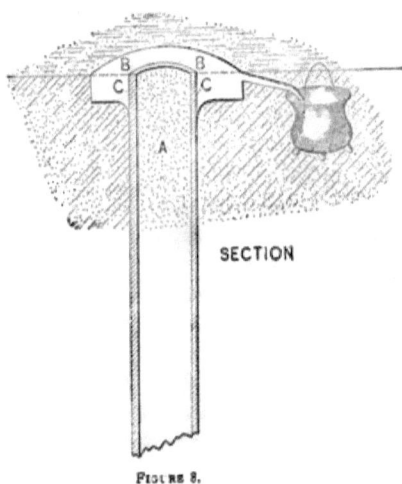

FIGURE 8.

these bends are turned in more than one direction, it follows that the seams can only be made in short sections.

Sometimes heavy lead pipes are fitted and joined together with flanged ends and belts, and although lead pipes of a light section can have flanges tafted, they are stronger with ends cast on them. This is done by filling the end of the pipe for some distance with wet sand, as at A, Figure 8, and then standing it upright with more sand on its outside, with the necessary space indented to the size and form of the flange required. The lead must be red hot and free from oxide, or what is commonly called "dross," and must be poured as quickly as possible until the end of the pipe is completely fused, the superfluous molten lead flowing away as before described. Even after it is supposed that the union is complete, the pouring should be continued for some little time, as it frequently occurs that the part which can be seen at B is fused long before the part at C, and hence the pipe is weak just where a strain may be brought to bear when making a bolted connection. It is always necessary to cast these flanges a little thicker than required, as the lead shrinks in cooling, and leaves the surface so rough that it requires trimming. When pouring the lead on to the part to be united, good-sized ladles should be used, and great care must be taken not to disturb the sand by pouring from too great a height; doing this also cools the lead. For making large work it is sometimes necessary for two or three men to be pouring at the same time, and others fetching more melted lead to take their places in pouring. There must not be any delay, as the lead gives up its heat to the surroundings so quickly that it is soon set.

The earliest record of any patent for making lead pipes appears to have been granted, A.D. 1741, to James Creed, who invented a machine "for cutting sheet-lead into any required breadth, for the making of water-pipes of any diameter and strength, * * * and the slips of lead so cut are turned up into pipes by means of rollers of wood, iron, etc., * .* * and at the same time the edges are scraped fit for soldering by a tool fixed on the frame."

There must have been some improvement after this in the making of lead pipes, for in the year A.D. 1790 a patent was granted to John Wilkinson, whose specification says: "I cast the lead in lengths, as is practised in the common way. This is put upon a polished rod or round mandrel of iron, or any other metal, such mandrels being made of different lengths and diameters, according to the size that is wanted. This rod or mandrel, with the cast-lead upon it, is put repeatedly through or between rollers with grooves of different sizes, according to the external diameter required, and extended to the length or thickness ordered, or drawn through gauges or collars of different dimensions upon said mandrel, each succeeding collar being less than the former, etc."

About seven years after the above patent was granted, a great stride was made by Joseph Bramah, the inventor of the valve water-closet, who took out a patent for preserving and drawing off liquors. Among the instruments described are "sundry tubes," and a method of making them.

On referring to this sketch of the pipe-making machine, Figure 9, it will be seen that it is literally a "lead-squirting machine," a title for which, I think, we have to thank the Americans, whose laconic description at once conveys to the mind the process, and which requires very little more explanation. It has been stated that Bramah never made pipes with his machine, that being impossible, but the idea originated with him, and other people have only improved upon it.

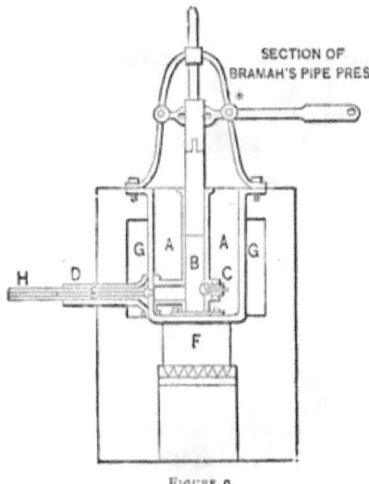

FIGURE 9.

In the drawing,—A is an iron melting pot; B, an iron or brass pump; C, the suction valve; D, the mould or tube; E, the core or mandrel; F, the fire; G G are flues; and H is the pipe as it issues from the nozzle of the mould.

The modern way of making drawn-lead pipes, from the smallest size used up to 6 inches in diameter, consists in filling a cylinder (which is kept heated by a fire and the necessary flues) with melted lead.

The bottom of the cylinder, which is movable and fits tightly inside the cylinder, is attached to the piston of a hydraulic ram, fed from an accumulator charged by pumps worked by a steam-

---

NOTE.—Figure 9 is incorrectly drawn. A slotted guide should be shown at * instead of the pin or bolt.

engine. The cylinder being charged with lead, and the engine started, the bottom is forced upward, causing the lead to escape through the die situated on the top. As the pipe issues from the orifice, it would naturally fall sideways and "buckle"; to obviate this, a cord is attached to the end of the pipe, and continued over a pulley fixed above the machine and kept taut by a man having hold of the other end. This is for pipes made and kept in straight lengths. Small-bore pipes, which are made in long lengths, are slowly wound on a drum by a man, who turns the drum fast or slow as the pipe issues from the machine.

In addition to pipes of a round section, there are manufacturers who have machines for making them of a square section. This square pipe is perfectly true and even in substance, and there is no reason why any other shape should not be made, say, for instance, with moulded front running parallel with the length. Some architects prefer lead rain-water pipes to iron ones, but they must either be content with plain round pipes or else pay rather heavily for hand-made ornamental ones, which often prevents their being used.

Most pipe-machines have the mandrel or core attached to the bottom of the cylinder, as shown at Figure 10. This mandrel gets bent at times, so that it is nearer to one side of the die or orifice than to the other, with the result that the pipe, as it issues, is thicker in substance on one side than the other. As the strength of a chain is only equal to its weakest link, so the strength of a pipe is only equal to that of its weakest side. Some makers, to obviate this, and get the pipe perfectly true and even in substance, put a bridge or guide-piece inside the cylinder some little distance from the top (see Figure 11, which is a section on C C, Figure 10), so that the mandrel may pass freely upward, but

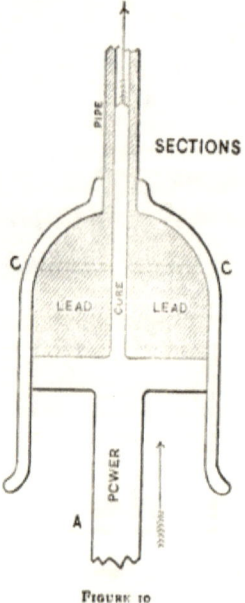

FIGURE 10

cannot get moved sideways. By this means the pipe is made more even, but it will not withstand such a bursting pressure, on its inside, as the other pipe will. The reason for this is that the lead when passing the guide-piece gets separated into two portions. These join together again afterward, but as the lead begins to lose its heat at this point, and is not always fused, it is only joined by compression; so although the pipe, as it issues from the die, looks perfectly even and in good condition, it will often be found to split when the plumber drives a tan-pin in the end.

FIGURE 11.

The amount of force required when making lead pipes is sometimes equal to a pressure of thirty-five to fifty hundred-weight on the square inch, according to the size being made; but it is easily understood that there is a great loss of this power by friction. To save this loss of power, a machine was invented to minimize the friction. To gain this object, the bottom of the cylinder, and the mandrel to it, was firmly fixed, and the die, which was attached to the *top* of the cylinder, was forced downward, the lead escaping upward through the die. The sectional sketch, Figure 12, will illustrate what is meant.

When pipes are made by the machine, Figure 10, the enormous power brought to bear compresses the lead as closely and compactly as possible, but it is doubtful if the machine last described does this so effectually.

Sometimes lead pipes, as they issue from the die, are passed through molten tin and a resin flux, by which means the lead becomes coated with tin. There are other patent means by which a tin pipe is covered with a lead pipe.

Lead-encased tin pipe is made in

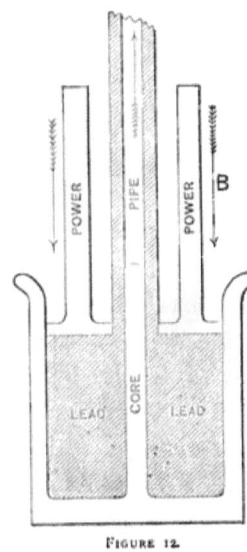

FIGURE 12.

the same manner as the ordinary lead pipe, excepting that the lead "slug" is cast with a hole through its centre, in which is placed a core equal to the internal diameter of the pipe to be made. The space between the core and the lead slug, which is left large or small according as it is intended the substance of the tin pipe shall be, is filled up with tin. When the whole is heated and forced through the die it is found that the two metals issue in the proper proportions, are perfectly true in section, and, on being cut open and bent or twisted in any direction, cannot be separated.

The two metals do not mix and form an alloy, as would appear probable at first sight, and when the specific gravities are compared one would be apt to think that the lead would sink to the bottom, but it is found possible to make this pipe of equal substance both in cross and longitudinal sections.

CHAPTER III.

HAND-MADE PIPES.

WE HAVE already referred to drawn pipe and heavy pipes with "burnt" seams. There are other ways of making them, and I have specimens of strong 1½-inch service pipes, as Figures 13 and 14, with "wiped" soldered seams. There appears to have been two ways of shaving them ready for soldering, practised by two different masters, who are reported to have been the best in the trade—this was 70 or 80 years ago—and who appear to have had almost a monopoly. A great deal of hand-made pipe is used nowadays, but only of a large section, say, from 3 inches in diameter upwards. When preparing to make this pipe the lead

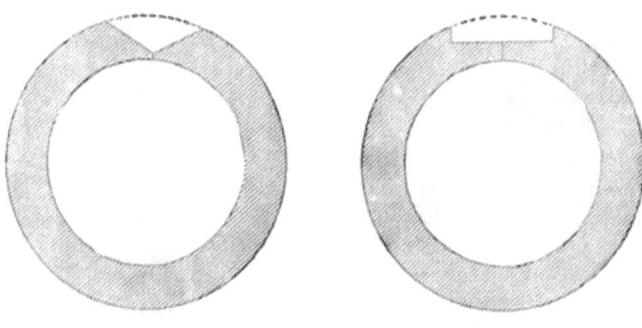

FIGURE 13.   FIGURE 14.

is cut into strips, generally 10 feet long, and the width equal to the perimeter of the pipe to be made. The lead is then dressed out perfectly flat. Most good tradesmen use a flapper, made out of a remnant of sheet-lead, to avoid making tool marks on the lead. The edges of the lead are then planed perfectly straight and parallel, and "soiled" for about ½-an-inch in width to prevent any solder that might run through from tinning on the inner side. The lead is then next folded round a wooden mandrel, this being generally a little larger than the intended pipe. Care must be taken not to make the edge of the lead ragged, and also to avoid using the dresser as much as possible. A good plumber can generally get the lead tight on the mandrel by rolling it backwards and forwards on the bench, and then taking the back side of his dresser, which should not have any roughness, and rubbing the edges of the lead down perfectly smooth. The mandrel is then drawn out, and the edges of the lead soiled on the outside for about 2 or 3 inches each side of the intended seam, and then shaved with a gauge hook, Figure 15, to the desired width. For small pipes, as 3-inch, about $\frac{1}{4}$-inch is shaved on each edge, which makes a $\frac{1}{2}$-inch seam, and for 4-inch the shaving is a little wider. Some plumbers can draw a $1\frac{1}{2}$-inch seam, but it is much easier to make it if smaller. The shaved parts are then "touched"—that is, a tallow candle rubbed on to prevent oxidation or tarnishing; neatsfoot oil is used

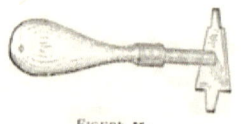

FIGURE 15.

sometimes instead of the candle. It is a good plan to rub touch on the soiled parts, as they sometimes have a little roughness to which the solder clings. This also prevents the water used by the "swabber" from softening the soil, so that it comes off.

Up to this part of the preparation the seam is gaping open; it is now pressed together and tacked. This is generally done with a red hot plumber's iron, filed quite smooth, and a piece of shaved lead, and the two edges of the pipe melted with the iron at the same time, the mate holding something inside the pipe to prevent a hole being burned through. Some plumbers burn these tacks about a foot apart all down the seam, and others only tack the ends of the pipe; whilst others again do not go to the trouble, but tie them with pieces of copper wire, or fix wooden clamps to prevent the ends of the seam gaping open so that the solder runs through.

When everything is prepared, the pipe, as described, and fixed perfectly level from end to end; the irons cleaned and heated to a bright red, but not too hot, which would oxidize the solder and convert part of it into dross, and so spoil the appearance of the seam; the solder heated to the proper degree, so that it can be poured on without burning holes through the pipe, and also carefully skimmed of dross; the "mate" with a bowl of clean water and a small sponge, or a piece of lead, or other sheet-metal, made in the form of a small cone, Figure 16, with a small hole

Figure 16.

in the point over which he can place his finger so as to let a small stream run at pleasure; the plumber begins by taking a ladleful of solder in one hand (usually the right, few men can use the left) and a hot iron in the other, and pours on the seam. As soon as the solder has heated the pipe sufficiently, the iron is drawn slowly and carefully on each side of the seam so as to melt off the superfluous metal, and cause the part remaining to flow until it is perfectly smooth on its surface. During the whole time that the iron is at work more metal is being poured on by the plumber as he walks slowly backwards.

As soon as about 4 to 6 inches of the seam is made the mate presses the sponge, or removes his finger from the orifice of the cone described above, and which is commonly called a "swab," and allows a small stream of water to flow, he slowly moving it backwards and forwards across the soldered seam. A great deal of practice is required to draw a seam with metal and irons. Beginners, as a rule, watch their iron so carefully that they forget to keep pouring on more solder, or they look to the ladle hand so much that they forget to use the iron. Some plumbers seem as if they never could use both hands at the same time, so that their seams always look patchy, as if made in short sections. Some men use a ladle with a small hole in the lip for pouring, so that the solder can only run as fast as necessary.

There are very few mates who can swab properly, and that is of as much importance as drawing the seam. If too much water is put on it cools the work so that the solder sets too fast, or, if the swabbing is done too close to the plumber it has the same effect, also causing steam to get into his eyes so that he cannot see properly. On the other hand, if the mate does not use enough water, or does not follow up quick enough, the solder cracks, and, at the same time, the part opens where the metal is melted, and some of it runs through and hangs down inside, which, if not removed afterwards, causes obstructions round which passing objects can cling. He must also pass his swab backwards and forwards across the seam, as if he allows the water to fall on it perpendicularly a depression is left on the surface of the solder.

The spare solder on each side of the seam sometimes clings to the sides of the pipe. This should be pushed off with a piece of wood so as not to scratch the soil. If this is not done a channel is formed down which the water can run, and spoil the plumber's heat, and if he should happen to pour any solder on this water its sudden conversion into steam might cause the metal to blow and do the men an injury. The seam should be swabbed at intervals of not more than 2 inches, and when it is done regularly it looks very nice when finished, and even when the pipe gets old and the seam as black as the lead, the swabbing can still be seen.

When preparing the seam it should be shaved rather deep so as not to leave a thin edge to the solder, Figure 17, and when soldering care should be taken not to get too much heat on, or the seam will not stand up bold and full. If the seam is badly

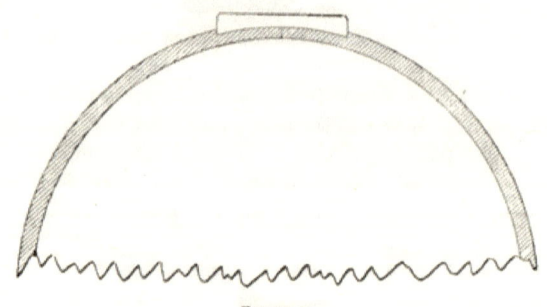

FIGURE 17.

prepared, or if the soil gets scratched by rubbing the iron too hard upon it, it is necessary to cut and trim the edges of the solder, but this should be avoided as much as possible. If all grease is removed off the lead, and care taken with the soiling, there is no necessity to touch it up afterwards. If the pipe is to have a "wiped" seam it is prepared in the same manner as for drawing, but it is generally shaved a little wider. The plumber pours as for drawing and wipes with his left hand, holding his cloth between his thumb and two forefingers, and generally using a small and rather thin cloth. The mate swabs as for drawing, but in this case a sponge is best to use, as it can be lightly passed over any black, dirty-looking marks which the cloth sometimes leaves on the solder. Some mates can make the seam look smart by swabbing on alternate sides which gives a pretty appearance, Figure 18.

In some parts of the country, instead of drawing seams with irons and common plumbers' solder (of two parts by weight of lead and one of tin), a finer kind of solder is used (containing a

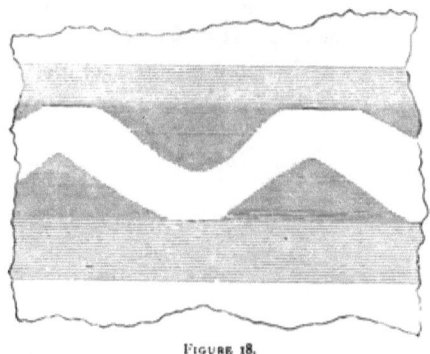

FIGURE 18.

greater proportion of tin), and a copper bit is used, but this is generally looked upon as very inferior work, because the seams are generally made small, and by inferior tradesmen, who, perhaps, have not sufficient practice to draw seams in the manner described above.

Pipe with soldered seam would last but a very short time, that is, when in contact with some kinds of acids; and even when used as soil pipe the soldered seam is generally the first part to go, so sometimes the pipe seam is "burnt" up. This is done with the aero-hydrogen blow-pipe, and is commonly known as "patent burning;" some call it autogenous soldering. This is very rarely done, as very few plumbers know how to do it. Nor is burning likely to come into general practice, especially at the present time when pipe-makers have got their machines to such perfection that they guarantee pipes of equal substance throughout. Neither is it always convenient to carry a burning-machine about from job to job. Some men don't care much about using it, there being some little risk with the gas, and a great many say that after using it for some time they cannot very well see to read small printed matter, showing that it has an injurious effect on the eyes.

When preparing the seam for burning, the lead is turned on a mandrel in the same manner as described for soldering seam pipe, excepting that it is not necessary to soil any part, no flux is required, and a strip of clean lead is held so as to melt on the seam instead of solder. Most plumbers prefer the seamless machine-made pipe, as it is much easier to bend.

CHAPTER IV.

PIPE BENDING AND ELBOWS.

To PROPERLY bend a piece of pipe it should be heated in the throat and kept cool at the heel or outside of the bend. All bends should be made on the bench and not in their position. A great many men think that to bend a $\frac{1}{2}$-inch or $\frac{3}{4}$-inch pipe before fixing is a waste of time, as it can so easily be bent in its place. This is a great mistake, as when bending around corners, the pipe generally buckles, and so contracts the water-way that a smaller-sized one with the bend properly made would allow as much water to pass through in a given time as the larger pipe with the obstruction. This applies to either waste or service pipes; in the former case the bend should be larger rather than less in diameter, for if a waste pipe should get choked by any means so that the force pump or a cane will not remove the obstruction, and the pipe has to be cut open, the cause of stoppage will be found at the bends.

Before describing the usual modes for pipe bending, just a few words on the way that the pipes get ill-used. In the first place, lead being very soft, it is rarely that pipes made of it come into the plumber's hands in the condition in which they leave the mills. Large sizes are generally laid on their sides and packed over each other, so that the top ones bruise those beneath, especially when they have to travel over rough roads for any distance; and even when conveyed by rail and securely packed with straw, sawdust, or other padding, they often get flattened by carelessness in moving the case about. Smaller-sized pipes, when made into coils of 30 feet and upward, are rolled about, so that the outside parts get flattened by the weight of the rest pressing upon them, and this evil is aggravated if the coil is rolled over any inequalities, such as stones, &c. To get these bruises out of the pipe is the plumber's first care, as he cannot do so after it is bent or in its position. Sometimes this forms a considerable item in the day's work, and the man has to work hard and yet get no credit for it. The simplest way to get the pipe into shape again, is to have a short tapering wooden mandrel, with one end the same size as the pipe and the other smaller and rounded. This is driven through the pipe, and should be followed by another one made parallel. As this is being driven through, any projections on the outside of the pipe should be dressed down, either by a soft wooden dresser or a lead flapper, and in such a way that as few tool marks as possible are left to disfigure the pipe.

Small-sized pipes, as a rule, are much thicker in proportion than large ones, so that when they get bruised on one side there is generally a corresponding bulging on the other; this bulging can generally be dressed in, and the pipe made to return to its original shape.

It is nearly always found necessary to straighten pipes, and this should be done on the bench as far as possible, especially when they are going to be fixed in a position where they can be seen. If this is done in their place they frequently get bruised on the back side, and sometimes the plumber tries to knock them straight, with the result that they get full of bruises and tool marks and are otherwise disfigured. A piece of soft wood and a hammer are best for straightening lead pipes so as not to leave any tool marks. When pipes are going to be laid in a trench or beneath flooring, where long easy bends will do, it may not be necessary to spend much time on them; but it matters not whether they are service or waste pipes, or pump suctions, they should always be fixed as straight as possible, and all bends must lay flat or they may become air-bound.

If a small pipe is to be bent cold, it is sometimes a good plan to slightly flatten the sides the reverse way to which it is going to be bent, and then make the bend, taking care not to make it too sharp so as to cripple the throat.

When bent cold, the usual result is that the pipe becomes reduced in thickness at the heel of the bend, B, Figure 19; at the same time it gets thicker in the throat, A; whereas, if the throat is heated first and the heel kept cold, we get it much thicker at A, and B is scarcely affected at all, but remains its original thickness. If the water-way becomes reduced at the bend of a small pipe, and it is some distance in from the end, there is no means of getting it out; but as a rule small pipes can be bent to a larger radius in proportion to their size than large pipes, so it is scarcely ever necessary to cripple them by making them too sharp. If the bend is near the end of the pipe the throat can be worked up with a hammer and bent piece of round iron called a bolt. When starting to make a bend of this kind the bolt is generally used to pull the ends of the pipe round, with the result that the pipe is almost invariably cut with it. Sometimes the piece is torn off at A, Figure 20. After it is pulled round, the bolt has its end pushed under the buckle, B, and is then hammered so as to force it outwards. If the end of the pipe is pulled too far it has to be pushed back again, as there is no room to get the bolt in; or, if the bolt is driven in with the hammer, the lead is worked into a buckle instead of being distributed sideways into the checks. On looking at the sketch, Figure 21, it will be seen that the plumber cannot get his tool under the buckle shown by dotted lines, so that part, instead of being worked

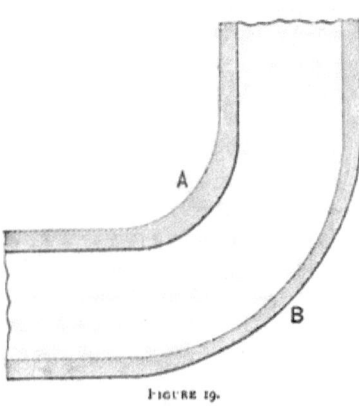

FIGURE 19.

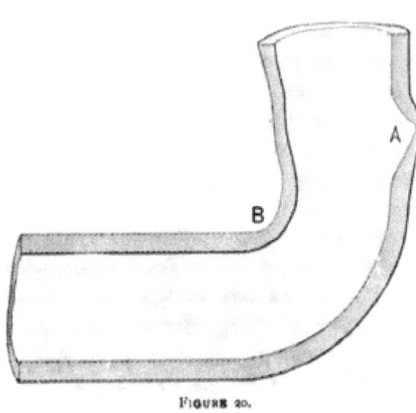

FIGURE 20.

up to G, is driven into lumps at D. A sharp bend cannot well be made at one bending, so it is much better to make it in two.

Some plumbers have steel bolts, but a good tough iron one is better, as the hammer does not glide off so easily, and neither does it spoil the hammer-face so soon. The end of the bolt should be quite smooth and perfectly round, so as not to make indentions inside the pipe, and should be as large as can be got

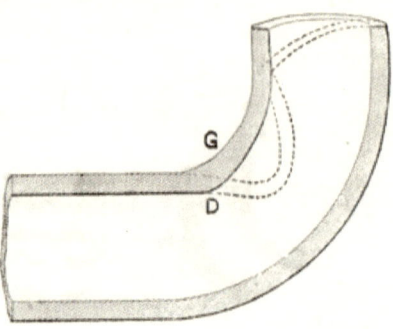

FIGURE 21.

into the pipe. If the bolt is a good length it does not jar the hand and hurt the wrist so much as when a short one is used. Sometimes a joint is made at right angles to the pipe at the extreme end, and is commonly called a knuckle-joint. To prepare the pipe for this, a piece is generally cut out of the side, as shown at Figure 22; the part at A is worked up

FIGURE 22.

with the hammer and bolt, and B bent upward, which causes the sides to bulge out. These sides are dressed or worked up (with a small hand dummy held inside if the pipe is large enough for it) so as to draw the lead out and prevent it becoming too thick, so that when the soldered joint is made, it may not look heavy and clumsy. When made, the

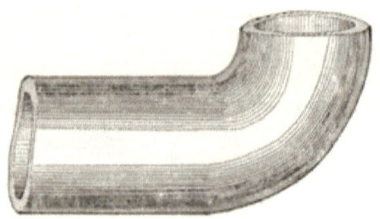

FIGURE 23.

bend presents the appearance shown in sketch, Figure 23. The foregoing description applies to bends made on pipes varying from ½ inch to 2 inches in diameter, but is sometimes applied to larger-sized pipes.

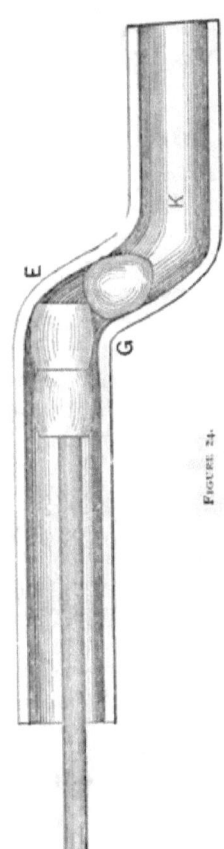

Figure 24.

When 1½ or 2-inch pipes are used for ventilating purposes, as a rule they are of light substance, although some people make no distinction between a waste and an air pipe, and use the same strength for each. But when a light pipe has to be bent to a small radius or sweep, it is very difficult to do so without crippling. As the bolt can't reach the bend, and the pipe is too small in which to work a dummy, there is only one recourse, and that is by driving through a series of boxwood balls, commonly called "bobbins.' A small one is sent first, and each successive one is larger, until the last, which is of the same diameter as the inside of the pipe. Short pieces of wood, rather smaller than the pipe, are driven after the bobbins, and these are called "followers," and the whole are forced forward by a series of blows with a wooden rod if some distance in, but if not far from the end a short piece of rounded wood is pushed in, and driven with blows from a hammer or mallet. Instead of forcing the bobbins through in the manner described, an iron or lead ball is sometimes allowed to fall on them. This necessitates the pipe being placed upright, so that the ball may fall and so generate sufficient force to drive the bobbins onward. The pipe has then to be turned the other end upward for the ball to roll out for using again. This is very tedious work, and it is very rarely that a good plumber adopts this system of making bends. They (the bobbins and followers) should never be used excepting by a good tradesman, and he, as a rule, can make his bends without their aid. Figure 24 shows that the whole of the force required to drive the bobbin round the bend is spent on the heel at E, so that instead of forcing the throat, G, outward, the bobbins strain the outside so that the lead is stretched to the utmost limit, and sometimes a large hole is made. A good tradesman, who knows how to use bobbins properly, will keep dressing the part at E with a soft hornbeam dresser, as his mate drives them through, and he hits good smart blows so as to force the part at G outward; a box dresser used for this would cut the lead. By warming G it comes out much easier. In referring to what was said above, when describing bends made with the hammer and bolt, and on looking at the position of the bobbins in sketch, it will be noticed that if the plumber is in too much of a hurry and pulls his bend around too sharp, that the lead will be driven into a buckle just below G. When this occurs, no more time should be wasted upon it, especially if the plumber has already made the bend, K, so that he cannot take them out and drive from the other end. Even when the bobbins are past the bend, a considerable number of followers must be sent after them, or else from the other end, so as to drive them back again and out of the pipe.

It does not matter how skilled the plumber is, he cannot always depend upon his followers going as he wishes. If he has them

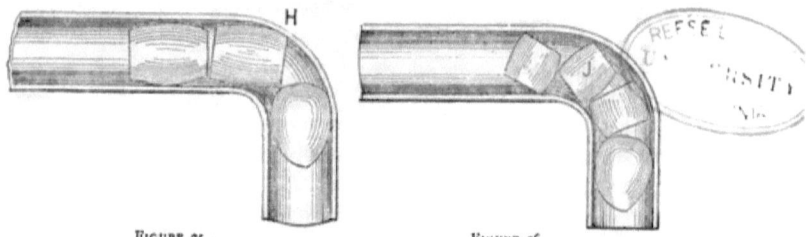

FIGURE 25. FIGURE 26.

too long they will get jammed in the bend, making it thin at the heel at H, Figure 25, and if he has them too short they get angled, as those at J, Figure 26. If they fit too tight the plumber cannot get them out. They ought to be so small that they will fall out of the pipe when it is held up; and then again, if they are too small they wedge together inside the pipe, and the more they are driven the tighter they get. Or perhaps one bobbin will split into halves, and the only way to remove them is to cut a slit in the pipe which has to be soldered over afterwards. When a plumber has made a bent pipe with bobbins and followers he should be careful to test if the pipe is clear of them, as it is no easy matter to get them out when the pipe is fixed, especially after water has passed through and caused them to swell. If three or four bends are to be made on one piece, say a 10-foot length of pipe, all the above evils are aggravated, and it requires about a bushel-basketful of these things to force the bobbin, A, Figure 27, through the whole length, and at every bend and turn in the pipe, one of

FIGURE 27.

them is sure to go wrong. On the whole, this way of bending lead pipes is not to be recommended, and should not be practised more than can be avoided. Another way of using bobbins is: have them perforated and strung on a strong cord. This is attached to a winch, and the bobbins are dragged through the pipe, pushing out all bruises, &c., as it passes. If three or four bends are in the piece of pipe, care must be taken or the cord will sometimes cut through the throat of the bends

It has been stated that it is a good plan to make the throats of the bends hot, as the buckles work out much easier. There are various ways of doing this: sometimes wood shavings are burnt under the part to be heated, or inside if the pipe is large enough; or if gas is convenient, the pipe can be held over a flame, or a flexible tube and nozzle or jet can be used. To test the heat of the pipe—which should never be allowed to get hotter than about 250° to 300° Fahr. (lead melts at about 617° Fahr., and is brittle at a much less heat than that)—pass a wet rag or sponge over it, or drop water on. If this water boils and dances about in globules, the pipe is about the right heat, but if the ebullition is so violent that the water will not stay on, but jumps off, so to speak, it is

PIPE BENDING AND ELBOWS. 47

too hot, and the lead will be found so brittle as sometimes to break, especially if the pipe is not made of pure but out of old lead, with which is, perhaps, mixed other metals, such as tin from old solder, zinc, antimony from old type metal, &c., &c. A much better way to heat the pipe for bending is to pour a ladleful of melted lead on the part, taking care not to have it too hot, and letting it lay there for a few seconds. This heats the pipe just where it is required, and does not make it so black and disfigured as when a flame is used to heat it. It is not a good plan to card-wire a bend to take out the tool-marks; a little water and fine sand (not coarse grit) is all that is required to get off any stain marks; the tool marks should be planished out with a smooth-faced boxwood dresser. A good plumber does not make more tool marks than he can help, and works them out as he proceeds. There is a certain amount of tact required when making bends, as sometimes a soft dresser is required, and at other times a hard boxwood one.

Pipes 3 inches and upwards in diameter are bent rather differently, in some respect, to those of a smaller size, and the throat, which becomes depressed in the act of bending, can be worked out again from the inside with a long rod with a lump on the end, commonly called a "dummy." Some men prefer, for a dummy, a rod of $\frac{1}{2}$ or $\frac{5}{8}$-inch iron, with an iron bulb like a plumber's soldering iron on the end, and an eye or loop turned

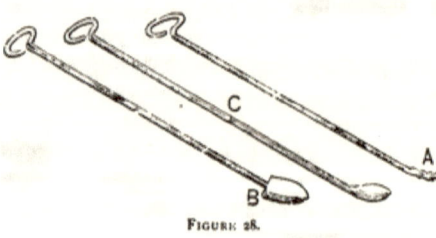

FIGURE 28.

on the other end as C, Figure 28, so that it can be turned sideways when required. Very little command can be had over the plain round rod, which slips and turns in the hand when using, and the hand also becomes cramped by long usage of the dummy, so that it cannot be used in a proper manner. As shown in the sketch, the dummy is in the position necessary for working up the throat, but to work the cheeks out it should be turned sideways, so that the thick part of the bulb is used, and not the point, as this would make indentions inside the pipe, and each of these would have a corresponding bump on the outside, entailing extra labour to work them out so that they should not be seen. The reason that iron bulbs are liked is that they can be placed in the fire and heated, and so help to keep the lead pipe warm during the operation of dummying; they also keep their shape better than any other kind, and can be bent or straightened as required.

If a plumber always had to carry the whole of the tools he might want at his work, he would require a strong man indeed for his mate, for as dummies require to be various lengths in the shaft or handle, and as large pipes require them to be heavier than small ones, it follows that these same tools would be a fair load, without any others, for a man to carry, so that as a rule they are rarely used made of all iron. Should one be required at a job, it is generally made out of a piece of iron rod, and if that is not to be had, a piece of $\frac{1}{2}$-inch or $\frac{3}{4}$-inch iron gas barrel or tubing. Iron rod is the best, because of being able to make the hand loop on the end. The other end can be jagged or cut with a chisel, and then tinned as at A, Figure 28. To cast a bulb on the rod a mould is made in moistened sand, or a piece of lead pipe is soiled inside, worked to the required shape and then buried in sand. The jagged end of the rod is then held in the proper position in the mould while molten plumbers' solder is poured in. The mould is then taken off and any roughness filed off the bulb, so as to make it smooth. If the dummy head is cast in sand there is generally a square shoulder left at B. Sometimes this is cut off with a hammer and chipping knife, but it is much better to melt it off with a heated iron or copper bit, as being less liable to loosen the head on the handle, and then made smooth with a file, as any roughness would injure the inside of the pipe. Gas piping at the best is only a make-shift dummy rod, as in spite of the end being tinned, the bulb of solder will soon work loose, and become almost useless. Gas pipe soon breaks off against the bulb by frequent bendings and straightenings to suit the angle of the bend being made.

The best way to commence making a bend in a large size pipe—say 3-inch, 4-inch, or 5-inch, and of the ordinary substance as six pound, seven pound, or eight pounds—is to slightly bend it

PIPE BENDING AND ELBOWS.

cold, and then *drive* in (not work) the bulges at the sides with a soft dresser; then heat the throat, keeping the heel of the bend as cool as possible by wetting with cold water. Slightly sprinkle the throat with water, or expectorate on it, to test its heat as before described. Next get astride the pipe and take the back side of the dresser, or something similar, with a piece of carpet or other padding, and press on the place where the bend is to be made, the mate at the same time lifting up the end. The pipe is then quickly laid on its side on the bench and the bulged part driven in and toward the heel. Figure 29—A shows how a section of the bend looks when first pulled up, and B when it has been dressed as described. The

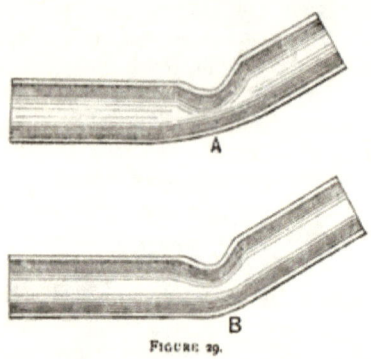

FIGURE 29.

illustration B shows that this driving in of the bulged part makes room for the dummy to work. The pipe is now again heated in the throat and a block placed, as shown at Figure 30, to prevent it becoming straight again. The mate then begins with his dummy to work or boss up the throat; he starts first to get up the centre part, and if he is not carefully watched he sometimes allows the

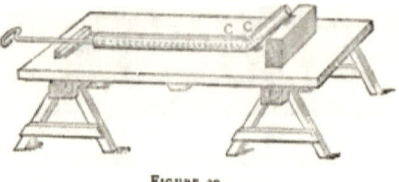

FIGURE 30.

back stroke of his dummy-head to knock against the heel of the bend and so make that part very thin. A good mate never allows his tool to touch the back side, but keeps the weight of it in his hand, so to speak, and gives a succession of quick, sharp strokes upwards.

All this wants doing as quickly as possible while the pipe is hot, but it should not be hit very hard with the dummy, as this would result in the lead being driven up into buckles instead of being worked into an uniform substance. While this is being done it will be found that a crease will form at C C, Figure 30. The plumber drives this away, or works it in with a boxwood dresser with a rounded face, at the same time as the dummy hits it inside. The pipe should be frequently tilted sideways, so that part of the superfluous lead should get worked outward into the checks or sides of the bend, instead of gathering in substance in the centre. The rounded part of the bulb should be used as much as possible, to avoid making the inside of the pipe rough, and unnecessary tool marks on the outside. Some plumbers can make a right-angled bend on a 4-inch pipe at two heats (a bend has been made in one heat , but it is much better to have three or even four heats, and on cutting open this bend it will be found to be the best and the most equal in substance.

After working out all the bruises, and, as far as possible, the tool marks after each bending, the pipe is pulled up again, and the whole operation described above is repeated until a bend of the required angle is made. At an exhibition of plumbers' work, held at Kensington some time ago, were about eight or ten lead syphon traps made out of 3½-inch lead soil pipe, shaped as in sketch, Figure 31, made by different men, and I don't think there was one that was any less in substance at X than anywhere else, thus showing what can be done with no tools but dummy and dresser.

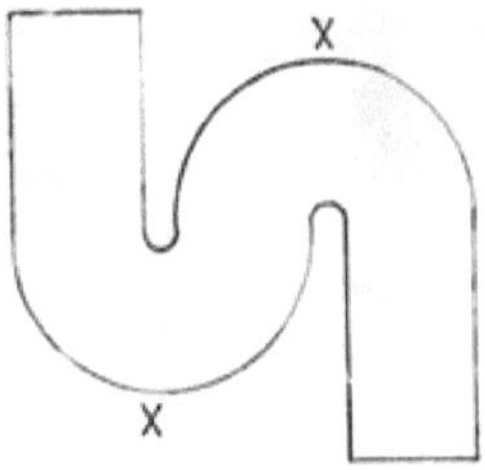

FIGURE 31.

CHAPTER V.

PIPE BENDING AND ELBOWS—*continued*.

WHEN, in spite of all the pains taken to properly make a bend on a soil pipe, it is found that the heel is thinner than the ordinary substance of the pipe, the throat can be worked or dummied out a little more, and the heel dressed in so as to thicken it. It is possible to so dummy up the throat of the bend as to make the lead as thin as a wafer at that part, especially when the pipe is pulled up too much at a time, so that there is only just room for the dummy to work in the centre after the cheeks are flattened. When this occurs it will be found that the lead has gathered in a hard, thick mass about half-way between the throat and the cheeks. It is a good plan to make bends a little larger in diameter than the rest of the pipe, and then, when as many tool marks as possible have been worked out with the dresser, use strips of six or seven-pound lead about 1½ to 2 inches wide as flappers, keeping one piece, which will bend itself by using, for the throat, and another piece for the heel of the bend, and so work in the surplus size to that of the pipe itself. Figure 32 is a good shape for a bending dresser, with the back (or top) side a little more rounding than the face, so that the back can be used for the centre of the throat of the bend, and the face for the hollow of the sides.

FIGURE 32.

For the back and sides an ordinary flat-faced dresser can be used, but it should not have any sharp edges, as those for lead laying, and it should be used lengthways as much as possible, so that any marks made by the grain of the wood may correspond with those on the pipe made by the die of the pipe-making machine. When a bend is properly made, the die marks will be found as illustrated at A, Figure 33,

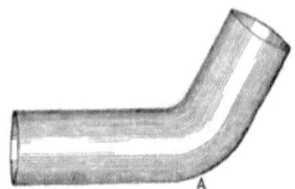

FIGURE 33.

showing that instead of the metal being driven into a body in the throat it has been distributed sideways.

It is generally acknowledged by most plumbers that a trap to be kept clean must not exceed in diameter the size of the inlet and waste pipes, so that when a 1-inch waste pipe is used the trap should be of the same size. To make a 1-inch S trap is a very troublesome bit of bending. The first bend, when it is like a U, is all right, because it can be reached with a bolt and hammer from both ends to work up the throat, but when the second bend has to be made these tools are useless. The best plan is to make the U first and let that be the outgo end of the trap, and then make the other bend, *being careful to always keep the sides so flat that the distance between the throat and heel is the same as the diameter of the pipe itself*. By the time this is done the waterway will be very much contracted, but one object is gained—the substance of metal in the heel is kept nearly, not quite, as it was originally, and the throat perhaps has buckled a little. Now cast a flat lead

FIGURE 34.

cake with a sinking equal in depth to half the size of the outside of the pipe, and the same width and shape as the trap, as shown at Figure 34. Cast another one, so that when they are put together there is a space for the trap to lie in. These castings should be of a good thickness and weight to resist any movement; weights can be placed upon them afterwards, or a wooden strut from the ceiling can be used so as to keep the flask quite tight and closed. Place the partly-made trap inside the flask, fasten down tight, and drive through the trap a small box, or other hard wood, well-greased bobbin, then a larger one, and so on until the pipe is the same bore throughout. These bobbins will have pushed out the flattened checks and any buckles that were in the throat, and at the same time the heel will have been affected very little. This last bend should be the bottom of the trap, as it is usual to solder a brass cap and screw in that part, and if the lead should have been reduced in substance, this would strengthen it.

By using heavier pipe, and extending the bends, the trap can be made very well without going to the above trouble, but traps made in this way do not look so workmanlike, and hold more water. This is a consideration where only driblets pass down. Traps with a long raking outlet, and not well supported, are apt to drop at the outlet and lose their seal.

Sometimes, when it is necessary to make bends of a large radius on a 3-inch or 4-inch pipe, it is a good plan to fill it with sand. First of all, fit a wooden plug in one end of the pipe and drive in four or five clout nails to prevent its coming out; stand the pipe on its end and gradually fill with sand, free from stones or shingle, and keep ramming the whole time so that it is tightly packed. When the pipe is full, fit in another plug and nail as described for the other end. Lay the pipe with the ends on blocks or trestles, then carefully, with the naked hand, press downwards where the bend is required; keep the hands continually moving to a fresh position, so as to get the sweep gradually, and not by a series of bends. The sides will be found to bulge outward a little, but they can be *pressed* in by laying a soft board on them, and using a lever, so as to squeeze them in. The pipe being full of sand, as the sides go in the throat and heel go out, and the bend will look clean and smart, and free from tool marks when finished.

If the bend is required to what might be called a medium radius, and with straight ends, it is a little more trouble. The pipe has to be loaded as described, and then lashed on to a board or plank with strong cord or a small rope, with pieces of hoop iron arranged something like a cart spring between it and the pipe, Figure 35, so as not to cripple the pipe, and also insure its bending

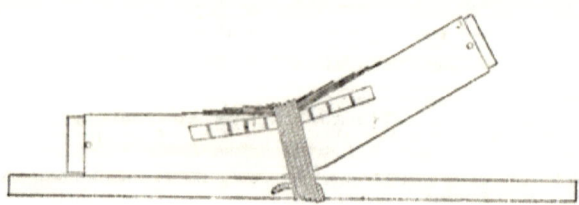

FIGURE 35.

in the required place. The plumber, after lashing and arranging the pipe, stands on the plank and lifts up the end of the pipe a little. If he tries to do too much at a time he will most likely pull it in two or tear it across the outside of the bend. After a slight pull he must unlash it and lay it on its side, and dress the bulged part in, using felt or a piece of carpet, or sometimes his cloth cap, if he wears one, to prevent the dressers marking the pipe. He then relashes it as before, taking care that it is not fastened in exactly the same place, but an inch or two on one side of it, and gives another pull up, and then dresses the sides in again, and so repeats the process until the bend is completed. These bends can be made to look clean and smart, but they are usually very thin on the back side. If they are going to be fixed thus, ⌐, with the bend upward, it does not much matter, but when they are to be fixed with the bend downwards, thus L, everything that comes down the pipe helps to drive the bottom, which has already been reduced in substance, downward, and in time perhaps a hole is made. Water bends are not good. If the pipe is filled with water, and the ends soldered over, we still get a weak part on the outside of the bend, so that we may come to the conclusion that those made in the ordinary way, by heating and contracting the lead in the throat with the dummies, are the best, as the heel or outside of the bend can be kept the original substance, and if required can even be made thicker by a good plumber, for what is the good of having seven or eight-pound pipe, if in certain places it is only equal to five or six-pound lead? It might as well be all that substance throughout, and besides there would be an economy in materials.

It has been stated that some hand-made traps made out of ordinary 3½-inch soil pipe were exhibited by working plumbers at South Kensington. There were also others made, as in Figure 36, in all sizes of pipe, by two working master plumbers, who stated that they always used them, made out of 4-inch pipe, for fixing under water closets. There were several remarks made as to the merits of these traps, and also as to the time it takes to make them. Opinions were divided as to the S, Figure 31, but a good many thought the knot was easier to make, for the reason

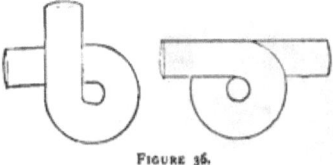

FIGURE 36.

that the bend is continued in the knot, while in the other it has to be reversed for the second half, by doing which the first bend is partly pulled open again. It was considered by some that a light hand dummy with iron handle could be so bent as to easily reach any part of the bend in the knot, whilst there is great difficulty in doing so with the S. One man had the misfortune to get a bruise in his trap that he could not reach with the dummy, so he had to resort to the old dodge of soldering on a copper wire loop, and pulling the bruise out, after which the wire was unsoldered, and the marks cleaned off the lead trap.

Some men who do not have much practice in bending so as to become experts, are in the habit of making bends in two halves and soldering a seam on each side—Figure 37 is one-half. These bends can be quickly made, and sometimes it is convenient to do them so. For instance, perhaps a bend is wanted to connect the eaves gutter of the house to the down pipe, and be made to fit around the cornice or sailing course of brickwork. To make these bends the lead is cut out

FIGURE 37.

the proper width and rather longer than required, the edges are then planed straight and the lead dressed rounding on a wood mandrel or on the plain part of the iron rain-water pipe if no mandrel is to be had. It is then bent as shown in sketch, which causes the sides at that part to open out flat. The *throat* is then worked in at A, so that it is hollowed in the same way as the straight parts, taking care not to make it too sharp and get it into a hard, thick mass, but keeping it well rounded. Some men get working up the *sides*, with the result that they are very much reduced in substance. It is much better to have the throat thicker than the sides thinner. When this half is made, the other half is rounded and then placed by the side of the first, to mark the proper distance; it is then bent the reverse way to the other, which causes the sides to open out. These sides have to be worked in, holding either a hand dummy or a mallet head inside so as to draw out the lead into a substance equal to the rest of the pipe, instead of getting it thicker.

If for a swan-neck or double bend, the last piece is then bent and worked—as before described for the first piece—to form the throat of the second bend, after making which the other half is bent and fitted to it. The edges of the parts that have been worked are then rasped to fit each other quite closely so that no solder will run through; they are then soiled and shaved with the gauge hook in the usual way and soldered. The edges of the inside of the pieces should be soiled so that should any solder run through it will not tin to the lead, and so can be easily knocked or melted off with a hot iron or copper bit. Before making these bends, the required shape should be set out on the bench with chalk, so that the work can be laid on it to mark the distance. A good many men, the first time they attempt them, make them too short between the bends.

Ship plumbers have very awkward bends to make sometimes. They generally use much heavier pipe, that is, of thicker substance than that used by plumbers on a building. Although bends made of heavy pipe do not cripple so much in the throat, it is harder work to make them. Some pipes have to be made tapering, or sometimes a cone with a bend on it is required. These are generally made in two pieces and then burned together, as described in an earlier paper. If they are well burned the seam can be cut off and cleaned up, and then burnished with the back of the boxwood dresser, so that when finished the work has the appearance of being made of one piece of lead.

There are several ways, besides those described, of making bends on soil pipes. For instance, several small V pieces can be cut out to form the throat, and a wide soldering wiped over them, or else solder them with a copper bit and clean them up afterwards, so as to make them look like a properly made bend. A slit can be cut in the pipe where it is intended the throat of the bend shall come; the slit is then pulled open and the pipe bent, which causes the sides to bulge out. This bulging is then carefully worked in and closed, and a wiped seam is made over it, or it is soldered with fine solder and cleaned off. The above ways of making bends cannot be compared, either for strength or appearance, to those made in the ordinary way with dummy and dresser.

In the shop that I was apprenticed in we used to bend some of our soil pipes; at the same time, we used to make as many elbows as bends, but now-a-days a man would be put down as an indifferent tradesman if he was seen making an elbow. Although I do not advocate them, one made in a proper manner is a very good substitute where, for want of the proper tools and convenience, a bend would be difficult to make.

When a man is going to make a bend, the ordinary way is for him to first of all make a template, or set a bevel to the required angle—he does the same for an elbow; he then cuts a V piece out of the side of the pipe where he requires it, and then bends it up and places his bevel to see if he has cut it right—he makes the remark that it is better to cut twice than cut too much the first time—and so, working by rule of thumb, he cuts another small piece off and tries again. Sometimes he has to cut two or three times before he gets it right, and then, probably, by bending the pipe so many times, he finds that he has nearly torn or broken it in two at the heel. By setting out the bend with chalk on the bench or floor, all this could be avoided. Consider the illustration, Figure 38, as representing the required angle—the lines are supposed to be the outside of the pipe. Now, if the pipe is laid between the lines E F, the throat G, can, by placing a small set square against it, be marked on the pipe; make a mark on the back side, represented at H; now lay the pipe so that it is between the lines K L,

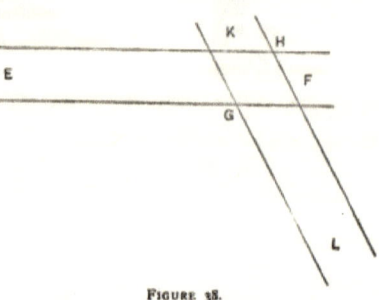

FIGURE 38.

and so that H is in its original position; transfer G on to the pipe again. Now, if the V piece, as represented by the three marks on the pipe, G H G, is sawn out, and the pipe bent so that the opening would be closed, it will be found to be the required angle, and if it is going to be soldered with fine solder and copper bit, that would be all that is required; but if the joint is going to be wiped, a piece must be left on—as shown by dotted lines, Figure 39—which must be slightly opened outward for the other side to enter. The back side of the pipe should not be cut, but should be left as shown at M. After the piece is cut out, the pipe

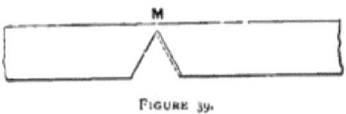

FIGURE 39.

can be soiled and shaved (before it is bent) as it lays on the bench; it should then be touched and bent round and laid on the lines, so as to ensure its being the right angle. It will be found that by bending the pipe, the part M will have bulged out a little; this must be worked in by a few gentle taps of the dresser, and if the extremity of the heel has flattened, that must be worked out from the inside with a dummy; or this flattening can be anticipated, and it can be worked out with the mallet head or hand dummy before the opening is quite closed. If both sides of the opening are worked out a little in the throat, the elbow when made would be more like a bend in shape, Figure 40. After the pipe is placed at the required angle, and before it is moved off the lines, it should

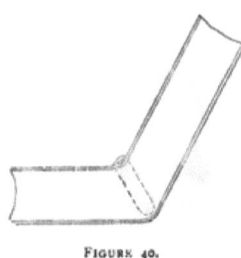

FIGURE 40.

be "tacked." The best way is to pour a little melted plumbers' solder on the upper side of the part to be soldered so that it "tins" to it, and let it remain until set. The elbow is then turned over and placed on two pieces of 1½ or 2-inch boards, long enough for the pipe to roll on without slipping off, and high enough for the spare solder to drop clear and out of the plumber's way, Figure 41. The plumber then begins to wipe the joint as it lays, and as soon as he has done 2 or 3 inches his mate swabs

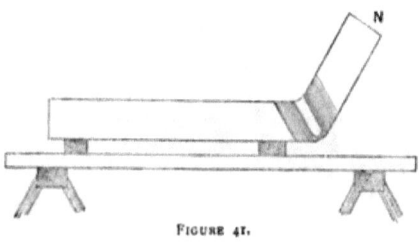

FIGURE 41.

it with cold water, taking care not to cool it too much or to go too near the part the plumber is working upon. The mate then raises the end of the pipe marked N in a slow and careful manner, and slightly presses it toward the throat so that it shall not open. As he slowly moves it the plumber keeps wiping until he has got right round.

Some plumbers make this joint at two heats by pouring solder on one side and wiping it, and then by turning it over and doing the other half. When the joint is wiped at one heat and no iron used, the cloth should be lightly pressed on the part that is finished while pouring on more solder; if this is not done, the tin out of the melted solder runs back and sets in little bright projections on the surface of the joint and makes it look very unsightly. Sometimes a little V piece, P, Figure 42, is notched on the entering side of the opening before closing the elbow for soldering, so that it may pass outside of the other, and so prevent the bend being pressed into a sharper angle while being rolled round and soldered, but, with care, this can be done without, and so avoid leaving an edge inside for hair or anything of that kind to cling to, and also a possibility, if it should be cut a little too far, of solder running through.

Figure 42.

The pipe should always socket in the direction of the current, and if the V piece is cut out as directed there would not be any such obstruction as illustrated at O, Figure 43, which shows it socketed the wrong way, in addition to the obstruction formed by not cutting out sufficient in the throat.

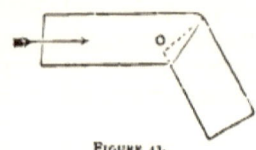

Figure 43.

If the elbow is made on a length of soldered seam pipe, the seam should always be arranged to be at the side in preference to either the throat or the heel of the elbow.

CHAPTER VI.

PIPE BENDING AND ELBOWS—*continued*.

ONE of the most necessary things for a plumber to have a knowledge of is geometry. He may not have time to take up Euclid to make a thorough study of it, but that is no reason why he should not have an elementary knowledge sufficient to be of use to him in his trade. For instance, instead of making templates for his bends, he can set them out as described in last chapter. If he wants to make a taper pipe out of a piece of lead, he can cut it out and prepare it for soldering before bending it into the required shape. He can also cut out pieces of lead to fit any part of a roof, or a spire, either when conical or octagonal, or any other number of sides. This part in geometry is called the development of surfaces, and is really very simple, although generally looked upon as very difficult. He can also take a piece of sheet lead and cut it out the required shape for an elbow, and then soil, shave and fold it up ready for soldering without any previous fitting.

Supposing that he wants to make an elbow out of a piece of lead, at an angle of 120°, and the pipe to be 4 inches in diameter. He first of all cuts the lead out the required size. If the elbow is to be, say, 12 inches long each side of the heel, we have one dimension, so the lead must be 2 feet long. The width we arrive at by multiplying the diameter, 4 inches, by 3·1416, but for those who do not clearly understand decimals, multiply by 3$\frac{1}{7}$. So we get 4 inches by 3·1416, which gives 12·5664 inches, or a little over 12$\frac{1}{2}$ inches; so to make the required elbow we want a piece of lead 2 feet long, and, say, 12$\frac{5}{8}$ inches wide. Now, take a pair of compasses with a pencil point, and set them to a radius of half the diameter of the intended pipe, viz., 2 inches, setting them a little full to allow for thickness of the lead, and describe a circle on a clean board with the face well chalked, so that the lines will be clear and distinct (or if no board or pencil compasses are to be had, a remnant of sheet lead flapped out smooth will do as well, and use a pair of pointed compasses). Now lay down the compasses without shifting them, and draw two diameters at right angles to each other, as A B and C D, Figure 44. Take up the compasses, and with A B C D as centres, cut the circle as shown, which will divide it into twelve equal parts. Now draw the pipe full size, and at the required angle immediately beneath the circle, as shown in the illustration, Figure 44, and project the divisions of the circle on to it, as shown by dotted lines.

On looking at the illustration it will be seen that these dotted lines appear to be closer together at the sides than in the centre, but on looking at the circle they will be found equi-distant. So for distances apart we must measure off the circle (or, what would be called in geometry, the sectional elevation), but the distances on what would be called the plan, as shown by the dotted lines, represent the correct length or distance from the end to that part of the pipe cut by the niche, or mitre, *a b*. Now take the piece of lead which has been prepared, and divide it lengthways with a chalk line into twelve equal parts, by first setting the compasses to the distances on the circle, and stepping it across the end, as shown by Figure 45. The compasses should be set a little full to allow for the rise of the circle between the points. Now, as a rule, the seam of the pipe should be on the side, and as D is represented in that position, we will take that as the seam, so that D D on Figure 45 will be the parts that will come together in the seam. Take the compasses and put one leg on D and the other on the other

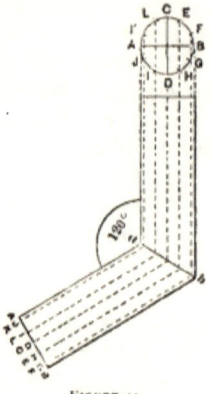

FIGURE 44

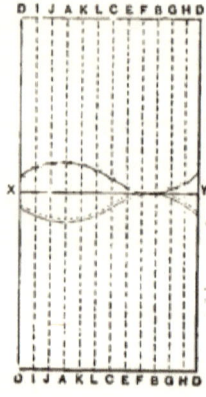

FIGURE 45

extremity of the line cut by the niche on Figure 44, and transfer it to the line D D in Figure 45 on *both* sides of the lead. Measure off the next line I, and transfer it in the same manner; then J, and then A. Now, supposing that the plan of the pipe was transparent, the line J also represents K, so that distance must be transferred on to line K in Figure 45. I also represents L, and D, C, and so on. To save time and avoid altering the compasses as much as possible, it will readily be seen that two distances can be marked each time in some cases. After these distances have been marked off, they can be joined together by freehand, and this represents one-half of the intended elbow.

Instead of repeating these measurements from the other end, a line X Y can be drawn across the piece of lead at right angles to where the line B in Figure 45 is cut, and then by placing one point of the compasses on it, and the other point on the lines as marked, the distances can be transferred to the other side, and then joined by freehand as described before. Draw a line parallel and about ⅜-inch away on one of the sides for the other one to socket into when folded up. Cut out the piece of lead enclosed by these lines, soil, shave, and fold it up, and it will be found to fit and be the required angle. It would be much better and fit truer, if twice the number of longitudinal lines were used, but they are not shown in the illustration, as it must, of necessity, be drawn to a small scale. The foregoing directions will be more readily understood if it is imagined for the time that instead of the pipe being round, it has the same number of sides as denoted by the longitudinal lines. This will, perhaps, be made plainer by studying the following illustrations, which show how a piece of lead would require to be cut to form a double elbow made out of square pipe.

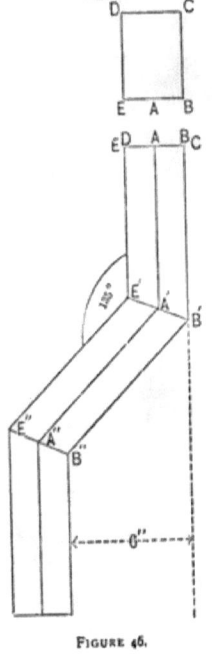

FIGURE 46.

PIPE BENDING AND ELBOWS. 63

Supposing that this double elbow is wanted for a rain-water pipe and to fit over the plinth or projection round the base of a building which stands forward 6 inches beyond the superstructure, that the ends of the elbow are to be 9 inches long, the pipe itself 4 inches by 3 inches, and the solder seam to be in the centre of the side. First of all, set it out as before advised to the required angle, say 135°, see Figure 46, and measure the size of the piece of lead required. It will be found to be 2 feet 4¾ inches × 1 foot 2 inches; after it has been squared, and the edges planed straight, make chalk lines down it as shown by lines in Figure 47. As the seam is to be in the centre of the side, the first line will be 1½ inch in, which is equal to half the side of 3 inches, the next line should be 4 inches from the first, for the front, the next 3 inches from the second for the other side, then 4 inches for the back, 1½ inch will be left to complete the other side. These four lines represent the four angles of the pipe. It is a common practice when making square pipe to run the point of the shavehook down these lines, to plough a piece out of the lead, so that the angles turn up true, and the arrises are sharp without any trouble dressing them up. Although when finished, the pipe or elbow looks neat and free from tool marks, still the practice should be condemned as the lead is much weakened, and it is a common occurrence for the angles to split because of this.

After the piece of lead has been lined out as in Figure 47, measure the distance A A' on Figure 46, and mark off the same distance on A A', Figure 47, on both sides of the lead. Then measure B B, as seen

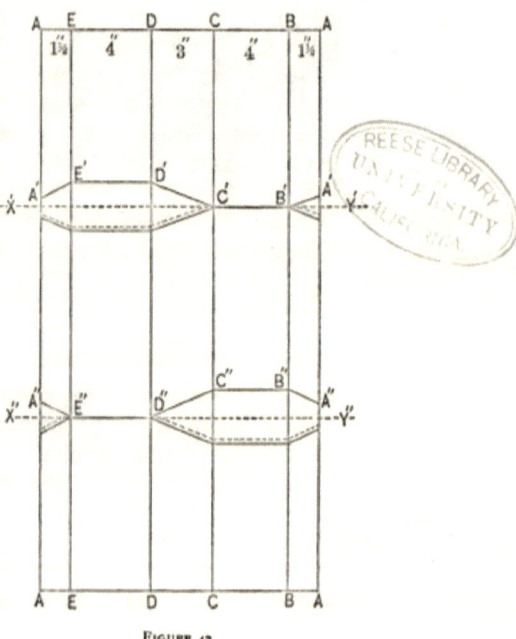

FIGURE 47.

in Figure 46, and mark off on B B', Figure 47. C C' is the same as B B', so a line can be drawn across at right angles. Now measure off and transfer E E', and as D D' is the same as E E' a line can be drawn across these also. Draw lines connecting the points, and one half of the elbow is shown. Now draw the line X' Y' and with the compasses transfer the distances already marked, on the other side of it; draw a parallel line about ⅜-inch away for the other side to socket into, and one of the elbows is set out. Now with the compasses take off the distance B' B" on Figure 46, and measure B' B" on Figure 47. Without altering the compasses, measure all the other points on their respective lines. Draw X" Y" and transfer as described for the other; allow a piece for socketing; cut out and fold up. In this case the soiling and shaving should be done after it is folded up, as in working the angles up to sharp arrises it would get scratched and tarnished and so require re-doing. Some of my readers, no doubt, will say—Why all this trouble to make a double elbow? Would it not be much easier to make the pipe first and then saw out V pieces and so get the required piece of work? I answer, yes! but my reason for giving the information is to get apprentices and others who now look upon geometry as dry work to take an interest in it, and see the value of it, and so be able to work to rules instead of by guesswork.

It is an excellent plan to draw the illustrations here given on cardboard, and then cut them out and fold them up, and then others of a different kind can be attempted, so as to get to thoroughly understand the principles and apply them in the workshop.

Before taking leave of this subject, we will consider one more problem, viz.: how to cut out a piece of lead to make a double elbow to fit in an angle of a building, to get over the plinth as described above, the pipe to be 4 inches and square, the ends to be the same length as for the last one. In the last problem we assumed the angle of the elbow to be 135° to fit the top of the plinth.

If we were to make the elbow we are now considering at that angle it would be found not to fit the angle of the same plinth, and a moment's thought will lead to the conclusion that the angle

PIPE BENDING AND ELBOWS. 65

must be more acute or sharp. To find what the angle should be
we must first find what is the
distance that one angle of the
building is from the other, and
to do this we must first draw
a plan of them. Let A A' A,
Figure 48, represent the face of
the wall of the building, and
B B' B the face of the wall
which constitutes the plinth. If
they are drawn 6 inches apart
to represent the projection we
can then measure B' A', and
so get the distance those two
points would be if horizontal
or on the same level. This
must not be accepted as the
length between the heel of one
elbow and the throat of the
other; to get that, and also the
angle of the elbows, we must
next proceed to set out an
elevation showing the face of
the two walls, and the top of
the plinth at an angle of 135°.
If the distance B' A' in Figure
48 is now drawn in front of
the part representing the plinth,
Figure 49, as shewn by thick
line and the top end connected
to angle A', we get the required
distance and angle at the same
time, and this forms the back
corner of the pipe. We next
proceed to draw a section of
a 4-inch square pipe with its
diagonal at right angles to the
line A A', Figure 49, as shown

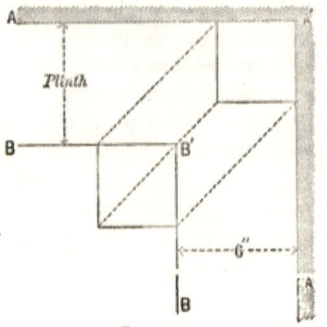

FIGURE 48.

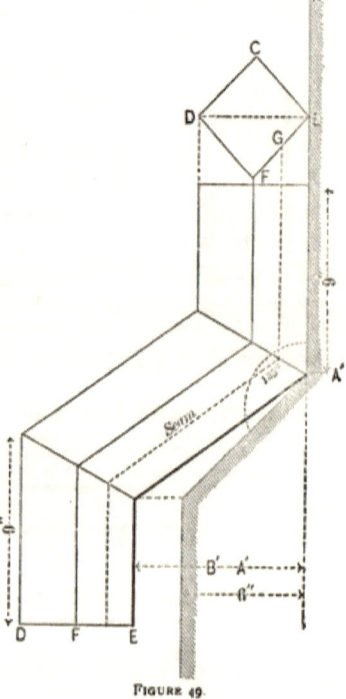

FIGURE 49.

F

at C D E F; project the front and side angles as shown by dotted lines drawn parallel to those representing the back corner of the pipe, as stated. Now set out the piece of lead, which should be 16 inches wide, as follows: 2 inches from one edge draw a parallel line, then three others each 4 inches from the one preceding, which will leave 2 inches to complete the other side. Supposing the seam is to be in the centre of the side E F, Figure 49, mark off the distance E to where it is intersected by the mitre on E in Figure 50, then transfer C in the same manner, then D, then F, which is the same as C, then mark on the edges of the lead the distance from the end to the mitre of the part forming the seam; draw an X' Y' through E and transfer the distances on the other side as before explained. Measure the distances for the other elbow, as explained for Figures 46 and 47, and on folding the lead it will be found to be the required angle and to fit in the position for which it is intended. If the angles are going to be wiped, the usual ⅜-inch must be left on one side for the other to socket into. If the soldered seam is not wanted to be seen, it can be soldered on the back angles, where it would be out of sight, or in the case now before us, it could be soiled and shaved and soldered inside either with a copper bit and fine solder, or with ordinary plumber's

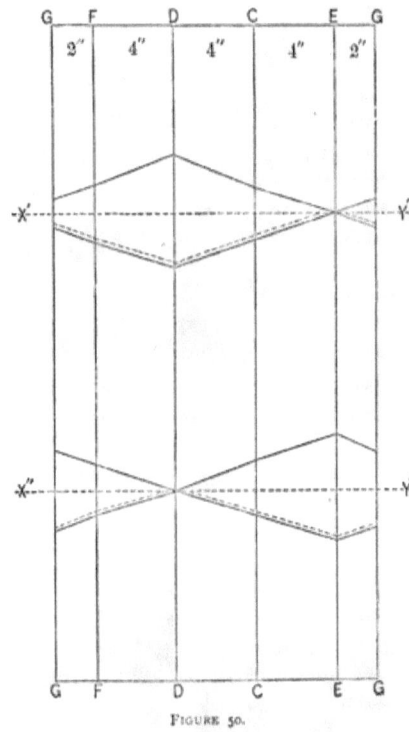

FIGURE 50.

solder, and "wiped" out. If the ends of the elbow were longer this would be more difficult to do, as a man could not get his hand inside; but in some cases, where the pipe is not too long, a cloth can be fastened on the end of a piece of wood, commonly called a "cat's paw," and wipe out the angle in that way.

It is possible to make bends on round pipes and then work them into a square section afterward, as illustrated, Figure 51, on looking at which it will be seen to have bends, in distinction to what we have hitherto called elbows. It is not at all difficult to make them. For a 3-inch square pipe a piece of 4-inch round pipe should be used, and great care taken in making the bends to keep the lead as equal in substance as possible. After making the bends on the round pipe a small hand dummy should be used to slightly raise the parts intended for the angles, and then the sides dressed in and the arrises worked up on the outside. It is necessary to make a chalk mark on the pipe where the angles should come, as in the absence of these guides, it will sometimes

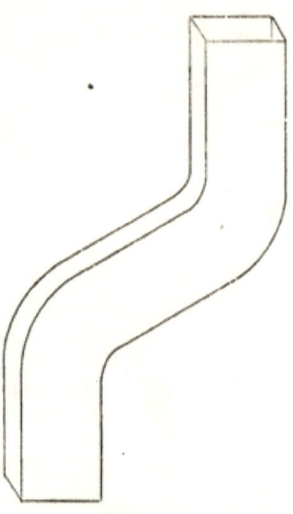

FIGURE 51.

occur that the bend gets twisted, and when this happens it requires a good deal of trouble to get it into the proper shape. These bends can only be made when a short piece of pipe is required, so that a small hand dummy will reach any part. A good many similar bends are made by cutting out two pieces of lead the size and shape of the sides of the required bend, and then bend a piece of lead to fit the front and "wipe" the two inside angles forming the front; afterwards bend and fit a similar piece on the back side and solder the two back angles on the outside. If the bend has a moulded front it is much better to have a wooden block made the shape of the required bend and the same size as the pipe. Cut out a piece of lead equal in width to the front and two sides of the block, allowing about 1 inch extra on each side for turning round the back side for nailing and fixing. Lay this piece of lead on the block and work down the sides; if there is any moulding to work, the lead must be tightly fixed, or else it will continually keep rising, and it is important that when once the lead is worked into the moulding it should be kept there, as if it gets moved ever so little, and a fresh chasing made, there will be so many tool marks made, as to be almost impossible to get them out again; and, in addition, the lead generally gets considerably reduced in substance if worked two or three times. After the bend is worked, take it off the wood block, trim the edges, and bend and solder a piece on the back side.

CHAPTER VII.

JOINT MAKING.

A GREAT many young plumbers, by dint of perseverance, succeed in wiping a very fair joint long before they know how to properly prepare the ends of the pipe. This arises from their being allowed to take a short piece of pipe, soiling and shaving a part in the centre, and wiping the joint on that, as if building the solder on was the only thing of importance. To prepare the ends of the pipe properly is of very great importance, and the soundness of the joint depends a great deal on this. In addition, care should be taken not to reduce the waterway, and also not to make the pipe any weaker at the joint. The ends should also be fitted very tightly, or else the solder will run through when the joint is wiped horizontally or underhand and lie in a body inside the pipe; or, if it is wiped upright, the solder that runs through falls down inside the pipe, like a lot of threads or fine ribbons, and lie in the pipe until the water is allowed to flow through. This causes the ribbons to get washed into the valves, causing them to leak by clinging round the working parts. First of all, when opening the ends of pipes, a box or other hard wood "tan-pin" should be used. The common way to open a large pipe is to dummy the end out, and for small pipes the end of the small bolt is often twisted round inside the end of the pipe to open it. Neither of these ways open the pipes true and even, and when the bolt is used the pipe is made rough inside, and also weakened at A, Figure 52. Both ends of the pipes should be opened. The advantage is, that it does not matter which way the current may be running, there are no sharp edges to check the free flow of water. If only one end of the pipe is opened, the entering end will sometimes curl inwards, especially if the joint is made upright, as the weight of the top piece presses upon

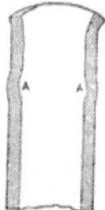

FIGURE 52.

that part, as at B, Figure 53; whereas, if both ends are opened this is avoided. One end must be opened wider for the other to enter. After the ends have been opened as described, they should be rasped or filed down to a thin edge on the outside, and care should be bestowed on this, especially for the entering end. If this is made as thin as possible it will give a little, and adjust itself to any little inequality on the inside of the other, or socket, end. Very few men take any pains with this, and most are satisfied with just rasping off the outer arris, or they sometimes whittle it off with a pocket-knife. When it is found that the ends do not fit tight, and there is a risk of the solder running through, the top edge of the socket end is closed round the entering end. This is a great mistake, as, if the ends are properly fitted, the socket end can be left open, as at C C, Figure 54; and if it is shaved inside the solder can enter and flow round, so that should none at all be left on the outside the joint would still be water-tight. If this was more often attended to there would be fewer *sweaty* joints.

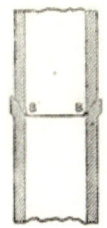

FIGURE 53.

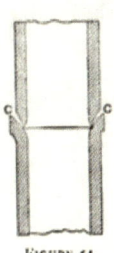

FIGURE 54.

After the pipe ends have been prepared so far, they are often rubbed with a piece of card-wire to take off the grease. This is a mistake, as a portion of the lead is scratched off, and, although not much harm is done, still the pipe is unnecessarily weakened. If the ends are well chalked and then wiped with an old rag—which must not be greasy, or a handful of wood shavings, that is all that is required (unless it is old pipe and full of grit on the surface, when the card-wire is necessary). After the grease is "killed," the ends should be soiled and dried. Some plumbers shave the ends and soil them afterwards. This causes the pipe to be weaker at the ends beyond the joint than anywhere else, and the joints cannot be wiped quite so clean at the edges as when the pipe is soiled first and shaved afterwards.

After the pipes have been soiled and dried the next thing is to shave them. If this is not properly done the solder cannot alloy with the lead, and the joint will leak. To shave a pipe properly the shave-hook must be sharp and must be held firmly.

It is important that the blade be firmly riveted or fastened on the shaft or handle. If these points are not attended to the lead is only scraped, leaving a dull face on it, or else the shaving gets on the edge of the hook and so prevents it taking off a clean shaving. Sometimes a false face is left on the lead, to which the solder adheres, leaving a space beneath through which the water can afterwards find its way. Although it is advisable to use a sharp hook and with a firm hand, it does not follow that unnecessary pressure need be applied to gouge out a deep incision in the pipe, as at D D, Figure 55, or that shaving after shaving should be taken off until the pipe is only half or perhaps a third of its original thickness, as shown at E E. It is not easy with small pipes, but for large pipes it is a much better plan, to shave *round* the pipe instead of lengthways. The hook for this, Figure 56, should have its cutting edge slightly rounding in its length to prevent the shavings clogging the edge, and it should have a small rounded corner for shaving the part that will form the end of the joint. It is a bad plan to use a sharp-pointed hook for this, as it makes the pipe very weak at this point, and there are very few plumbers who leave a body of solder right up to the end of the joint to fill up what the shave-hook has taken out. Very often a joint leaks because of a narrow strip of space not being shaved. If the shave-hook is passed round the pipe as well as lengthways, and the extreme end of the pipe carefully shaved so that the rasp marks are taken off—especially when they have become filled with soil, there is no liability of the joint leaking from that cause. Sometimes coarse rasps are used for preparing the ends of pipes; but a fine one, as used by cabinet-makers, is much better, as it makes the work smoother, and does not drag and distort the ends so much. During the operation of both rasping and shaving pipes, either for straight or branch joints, something should always be placed in the end or opening, as the case may be, to prevent the raspings or lead shavings falling into the pipe. If this was always done it would

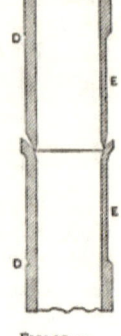

FIGURE 55.

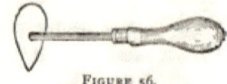

FIGURE 56.

very often save the trouble of taking the valves and fittings to pieces to clear them. Where valves have india-rubbers they often get cut and want renewing because of this precaution not being taken.

It has already been mentioned that pipes as a rule are greasy, and they should be chalked before soiling, to kill the grease. If this is not done and the soil is made too strong—that is, too much size used—it will generally peel off, but if it is made too thin it will rub off and get on the cloth, and so get worked on to the joint, making it look black and dirty. It is also difficult to make the joint, as the soil rubs off and the solder tins to the pipe beyond the required space, making it look unsightly in shape and necessitating its being trimmed up afterwards. This trimming-up is a source of weakness, as it is generally done with a pocket-knife, and in cutting off the ragged ends the lead pipe is sometimes nicked so that, should it require to be bent afterwards near the joint, this nick opens and the pipe breaks right through. When necessary to trim the joint, which rarely occurs with a good joint-wiper, the knife should be forced between the pipe and the superfluous solder, and so cut it off in that way. When joints have to be prepared on the bench, and then taken and placed in their position for making, the ends of the pipes should be tinned by pouring melted solder on the prepared ends, as if the joint was going to be made, and then wiping it all off again. This should always be done for large pipes, but for small services it is sufficient to "touch" them—that is, cover the shaved parts with tallow, and then wrap a piece of paper round to keep them clean and prevent any dust adhering to the grease. When the ends of the pipe are tinned before fixing they should be soiled inside and the socket end shaved when in its position, so as to insure its being quite smooth for the entering end to fit into.

When the pipe is to be fixed horizontally, a piece of board should be placed to catch the solder that falls during the operation of making the joint, or, better still, a piece of lead well soiled, so that it can be folded and break up the solder before it is quite set, and also act as a shoot for pitching it into the pot for remelting. But when the joint is going to be made upright, it is a common and slovenly way to twist wood shavings or paper into a band and

JOINT MAKING.

tie it on to the pipe just below the joint to catch the spare solder as it falls. This practice cannot be too strongly condemned, as, after the joint is made, when using a red-hot iron to melt off the rough ring of solder that is formed, the shavings or paper are set on fire; the solder being too hot to take hold of, and the shavings all ablaze, they are allowed to fall, at the serious risk of the house or building being burnt down. In addition there is always a great waste of solder through this inefficient way of securing what falls when making the joint. This could be avoided by cutting out a piece of six or seven-pound lead as a collar, Figure 57, with two ears, so that when placed round the pipe it will fit as a saucer or tray, and if bent round tightly, and the ears bent one inwards and the other outwards, they will lock into each other and so make the whole thing self-supporting. Where the collar is lapped there is a small space left at the bottom where the solder can run through; but a sprinkling of dust will generally stop this up. After the joint is made this lead collar, which should have been well soiled, can be removed, and then two places melted in the ring of solder, on opposite sides of the pipe, when the remainder can be taken away in two halves.

FIGURE 57.

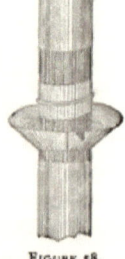

FIGURE 58.

For upright work the pipe should be soiled a good distance down from the bottom end of the joint so that the solder shall not tin to the pipe; and the collar should be fixed about 3 or 4 inches down, Figure 58, so that when wiping the solder shall not burn the little finger of the hand. The collar should not be fixed too low, as the caught solder helps to keep the pipe hot whilst the joint is being made.

Figure 59 represents a joint made by a good plumber who is very fond of thin cloths. He says he "likes to feel his joint." Although it may be perfectly symmetrical, it is not a good joint, as it is weaker at the ends than elsewhere. The pipe is reduced in substance when shaving, and in the joint spoken of this

FIGURE 59.

weakness should have been covered with solder, so as to strengthen it. If the joint had been shaved shorter, so that the metal covered the shaving, it would have been much stronger; but the joint would have looked short and out of proportion to the size of the pipe, as in Figure 60.

Thin cloths are a mistake for making joints, especially on large-size pipes. If the cloths are of a good thickness the joints

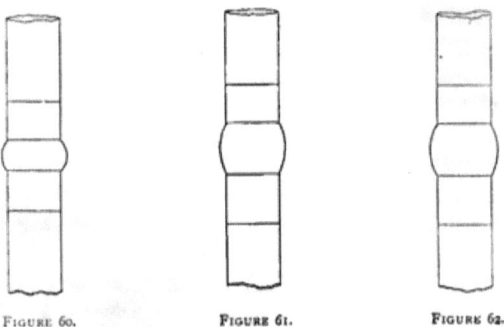

FIGURE 60.   FIGURE 61.   FIGURE 62.

can be wiped much truer, and there is not such a liability of bare places being left on the ends of them. The illustration, Figure 61, shows a properly-shaped wiped joint, suitable for a ventilation or light waste pipe where the pipe used is light in substance; but this joint would not be strong enough for a pipe of heavy substance for conveying water under pressure, and the joint should be stronger in proportion, as shown by Figure 62.

Very few men make any difference in the strength of their joints, but make all alike, irrespective of the use the pipe is to be put to; but a very little thought will prove that there should be a difference, although if an error is made it should be on the side of strength. This applies more especially to pipes that have brass-work for attaching to cisterns or fittings of any kind soldered to them, or stop-cocks, as in some cases they have to withstand a strain when being screwed up or turned. Brass-work is generally made as small as possible, to save the metal, and the ends are generally made so short that there is not much to solder to. The result is the parts that meet are not in the centre of the joint. A

JOINT MAKING. 75

glance at Figures 63 and 64 will explain what is meant. The joint should be made much heavier and a greater body of solder left on, to get a sufficient proportion over the part that requires it.

There is a great variety of opinions amongst plumbers as to the length joints should be made. Some are in favor of very short joints, whilst others prefer very long ones, even so far as to make them 4 inches long on a ½-inch or ¾-inch pipe. This is absurd, and they have been compared to "dirty tinned ends," and not joints at all, especially when there is scarcely any solder left on. Joints made in this manner frequently break in half when screwing up the union. A reference to the illustration, Figure 64 will show the absurdity of making joints in that way. The accompanying table is a very fair average length for all wiped joints:—

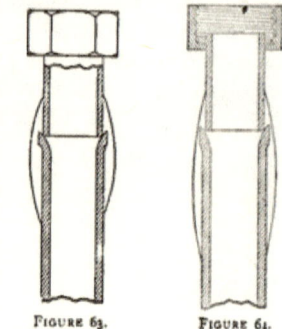

FIGURE 63.   FIGURE 64.

6-inch pipe,	4-inch joints.			1½-inch pipe,	3-inch joints.		
5 ,,	,,	3¾ ,,	,,	1¼ ,,	,,	3 ,,	,,
4 ,,	,,	3½ ,,	,,	1 ,,	,,	3 ,,	,,
3 ,,	,,	3½ ,,	,,	¾ ,,	,,	2¾ ,,	,,
2 ,,	,,	3¼ ,,	,,	½ ,,	,,	2½ ,,	,,

It is the one several good men work to, and the joints look fairly proportionate to the sizes of the pipes.

CHAPTER VIII.

JOINT MAKING—*continued*.

MOST men prefer certain ways of making a joint. One man will make all the joints that he possibly can in an upright position, another one will do almost anything so as to be able to make the joint underhand; one man never requires an iron, another one cannot make his joint without it; one man does not like a wood splash-stick to put on his metal with, as the smoke gets into his eyes, and yet another one says that an iron one cools the metal; one likes fustian for his cloths, another prefers bedtick. One man likes them fastened with pins, as they bend much easier; another one says the solder tins to the pins and pulls them out, or they get hot and burn his fingers, and he prefers to have them sewed. And so each man follows his own ideas; but I must say that I like fustian-cloths tacked with needle and thread, and iron splash-sticks. I don't mind which way the joint comes, either upright or underhand; and although I can make a joint either with or without an iron, I must say that I like the old-fashioned way of using an iron when making a joint, and my reason is that the joints do not so often sweat, because the solder can be used finer—that is, richer in tin. In addition, men who do not use irons pour on solder, in trying to get up a heat, until they have almost melted the pipe, and then when they have nearly finished the joint the heat is so great as to keep the tin melted, which partly separates from the lead and keeps running, the plumber meanwhile wiping round and round the joint, and finally wiping off this tin which should have remained alloyed with the lead as solder. The solder, having parted with a portion of the tin, is full of cellules,—and so the joint leaks; the plumber growls and says the metal is porous. This continually going round the joint each time leaves a ragged edge, which has to be trimmed off afterwards. So, for these reasons, the use of the iron may be considered an advantage. Not being necessary to get up such a great heat, the joint can be

JOINT MAKING.

wiped much cleaner at the edges. The two metals forming the solder do not separate to such a serious extent, and hence the joint is much sounder. The joint also looks cleaner, and if a black mark is left with the cloth, a sponge—dipped in a bowl of clean water—lightly passed over while the joint is still hot, will generally remove the mark.

After a joint is made, and while still hot, the usual way is to rub a tallow candle over it and wipe off the grease with a clean rag or pocket-handkerchief. Some think that in addition to cleaning the joint the grease gets into the pores and so prevents the sweating spoken of. If the joint is not sound without the aid of grease it can't be a very good one.

Years ago it was the practice to *overcast* all joints. This consisted in wiping the joint first and then rubbing the hot iron over it so as to leave a number of faces. The object sought was to close any pores in the solder to prevent the water from oozing through. A good many men used to do this to all joints, no matter whether they were on soil, waste, or service pipes. To pipes that did not convey water under pressure this was quite superfluous; but there was some excuse for doing so to main service pipes which had to withstand a great pressure of water. Men who were really good joint wipers used to overcast some joints, but only to water main pipes, or to the tail or suction pipes of lead jack-pumps—more especially to pumps, as, although the water could not leak through, air could be drawn in and so allow the water to fall back into the well or cistern, so that it was necessary to pump for a few seconds before the water would flow from the nozzle; so, to make sure, good tradesmen would very often overcast their joints, it being difficult sometimes to find out a defective one in a suction pipe, which would have shown itself by leaking if in a service pipe. A good many country plumbers treat their joints to pumps in this manner at the present day, but the practice has died out amongst town tradesmen. Figure 65 shows an elevation, and Figure 66 a section through the centre of an *overcast* joint.

FIGURE 65.

FIGURE 66.

Lead joints have been made on lead service pipes. One from Hampton Court Palace is shown at Figure 67. Instead of the

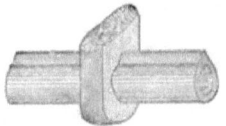
FIGURE 67.

ends of the pipe being soldered together, a quantity of molten lead was poured on, and it presented the appearance as shown. From the rough appearance of the inside of the joint it could be seen that the pipe ends were filled with sand, and then, doubtless, they were butted together and buried in more sand, leaving a space round the part to be joined, into which was poured molten lead until the pipe itself was fused, the surplus metal running away over a weir in the same manner as was described for burning seams on heavy pipe. It has been suggested that common salt would make a good core, as it would answer perfectly, and when water was allowed to pass through, the salt would dissolve. Sand would probably get baked hard and so require some time for the water to displace it. Salt would not be so injurious as sand to any fittings that it might have to pass through.

When soil pipes or waste pipes are fixed in a chase in the wall it is a very good plan to make what is commonly called a "*block joint.*" This is a very easy joint to wipe, but it is sometimes improperly prepared. On looking at Figure 68, which

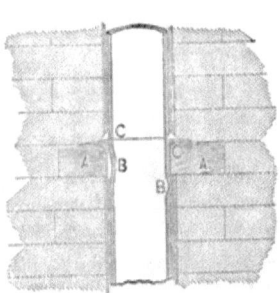
FIGURE 68.

shows a section of pipe and joint, A A represents a wooden (or *stone*) block, perforated for the pipe to pass through, and the ends long enough to lay on the brickwork each side. The carpenter or mason should be instructed to take off the sharp arris round the hole. A common way is for the plumber to taft over the pipe sufficiently to get enough to lay on the block for him to solder to. When the plumber opens the end of the

pipe, if the block fits tightly, the end grain of the wood cuts into the lead and so weakens it just where it should be the strongest, as the pipe is partly suspended at this point. The plumber, in

JOINT MAKING.

trying to get sufficient to solder to, very often gets the pipe buckled, as at B B, Figure 68; also a hard lump at C C; and the outer edge of the flange so thin that he can scarcely shave a face on it for soldering to.

Now, the block should always fit moderately close round the pipe so as to well support it, and the sharp arris round the hole on the wood, or stone, should be taken off, both on top and under sides; and instead of flanging the lead over, the plumber should cut a collar out of a piece of sheet-lead, and soil and shave and tin it before fixing. This collar should then be placed round the pipe on the block and the pipe tafted over it. The collar should be at least ½-inch larger than the outside of the pipe, so that it shall not cut into it when the "tan-pin" is driven in, as described above when speaking of the block; and the pipe should be tafted as shown in section at D D, Figure 69, which shows a raised bead instead of being made flat, and which prevents the lead being driven into a hard mass with a sharp corner underneath. Great numbers of taft joints are found to be broken under the part shown by the right hand C, Figure 68. This does not take place when the end is prepared as shown at Figure 69. On looking at this Figure it will be seen that the bead occupies a certain amount of space which would have to be filled with solder if the joint was prepared as shown in Figure 68, so that really there is an economy of solder when it is prepared in this manner; at the same time a fair amount should be left on, so as to get a good strong flange to support the upper portion of the pipe. Some plumbers, when they begin to work the taft over, have too long a length of pipe projecting. They work it partly over and then cut off a portion with hammer and perhaps a blunt chipping-knife, and so buckle and distort the end of the pipe that they cannot make a neat taft. The end of the pipe, if too long, should always be cut off with a saw, allowing ¾-inch for tafting, as that is quite sufficient for the purpose, and there is less liability of splitting or tearing the pipe.

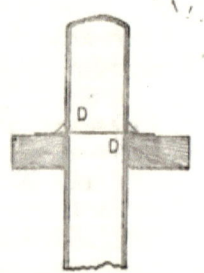

FIGURE 69.

This applies to this kind of joint in various other positions.

Say, for example, a trap has to be tafted over for soldering to the lead safe under a water-closet. If it is done as at A, Figure 70,

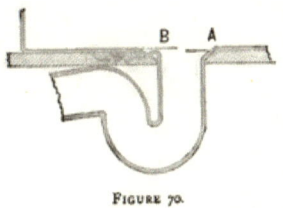

FIGURE 70.

when the heated solder is poured on, the lead expands and the edge curls upwards, and after the joint is made, the edge of the pipe or trap is sometimes found projecting above the soldering, and will often be found to open as the solder cools and shrinks. If the trap is tafted as shown at B, this could not occur. If these joints are properly prepared it is not necessary to use much solder, as little or no strength is required; a flat surface is left for the water-closet to seat upon, and thus requires less cement for bedding it.

The lead safe need not be dressed into the dishing on the wooden floor; by simply turning over the edge of the lead trap in the form of a bead—so that it does not stand above the level of the floor—it forces down the lead safe just sufficient to bury the edge of the trap and insures its being covered with solder.

In the case of soil pipes the flange should be shaved not less than 1½ inch wide, so as to get a good support for the upper length of pipe, as before stated; but when soldering a trap to a safe, ¾-inch wide is quite sufficient for all that is required. Now, a good many plumbers think a joint of this width too small and paltry-looking, so they endeavour to make it look bold by shaving it about 1½ inch wide. If the dishing in the wooden floor is this width, there is a great waste of solder to fill it up; and if it is made smaller than that, but the lead shaved the same width, it must necessarily be weaker, as the edge of the shaving has no solder left upon it to replace the lead taken off with the shave-hook.

When preparing these joints, a piece of board or a wood block is fitted into the trap, and a centre found so as to scribe the circles for shaving. Very often a pair of pointed compasses is used for this, with the result that where lead of a light substance is used for the safe it is, perhaps, cut half-way through. It is much better to have a pair of compasses with one end turned like a hook, with sharp edges. As it is not always convenient to carry a lot of tools, these can be made by bending a piece of sheet-copper so as to fit on one leg of the ordinary compasses, and then bending a piece of a steel watch-spring like a loop and soldering it to the copper thimble, as Figure 71. The edges can be sharpened on a stone, and they answer the purpose as well as those of a better description; they can be carried in the waistcoat pocket, and if lost or broken can be easily replaced. They can be made in the same way, but with a short, straight piece of the spring instead of the loop, and sharpened like a lancet-point, for cutting out paper patterns or thin leather washers for valve flanges—Figure 72: The steel can be soldered to the copper thimble with ordinary fine solder and copper bit, using, as a flux, hydrochloric acid saturated with zinc, commonly called "killed spirits."

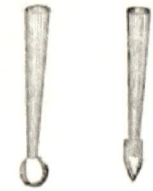

FIGURE 71.   FIGURE 72.

At a sanitary exhibition, held in London some time ago, were exhibited some joints made without solder on lead soil pipes. They were made by welting the ends of the pipes together, as shown in section, Figure 73. The advantage claimed for them was that as soldered joints are sometimes eaten away by the gases emanating from sewage, when there is no solder the joints cannot be affected in that manner. These joints will never come into general use, as to be effectual they must be made perfectly air-tight, in addition to being water-tight, and to get them so a mandrel must be fixed inside for the welt to be worked in quite closely, which cannot always be done when the work is in its position. To make them on heavy pipe would be rather difficult, and the lead would get very much reduced in substance when preparing the ends, and the joint would be, consequently, the weakest part of the pipe. Ordinary wiped joints could be made over them, and so strengthen them in that way. If joints are properly prepared, there is little or no portion of the solder exposed to the influence of sewage gases, so it will be seen that it is quite unnecessary to go to the trouble of preparing them in the manner described.

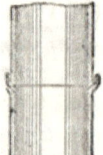

FIGURE 73.

When preparing pipes for making branch-wiped joints, most plumbers have a way of their own in opening the pipe into which

G

the other has to be joined. Say the joint is going to be made on a service pipe of any diameter,—some plumbers will bore a hole in the side with a gimlet or small auger; others will cut a slit in it with a sharp chipping-knife; others, again, cut an oval piece right out with the same tool. Sometimes a plumber takes the round side of the rasp and so reduces the thickness of the pipe as to make the lead so thin as to be able to easily open it afterwards. A V-piece can be sawn out, and on large-sized, heavy main pipes, a hole burnt through with a red-hot plumber's iron, so as not to distort or bruise the pipe out of shape. Some of these ways are very good and others bad. The great object is to get as much lead as possible to stand up for the end of the branch pipe to enter in such a manner that solder will not run through when the joint is being made, and so that there shall not be an obstruction to the free flow of water. In heavy pipes it is a good plan to reduce the substance of the lead a little, and then cut a short slit so that the sides can be pressed open with a chisel to allow the end of the bolt to enter. This bolt should be reduced at the end and bent so as to be able to work the lead upwards, which cannot very well be done with a straight bolt. The result of using a straight one is that the lead is so thickened up as to project inside the pipe, as shown at A A, Figure 74. If a bent bolt is used the hooked point can be placed under that part, as shown in section, Figure 75, and by hitting it at B with

FIGURE 74.   FIGURE 75.   FIGURE 76.

a hammer the evil spoken of can be avoided.

This bent bolt wants using carefully, or another evil sometimes takes place. Figure 76 shows a cross-section of a pipe opened ready for the joint, and C C a representation of what is actually the case with half the branch joints that are made—that is, the inside of the pipe is bruised and reduced in substance by using the thin end of the bent bolt to open the sides. This mischief is aggravated where these bruises are made below the centre of the bore of the pipe where no solder is put on the outside to strengthen the weakened portion. With a little care the *sides* can be opened without the bolt, and so avoid this. The water having to turn a sharp corner when the branch is made at right angles, the opening should always be made a good size, and the end of the branch pipe opened with a "tan-pin," and then the sides flattened to the diameter of the pipe, so as to present as little resistance as possible to the free flow. A glance at the section, Figure 77, will

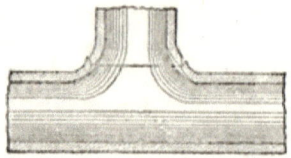 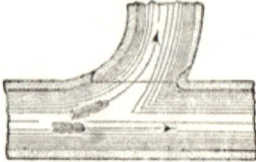

FIGURE 77.   FIGURE 78.

show what is meant. Figure 78 shows a much greater improvement, and it is no more trouble to prepare than the last one described. The hole in the main pipe having been opened much longer than is equal to the bore of the branch pipe, the branch pipe is bent as shown, sawn off at the proper angle, and fitted. As a rule, all pipes should socket in the direction of the current, but this is very rarely carried out with branch pipes for water services, and is of less importance if the point of connection is enlarged as described. It has been mentioned that if joints on pipes were properly prepared a portion of solder would enter the socket and flow round the spigot end; but a branch joint is not so easy to prepare so as to ensure its fitting so tight that solder cannot run through when the joint is being made, so that, as a rule, the outer edge has to be closed round the branch pipe. This is sometimes done with a hammer, which necessitates the marks made being shaved over again. It is a good plan to take a brad-awl or a point of the compasses to close the edge as tightly as possible by *pressing* it in.

For branch joints on soil and other large-sized pipes, of

moderate substance, a narrow strip is generally cut out of the main pipe about one inch less in length than the bore of the intended branch pipe, which should always be bent at the point of junction, as shown in Figure 78. The arrows should be reversed to denote the direction of the current in a waste or soil pipe. After the piece of lead is cut out and the sides forced open a little, a small handful of wood-shavings set fire to, and placed in the opening, softens the pipe so that the hole can be enlarged in a very few minutes with a small hand-dummy.

The reason for bending the end of the branch pipe is to prevent anything, passing down from a higher level, lodging in the end of it, and also to prevent the pipe being frequently stopped, as is often the case where small-sized soil pipes are used and a scanty supply of water is laid on to the water-closet, in which case an accumulation of soil and paper sometimes takes place in the branch or horizontal pipe. When this accumulation is forced onwards in a body, it projects across the vertical pipe and presses on the opposite side, as shown in section at Figure 79. Sometimes a brass cap and screw is soldered in, as shown, for easy access for removing this kind of obstruction; but this is

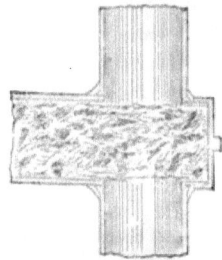

FIGURE 79.

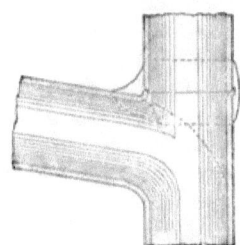

FIGURE 80.

generally an aggravation of the evil, as a ledge is formed for the end of the plug of paper, &c., to rest upon.

Branch soil pipes can be bent, as shown in Figure 80, and the vertical pipe from upper floors soldered into the top of the bend, so as to insure no stoppage taking place in the manner described. This is an excellent plan, but it entails a great waste of pipe, as the joints come at unequal intervals, and the work, when fixed, does not look so smart. As more joints are required, it means more labour and solder than when fixed in the ordinary way. At the exhibition, to which reference has already been made, were shown some "welted" branch joints. At the junction the vertical pipe was opened, as shown in section at Figure 81, with the double purpose of preventing a plug of paper, &c., resting against the opposite side, and also to prevent the discharges from a water-closet filling the pipe and so displacing the air that syphonage of water out of the traps will take place. Having never seen this principle in actual practice, no opinion can be expressed as to its efficiency for that purpose. Some people might demur and say that the enlarged part would get foul, as the water, &c., would splash about instead of passing over the inside surface with a scouring force sufficient to keep it from getting furred.

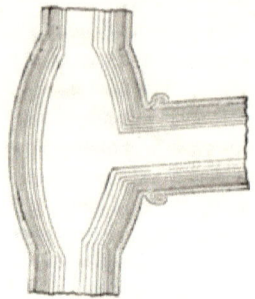

FIGURE 81.

The same precautions must be taken, when preparing branch joints for soldering, as already explained for straight joints. Some plumbers are in the habit of shaving their joints considerably more on the straight, or running, pipe than is required, as shown in elevation and section, Figures 82 and 83. Now, it does not

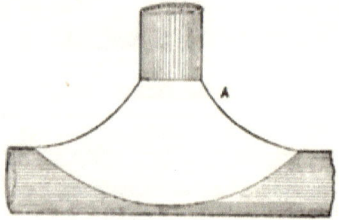

FIGURE 82.

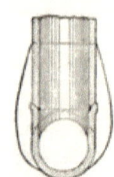

FIGURE 83.

matter what size the pipe is, it is never necessary for the solder to come below the centre of the pipe, and it is only a waste of metal to make them in that way. Neither does it add to their appearance; on the contrary, they always look dumpy. These joints are also more difficult to make, and require a thinnish cloth to make them with. If a board is placed close beneath the pipe, to catch the spare solder, in such a way as to keep the pipe as hot as possible, the little finger of the wiping hand frequently gets burnt. A great many men also shave the end of the branch pipe too long, with the result that when they make the joint no solder is left on the top edge. When wiping the hollow part shown at A, Figure 82, it is very difficult to leave the solder right up to the edge of the shaving.

The process of wiping a soldered joint can only be learned by practice, and it is almost impossible to describe in writing the various movements required. Even when learned, it is easy to get out of practice, and a good tradesman varies a great deal at different times.

CHAPTER IX.

JOINT MAKING—*continued*.

In London it is not at all uncommon for plumbers to be at work on the lead-work of roofs of buildings for several months at a time; so that afterwards, when they try to make a joint, they sometimes fail to do it at the first trial.

After the joint has been prepared, so far as fitting, soiling, and shaving are concerned, the first thing is to firmly fix it, so that it cannot move when being wiped. If an underhand joint, it is easy to lay it on the bench on wooden blocks, and then load the pipes with weights, or rolls of lead, or other heavy material that may be laying about. It is not at all a good plan to drive chisels or spikes, for fixing, into the bench, as it makes it full of holes for solder to run into, and also makes it so rough that pipes cannot be straightened, or lead dressed out flat upon it. Very few plumbers have a decent pair of compasses in their kit, through using them for fixing small pipes; the points get broken, the legs bent, and the joint destroyed by driving them in with a hammer. When underhand joints have to be made in their places, it is necessary sometimes to drive chisels into the wall to support the pipe; these chisels should be strong, so as not to bend, and the pipe should be firmly lashed to them to keep it from moving, and also to insure the ends from coming apart. If these precautions are not taken, the joint very often cracks right around, and this crack, being fine, cannot be seen until the water is allowed to enter the pipe, and a leakage is found. The same care is required when fixing a joint for making in an upright position. For small-sized pipes, which are thicker in substance, proportionally, than those of a larger size, it is sometimes only necessary to fix the bottom part, but for larger-sized pipes, especially those made in 10 or 12-foot lengths, the upper part should also be firmly fixed, so as to sustain the weight of the pipe.

In one case a man was fixing a stack of 6-inch soil pipes, with branches on each floor; above the highest branch it was reduced to 4 inches, and continued to the roof as a ventilation pipe; the pipes were fixed on wooden blocks at each 10 feet. The 6-inch pipe had its top end reduced to 4 inches and tafted back on a lead collar on the block. By some means, when tafting the end, the pipe became slightly larger, so that when the 4-inch length was fitted, it was found necessary to slightly enlarge the end for it to fit into the other one. The solder was splashed on and the joint just going to be wiped, when, lo! the 4-inch pipe disappeared down the inside of the 6-inch! If the top length of pipe had been properly lashed and fixed, this would not have happened.

It is a good plan to leave out the bottom length of all vertical stacks of pipes until the work is completed, so that any obstructions may be detected and removed before it is fixed. This was done in the above case, so that no harm was done beyond spoiling the length of pipe that fell through.

When fixing branch joints, the same care in firmly fixing must be taken, especially on large-sized pipes. The common way to prevent the branches of soil pipe from entering too far into the main pipe, is to cut tags, as shown at A, Figure 84. This is not a good plan, as there is a rough edge left inside for passing objects to cling to, and sometimes a little spur of solder will project through on the inside. A better plan, and one often practised, is to have a short wooden mandrel placed inside the main pipe to support the weight of the branch. Should this mandrel be forgotten until the pipe is fixed, there is great difficulty in getting it out afterwards. The best plan is to place two parallel pieces of 1¼ or 1½-inch board inside the pipe, and a little longer than the opening for the branch, and then force in two small wooden wedges in such a way that the end of the branch pipe shall rest upon them, and not enter quite so far as the inside face of the main pipe. A glance at section,

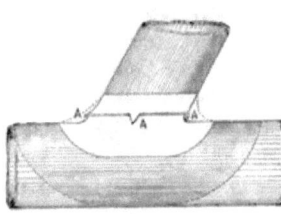

FIGURE 84.

Figure 85, will illustrate what is meant. The advantage claimed for this is, that a plumb-bob on the end of a chalk line would displace it, should it be left in, so that it would fall out at the bottom of the stack. If not too long, the pieces would pass any bends, but care should be taken not to have them too short, so as to avoid a possibility of their becoming angled, or getting cornerwise, and so jam in the pipe.

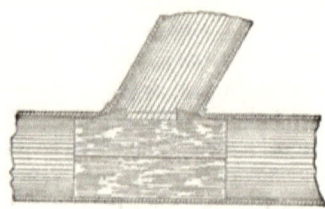

FIGURE 85.

For fixing brass unions, ferrules, or bosses, on the ends of pipes for making straight-wiped joints, splints are often used. Splints are pieces of common lath, split up into long narrow strips and tightly fitted inside the pipe, &c. When joints are fixed together in this way, they can be "*rolled*," that is, instead of passing the cloth round the pipe to form the joint, the pipe can be rolled backward and forward on two wooden blocks—Figure 86, and the cloth simply pressed against the joint to form it into shape. Some men will make this joint with an ordinary underhand cloth; others make the joint partly with this cloth, and then finish with a smaller one, which the mate keeps warmed ready. Others use the

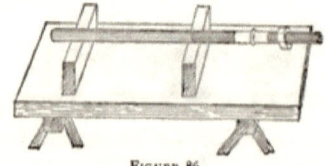

FIGURE 86.

small cloth only in this way: Take a ladle of molten solder, heated to the proper degree, in the right hand, cloth in the left, and pour on the joint, the mate slowly rolling the pipe toward the worker, so that he can keep patting the solder to keep it from falling off. When the heat is about right, lay down the ladle and take the cloth in the right hand, roll the pipe towards you, place the end of the cloth, which should be held with the thumb and forefinger only, on the pipe-end of the joint, and smartly roll it away from you with the left hand, pressing the cloth as hard as possible on the pipe; the end of the joint will be found to be wiped nice and clean. Smartly roll the pipe towards you, and then use the second or third finger on the other end of the cloth, and press it on the brasswork; quickly roll the pipe from you, and the other end of the joint is wiped clean. If, after doing this, the body looks too large and bulky, wipe the superfluous metal from right to left on the pipe, and wipe that end again, as before described. Now deliberately and carefully hold the cloth on the joint and press the ends hard on the two extremities; smartly roll the pipe away from you. If a good heat is got up, this can be done two or three times until the joint is as true as if it had been turned in a lathe. A slight pause can now be made, and the cloth should be looked at to see that there is no solder sticking to it. About this time the solder is almost set, and a light touch with the cloth will remove the small ridge that will be left at the last roll of the pipe. This will leave a slight dark mark, which a sponge of clean water, lightly passed over it, will remove. When making these joints, only two fingers are required, namely, those at the ends of the cloth. If the cloth is pressed in the centre, it will cause the joint to be too slender, and sometimes to split or open lengthways, so that it has to be re-made. Those men who like to wipe with the left hand, turn the pipe and joint the other way so that they can roll them with the right hand. The cloths used for these joints should be of moderate thickness, and if they are made of old fustian, which has ribs on the face, they score the face of the solder on the joint, giving it a smarter appearance than when only a small portion at a time is wiped, and thus leaving the cloth-marks crossing each other in all directions. The principal troubles a young joint-wiper has to contend against are to keep the solder from falling off, and to get the ends of the joint wiped clean, so as not to require trimming. To prevent it falling off, care should be taken not to get the pipe too hot, and also to keep it in motion, so as not to have one portion of the solder heated more than another, one part in a fluid condition and the other almost set. When making underhand joints, and the pipe is made too hot, the bottom side of the joint sometimes falls off, and, even if made, it is a common occurrence for it to have less solder on the bottom than elsewhere. To wipe the edges clean, they should be done first. Some young men are so anxious to mould the body of the joint that they lose their heat. The body or centre should

JOINT MAKING.

generally be left until last, as that part retains its heat longest. When wiping joints in an upright position, the top edge should be done first, and then, in some cases, the body. As the tin generally flows to the bottom edge, which is kept hot by the spare solder in the collar placed to catch it, it can be wiped last, but it should be done quickly, as the tin sometimes sets, leaving small projections which tear the cloth.

Plumbers, as a rule, scorn to make joints any other way than by wiping them, but there are scores of quacks who go about usurping the title of plumber, who make theirs with a copperbit, and could not wipe a joint if they tried. These people work cheap, and, to put it mildly, do not do the trade any credit. Now, joint wiping is not the only criterion of a tradesman; at the same time, if he cannot wipe a joint, what other branch can he do in a proper manner?

Figure 87 represents a section of a copper-bit joint as usually made. At a glance it will be seen to be very weak at A. This part is covered with solder when the joint is wiped, and so strengthened. Soil pipes, when joined in this way, soon break,

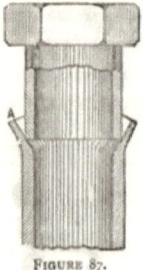
Figure 87.

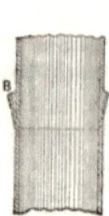
Figure 88.

Figure 89.

especially when a man thinks he will insure its being a sound joint by letting one end enter a good distance, say ¾-inch or perhaps 3 inches, into the other, and then proceeds to run a small bead of fine solder round the top edge of the bottom pipe, as shown at B, Figure 88. This kind of joint would be much stronger if made with a blow-pipe, but this is rarely done, as it makes the man's cheeks ache to get up sufficient heat for the solder to flow round and *down to the bottom of the socket*. There are several kinds of lamps by which this description of joint can be made, but when done it will not bear comparison with a good wiped joint. It is necessary to make them with a lamp under special circumstances; for instance, an ornamental pipe fixed in view, where a bulged joint would not look well; but in this case it could be strengthened by a band or other ornament, as shown in section, Figure 89. Care should, however, be taken that the socket is filled with solder, and the heat so great as *to insure its tinning to the lead right down to the bottom.* This cannot always be guaranteed unless the ends are prepared and tinned before putting in place. The ends of the pipe must be quite smooth and true; if care is not bestowed on this, it frequently happens that just as the joint is all but finished the solder runs through and down the inside of the pipe.

Ordinary copper-bit joints are not good for the kind of work that falls under the head of plumbing; they may do very well for gas-fitters, who, as a rule, use very light lead pipe when they use any at all, but although thousands, no doubt, are made, they are not so good as a blow-pipe joint. But there is another way of making copper-bit joints that really is strong—I do not know whether to call them Scotch joints or not, but the first I ever saw made was by a Scotchman.

Figures 90 and 91 show a section and elevation of an overcast copper-bit joint, and it will be noticed that it is strengthened

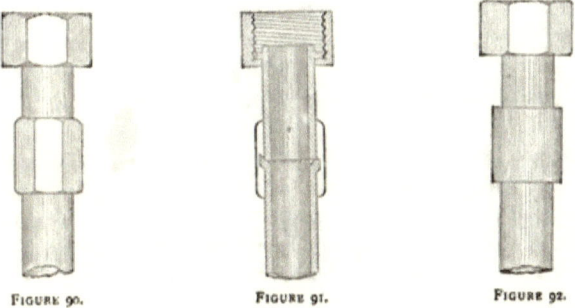

FIGURE 90.   FIGURE 91.   FIGURE 92.

on the outside, in the same manner as a wiped joint, by a body of solder. These joints can only be made in a horizontal position, when the parts to be united can be turned round, so that when one face is made or soldered it can be rolled over, so that another portion of the pipe, &c., is upward, ready for another section to be soldered. The proper way to make these joints is to solder the brasswork to the pipe by filling the socket-end with solder like an ordinary copper-bit or blow-pipe joint, to do which it should be placed upright; then lay the pipe on its side and build sufficient solder on the outside to make the joint. It should now be allowed to cool a little, as if allowed to get too hot the solder will run and the brasswork fall off. After this cooling has taken place, the copper-bit, with a well-tinned face, should only be sufficiently heated to just melt the solder, and not to render it so fluid as to run off the work. Hold the copper-bit firmly, and let it just touch the solder at one end of the joint, taking care not to bury the face in the solder so as to touch the pipe. As soon as the solder melts, move the bit slowly towards the other end, pause for a second or two, and then suddenly and smartly lift the bit upwards; if this is not done quickly, part of the solder will adhere to the bit instead of remaining behind and leaving the joint full up to the edge. Now roll the pipe, &c., to present another face on its top side, and serve that the same. These joints are very easy to make, and require no trimming at the ends if care is taken with them; they also look nice and clean, as each face presents a *bright* surface. These joints are suitable for brass unions to lavatory valves, which are generally too short to make wiped joints to, and they may be considered superior, and where the unions are short, as stated, they are stronger. If the reader will refer to what was said about long joints, and to the illustration given in an earlier chapter, I think he will agree that this is so. These joints can also be made to the waste unions of washhand-basins, where the whole weight of the branch waste, and sometimes the trap as well, is suspended from a frail piece of crockeryware; they have an advantage over the wiped joint, as they are much lighter.

After these joints have been made, they are sometimes filed and cleaned up so as to present the appearance shown at Figure 92. This is a waste of time, they are no stronger for it, and although some first-class tradesmen do this, it should be condemned quite as much as serving a wiped joint so, which is generally looked upon as showing want of skill in wiping.

On large-sized pipes, these joints, which are sometimes called "*ribbon* joints," can be made by holding the bit steady and rolling the pipe round, but the same attention must be paid to the heat of the bit, and care taken not to get the work too hot, so that the solder sets almost as soon as it leaves the bit; but half-an-hour's practice is worth more than pages of description.

About two years ago, a series of lectures was given by a master plumber in the rooms of the Society of Arts, and he showed a specimen of a welted joint. This consisted of a hard-metal nipple with grooved external surfaces; the two ends of the lead pipe, that were to be joined, were passed over it and then pressed together so as to be perfectly water-tight. This cannot be equal to a good old-fashioned wiped joint, which could be made almost as quickly, and which, when properly made, would resist any strain that it might be subjected to. There are patent moulds for casting soldered joints on lead pipe, which consist of placing iron flasks round the pipe, after the joint has been prepared in the usual manner, and then pouring in melted solder. The poor plumber's assistant, who would have to carry all the apparatus for making these joints, would be entitled to pity, and if a man cannot wipe a joint he cannot be called a plumber, and has no business in the trade. Fancy a firm, employing about 150 pairs of plumbers, having to buy these things! A complete set, for making joints in all sizes and positions, would cost much more than any master would be disposed to pay, and each man would require a set to himself, unless there were several men working together, and then there would be a waste of time borrowing and taking back; and even then there are hundreds of joints made where they could not possibly be used.

CHAPTER X.

PIPE FIXING.

There are several ways for fixing lead pipes. For services, the common way is to drive into the wall a series of iron wall-hooks. If the pipe is horizontal and fixed on the face of the wall, and not recessed or covered with plaster, it frequently presents the appearance illustrated in Figure 93. The lead pipe being very soft, and having no support between the wall-hooks, it hangs down, and at the same time the whole of the

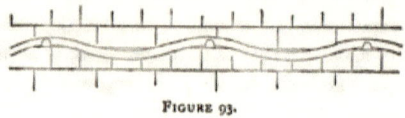

FIGURE 93.

weight is resting upon such a small portion that the hook cuts into the pipe and bruises it.

Sometimes, to save the pipe from being cut, a piece of sheet-lead is placed between it and the hook; or a small piece of plasterers' lath is used for the same purpose. Although this is a good plan, it gives the work a patchy appearance. A much better way is to have a wooden fillet nailed on the wall, with a hollow on its top side for the pipe to lie in, as shown in section, Figure 94. The advantages gained are, that the pipe is not cut, as predescribed, and it can be drained empty if fixed to a very slight inclination, as there are no baggy parts for water to lie in. This is of great importance, as, when out of use, the pipes can be emptied to avoid water freezing in them in cold weather. Another reason is, that when an intermittent supply of water

FIGURE 9

passes through pipes fixed on hooks, water will be retained in the lower parts, and those over the hooks become charged with air. Under certain conditions this will prevent water from passing through, and when it occurs, some plumbers try to remove it with a force pump. Other plumbers, who know the cause, will simply straighten the pipe to allow the air to escape. If it is very old pipe, it will sometimes leak where cut with the hook. To avoid this, the plumber will drive a common pin into the highest parts of the pipe, and so let out the air, afterward closing the small hole made, by driving in the sides with a hammer and the side of the point-end of a chisel. Larger-sized pipes, when fixed horizontally, generally have lead tacks soldered to their sides, which are then hooked to the wall. These pipes frequently sag down, and the tacks tear away from the pipe (or the pipe itself tears), so that it is much better to have the wooden fillet to support it, as described for smaller-sized pipes. This wooden fillet will also answer as fixing for a casing for protection against frost or injury.

When service pipes are fixed vertically with wall-hooks, unless they are driven in very tight, the pipe slips through, and the edges of both the shank and head of the hook cut into the pipe. If there should be a slight bend in the pipe, the hook will cut through the lead and the pipe will leak. If a waste pipe, through which hot and cold water pass at intervals, is fixed in this manner, this evil is aggravated by the expansion and contraction of the lead. Then, again, if the hook is driven in tight, so that the pipe cannot slip downward, the pipe becomes so bruised as to contract the water-way, as shown in section at Figure 95, so that a smaller pipe, full bore throughout its length, would allow as much water to pass through in a given time.

FIGURE 95.

Pieces of lead, placed between the hook and pipe as prestated, save this to a certain extent, but do not prevent the evil. Pipe-hooks are so narrow in the shank and hook that they cut the pipe more than wall-hooks.

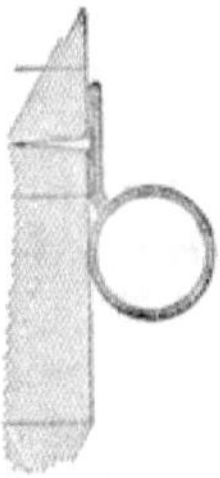

For fixing large-sized (say 4-inch soil) pipes, the usual way is to solder on "tacks," that is, pieces of lead, which should be thicker (rather than less in substance) than the soil pipe. These tacks are generally 9 inches square; one edge is soldered to the pipe, as shown in section, Figure 96, and then

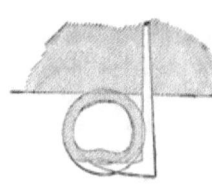

FIGURE 96.

wall-hooks are driven through the tack into the brick wall, and the other half of the tack folded back to protect the hook and also hide the ragged appearance. These hooks must be driven into the *horizontal* joints of the brickwork—if driven into the vertical ones the mortar would crumble beneath them when the weight comes to bear, and so let the pipe gradually slide downward. The horizontal joints in brickwork are generally about 3 inches apart, and as the tacks are 9 inches long, it will be seen that only two wall-hooks can be driven in to be of any advantage; if a third one were used, as at A, Figure 97, a very slight settlement of the pipe would cause it to be useless by the lead tearing away. Lead tacks should never be more than 3 feet 4 inches apart from centre to centre, but it is much better to have four of these tacks to each 10-foot length of pipe. In one case a stack of soil pipe, with the tacks 5 feet apart, had gradually slid downward and broken in several places. The work had only been done about ten years, but when we come to

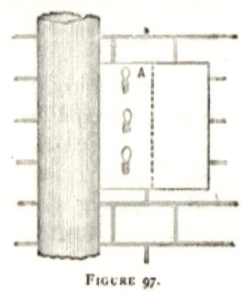

FIGURE 97.

think that three-quarters-of-a-hundredweight of lead (the pipe was 4-inch diameter, eight-pound lead) was suspended on four wall-hooks, the mystery is that it lasted so long.

Sometimes, where a lead soil pipe is fixed on an external face of a wall, the tacks are made of heavier lead, say nine or ten pounds per foot super, instead of the same substance as the pipe, and instead of common wall-hooks, specially-made nails, with flat stems and large heads, are driven in to support the pipe; instead of the tack being folded back the edges are trimmed, as shown in Figure 98, or to any other design that may be ordered. In this case the tacks should be put on in pairs for the sake of appearances, but where this is of no importance, it is better to fix them on alternate sides, and at equal distances apart. When tacks are soldered on in

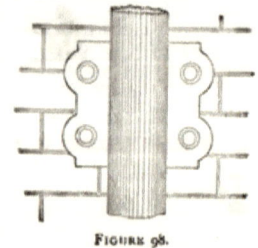

FIGURE 98.

pairs they are generally both done together, so that the seam includes a portion of the pipe and the two edges of the tacks. This is not a good plan, as the thickness of solder keeps the pipe from fitting closely to the wall, and the piece of pipe is liable to be torn out, so that really the strength of the fixing is only equal to one tack instead of two. Pairs of tacks should always be soldered on separately, as shown in section at Figure 99, so that

FIGURE 99.

there are two portions of the pipe soldered to instead of one. The tacks are generally "wiped" on to the pipe, and if the plumber uses a thin cloth, he generally wipes the edges of the seam so bare that the lead shows through. When this is done it is weak, as the lead, being reduced in substance with the shave-hook, should have those parts well covered with solder to make up the substance. It is a much better plan to *float* the seam by pouring heated solder on to the parts and about an inch on each side of the shaving until a heat is got up, and then have an iron heated and well cleaned from scales, and slowly draw it backward and forward over the solder until it flows and the surface is flat, as shown at A, Figure 100. If the solder is

FIGURE 100.

too hot, it will run off the seam. The edges should be made straight so as not to require trimming. After doing this, pause for a few seconds, and then smartly draw a knife across the ends of the soldering—while still hot—to cut them straight, and then pass the end of a rule or thin straight-edge under the tack, close to the pipe, to remove any spurs of solder that may have run through; have a sponge of clean water ready and swab the seam. All this should be done in much less time than it takes to describe. When one tack has been done and the spare solder removed, have a piece of brown paper ready with one edge pasted and stick it on the soiled part between the two seams — not on the solder, as it would make it look dirty—and then float on the other tack in the same way. If all this is quickly done, the two solderings can be made at one heat, although it may be necessary sometimes to have another iron hot and ready, as the first one

PIPE FIXING.   99

would have become too cool to cause the metal to flow nicely. On account of its appearance, this way of fixing pipes is very suitable for work when it is in sight, but when they are going to be fixed inside a building, and in a chase or other position where other pipes are run, or inside a casing where the projecting ears or tacks with the large-headed nails would take up too much room, it is an advantage to use what are commonly called "face-tacks." These are narrow tacks soldered on the face instead of on the back side, and a considerable amount of solder left on for the sake of strength, so that the tack will not tear where the nail is driven in, as is so common with the ordinary lead-tack. The solder being a harder material than lead it will withstand a greater strain. Figures 101 and 102 are respectively

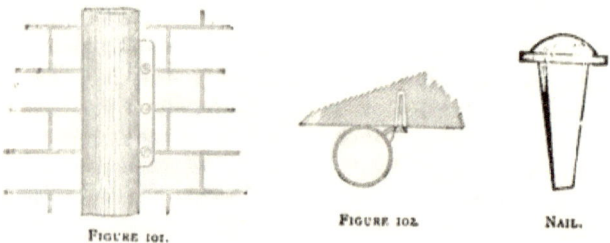

FIGURE 101.   FIGURE 102.   NAIL.

elevation and section of the kind of tack described. Some masters object to this kind of tack, because they think it takes more solder than the ordinary kind; but if properly done it does not take much more, and the saving in the size of the piece of lead will repay the value of the solder. When this kind of fixing is adopted, the ordinary nails used for fixing iron rain-water pipes answer the purpose very well, but if galvanized they are much superior and look better; they also last longer than plain iron ones, as those unprotected rust through where they are in actual contact with any lead; but this is only in the presence of moisture, or when not protected from rain. Sometimes iron nails with lead heads cast upon them are used, but they are not to be depended upon for any great length of time, as they soon rust through close to the head, so that it falls off. Where expense is not to be considered, copper or good gun-metal nails are to be depended upon, as these metals resist the influences of the atmosphere to a much greater extent than iron.

Almost all kinds of lead pipes can be fixed by face-tacks, such as *cold-water* waste pipes, main and ordinary service, and warning pipes. Some architects who find time to give a little thought to the plumbing work of a building, will have a chase or recess built in the structure in which to fix pipes, but they are very rarely made large enough, and more frequently so small that the pipes have to be fixed one in front of the other, with the result that those in front have no fixing at all. Perhaps the chase is crowded with lead pipes in this manner, and then the gas-fitter tries to find room for his mains, there is no other way for the gas pipes to go, and the fitter thinks that no harm is done if he bruises the soil pipe a bit to make room for them, or there may be just room for the pipe, but to get his tongs in to screw it up he has to make room by bruising some other pipe. Sometimes the iron hot-water circulation pipes have to be squeezed in as well, and, in addition to the evils enumerated above, these pipes so expand and contract that the sockets, rubbing and chafing against their neighbours, the lead pipes, soon wear a hole in one or other of them. If these hot pipes are fixed at the ends so that they cannot expand, they bulge sideways, and so indent the other pipes.

In spite of the chase being so crowded with pipes, it is not at all uncommon for the tenant of the house to find that a few speaking-tubes would be a convenience, and as there is no other way of taking them without destroying the decorations of the staircase, they have to go into the chase with the other pipes. Very probably it is found that more bells are required, either electric, pneumatic, or ordinary wire-hung bells, and each tradesman has to exercise his ingenuity to the utmost to find room for his fittings, all of them in this already crowded recess. After a period of time, perhaps, a leakage of water is found to be taking place somewhere in this chase, and its position cannot be seen, but it is so serious that it must be found and put to rights. The pipes that can be seen appear to be all right, therefore it must be one of those at the back. The plumber begins by cutting all the bell wires or tubes, and partly taking down the speaking tube and gas main, and then perhaps he can see enough to know that the leakage is taking place at a higher level, and so he goes up to the next floor and possibly succeeds in finding the actual position, but cannot get at it on account of the hot-water pipes. Of course he cannot disturb these until he has put out the kitchen fire and emptied the boiler or water-back. He then takes down the circulation pipes, and perhaps has to cut one or two lead service pipes before he can gain access to repair the one that is defective. The mistress of the house complains, the cook grumbles, and the master, in a towering rage, sends the poor plumber home in disgrace because he made so much fuss about repairing a broken pipe, which the master of the house, in his ignorance, says took only about an hour to do.

My readers may think this a rather highly-coloured illustration, but it is a fact. Some architects, who perhaps may have noticed this state of things, will have a good wide recess made to receive the work required, and also leave a margin for any pipes that may be found necessary at a future time, and so that the hot-water pipes may not have any effect on the lead pipes by being too close to them, which not only injures the lead, but perhaps warms the water in the pipes so that it is unpleasant for drinking purposes. In addition to the necessity of a wide chase, so that all pipes may be fixed side by side where they can be readily accessible for repairs, it is a good plan to line the back of the chase, either by covering the brickwork with cement and making the face even so as not to bruise the back sides of the pipes, and with wooden blocks at intervals for fastening them to, or else to fix boards over the whole of the surface (except where the soil pipe is to be fixed). Now, if the soil pipe had face-tacks soldered on and fixed with nails as described, the other pipes, which are generally smaller in size, could have face-tacks too, but they should be fixed to the wood back or blocks, according as the chase was prepared, with strong inch screws, so that the pipe could not slip downward or move in any way. By taking out the screws the pipe could be pulled forward for making any necessary repairs, or, if found necessary, a branch pipe made good to it, without the trouble of moving any of the other pipes. The hot-water pipes are useful in cold countries to prevent frost affecting the cold-water pipes, but they should on no account be allowed to be in actual contact with any lead or other pipe if used for conveying drinking-water. When pipes are fixed separately, it is much easier to protect them with felt, slag-wool, or any other substitute, from the effects of frost, as, if the back wall was boarded, they need only to be covered on the front side, and long strips of wood, with longish screws used to fasten the material as close to the pipe as possible.

CHAPTER XI.

RAIN-WATER PIPES.

A GREAT many architects are in favour of lead rain-water pipes, as when well fixed they last so much longer and do not split in the same manner by the effect of frost as iron ones. Iron ones also have their sockets or hubs frequently burst by oxidation; any water running down the outside of the pipe, enters the sockets and causes the inside to rust; this rust or oxidation swells with such force as to cause the iron socket to break. Sometimes a piece falls off and leaves a hole through which rain-water can escape and soak into the walls of the house, and very often where these walls are of porous stone or of brick, the decorations on the interior face are destroyed. Where wooden skirtings are used they become rotten, and any fittings or shelves in the vicinity have their contents brought into a mouldy, and very often decayed condition. Another advantage of leaden rain-water pipes is, that if a length should become injured in any way, it can be easily taken out, or repaired in its place, while an iron one cannot be repaired or a new length properly fixed without taking down some of the other pipes, unless it happens to be the bottom length. Iron pipes periodically want painting to protect them from oxidation, but this is never done on the inside—where it wants it most, nor yet on the back side, as the painter cannot get his brush behind. Lead pipes do not require any painting to protect them, and have a certain amount of value attached to them when old and past repair. There are thousands of stacks of leaden rain-water pipes fixed in London, mostly to houses that have been built a great many years, and when some of these houses undergo repairing, the best architects generally leave the old leaden pipes which are in fair condition, and perhaps take out one length at the bottom and fix iron, as this part often gets battered about and damaged by being knocked with brooms or passing objects. A great many new houses or buildings of a high-class description have leaden stack pipes; those lower in the scale have iron ones; and those very cheaply built mostly have zinc of about No. 9 gauge.

A great many of these old leaden rain-water pipes are fixed with iron wall-hooks in such a way that the fastenings cannot be seen. The lengths of these pipes are about 6 feet long, and the joints are the end of the upper pipe socketed into that below; consequently, there is no solder used in any way excepting to the seam when the pipe was made.

This fixing, Figure 103, is mostly applied when the pipe is in an angle of the building, and when there is not an angle convenient, a chase is left in the brickwork for the pipe to fit into,

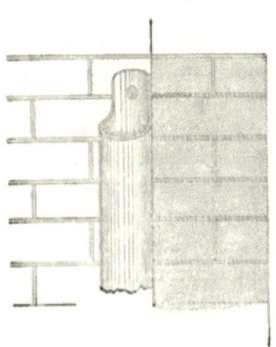

FIGURE 103. ELEVATION.   PLAN ENLARGED.

so as to be flush with the face. By doing this, more fixings, in the way of wall-hooks, could be driven in than if it were fixed on the face of the wall, in which case the whole weight of the length of pipes would have to be sustained on one hook, whereas, in an angle, two could be used, and when fixed in a chase, it is possible to get three in. These pipes sometimes have astragals or ornaments at the joints. A piece is cut out of the front, leaving a tag on the back side for fixing, as illustrated.

Another way of fixing leaden rain-water pipes was, and is, to solder tacks on the sides of the top or socket end. Then saw a piece of ½-inch or ¾-inch lead pipe longitudinally, and fold it around the larger pipe, and solder them together with fine solder and copper-bit; or sometimes these astragals are cast solid and put on. The plumber will often himself design a small neat moulding, and make the flask for casting these astragals. He generally proceeds first to paste a section of his design, drawn on paper, on a small piece of sheet zinc, which he then cuts so that the edge represents the required shape, as shown in Figure 104; this is then nailed upon a piece of thin board, cut to the same shape, but a little larger, and with another small, narrow piece of board nailed or screwed on at right angles. The plumber then gets a smooth piece of board about 18 inches long and 7 inches wide, and on one edge nails

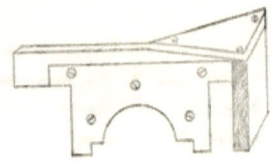

FIGURE 104.

a strip of wood for his mould to run against. The next thing is to mix some plaster of Paris and water to the proper consistency and lay it on the board, which should be previously wetted, and then, before the plaster sets, to pass the prepared zinc mould over it, so as to remove all, excepting a portion which will be found to be the required shape, as shown in sketch, Figure 105.

This should then be placed in a dry situation for the plaster to set and any moisture to evaporate. When thoroughly dry, block the ends with a piece of well-soiled lead backed up with sand. The sides will have been formed with plaster, the zinc mould being cut so as to allow for this. When all is ready, pour melted lead

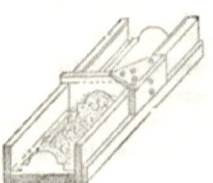

FIGURE 105.

on as quickly as possible, until it overflows the sides; care should be taken to level the plaster mould so that the lead may be the same thickness at each end. When the lead is set it can be taken off and the plaster will be left adhering to the board. The outer edges of this half flask should be made perfectly straight and smooth, and three or four conical-shaped holes made in the faces, outside the moulded sinking. This sinking should then be filled flush with plaster, or fine loamy sand, slightly moistened; when this is done and the lead covered with a coating of thin soil, place a band of well-soiled sheet-lead around the edges, well back it up with wet sand, and then pour on melted lead to a depth of not less than 1 inch. In all cases the lead requires to be poured on quickly, or the flask will be flaky, and before it has been used many times will become so rough that what is cast in it will require trimming and cleaning up before it can be used.

If these lead astragals or mouldings are made to a large size they should be cast hollow on the back. To do this, the first half of the flask should have the moulded sinking filled with plaster, and before it sets a portion should be removed by a piece of zinc cut to the required shape and fastened upon a small piece of wood, in the same manner as described before. Its shape will be as sketch, Figure 106. When the plaster is perfectly dry, prepare and cast the other half of the flask, as before described. Separate the two pieces of lead, remove the plaster, well soil the lead, fit them together, and fasten with clamps or hand-screws. If these are not at hand have some small pieces of board, notched to fit loosely on the edges of the lead flask, and wood wedges cut. These will answer the purpose almost as well as the screw clamps. Cut a pouring-hole in one end of the lead flask, allowing space enough for the air to pass out as the lead is poured in.

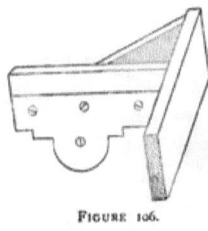

FIGURE 106.

When all is prepared, stand the flask on its end and pour molten lead in at the other end, taking care not to have it too hot so as to melt the flask. Some men prefer to have pouring-holes at the edges of the flask, but endways is best, and the casting better and not so flaky. These mouldings are simply bent around the pipe and soldered with copper-bit or blow-pipe lamp and fine solder.

If only a few astragals are needed, a carpenter can make a pattern of wood with which to make a print in moistened loam or fine sand, and then pour melted lead into it. By this means it is difficult to make the casting true, as it requires trimming and filing to make it an equal thickness its whole length. Sometimes a lead socket with ears and astragals is cast all in one piece and soldered to the pipe afterward. Figure 107 is an example of one for square pipe, but this is a trouble to cast, as the flask has to be in six pieces or it cannot be drawn away from the casting. The easiest way for an ordinary plumber to make it, is to cast a length and then niche and solder it together, and cast the two ears with a band all in one piece, to pass behind the pipe and solder on afterward. This makes it very much stronger, but as the weight of the pipe will sometimes cause the ears to bend, it is a good plan to cut off about three inches of the pipe on the front and two sides, and then flange over the piece left on the back and turn it into the wall and fasten with lead wedges and Portland cement.

FIGURE 107.

It very often occurs when a cast-iron rain-water pipe is fixed to a building, that a bend is required to a different length to any iron ones kept in stock, and as a special-made bend would be expensive, it being necessary to make pattern and core-box for it, it generally falls to the plumber to make one out of lead pipe. After making the bend, the plumber generally folds a piece of sheet-lead around the socket of one of the iron pipes and works it into the mouldings, so that when this dummy socket is made and soldered on the lead bend, it matches the iron in shape, and when the whole is painted, the difference in appearance is so slight as not to be noticeable. The ears can be made by cutting out a piece of wood with which to make a print in sand, and then run melted lead into it. These ears should be very strongly soldered to the bend, as when iron pipes are fixed so that the spigot end is tight on the bottom of the socket or hub, any expansion that takes place will force the bend downward, and perhaps break it away from its fixings.

CHAPTER XII.

LINING SINKS AND CISTERNS.

ORDINARY-SIZED cisterns are generally lined with six-pound lead sides and seven-pound bottoms, but for best work and large cisterns, seven and eight-pound is very often used. Small-sized cisterns are usually lined with lead of the same substance throughout, and in one piece, with the angle pieces cut out. The plumber generally opens out his piece of lead and carefully takes out the creases with a leaden flapper, so as not to bruise or indent it; he then marks out the part for the bottom, and then the sides and ends, allowing in all cases for the thickness of lead, so that it will not fit too tight in the wooden case or shell. At each angle is left about ½ or ¾-inch on the side that will form the under-cloak, as shown by dotted lines, A, A, Figure 108, and corners are left on, as at B, B, so that when the lead is turned over the top edge of the cistern they shall meet in a mitre.

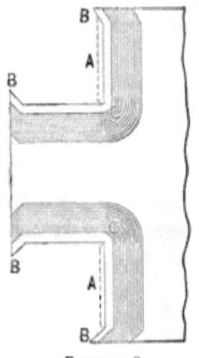

FIGURE 108.

The lead is now soiled about 3 or 4 inches wide around the parts to be soldered, and when dry they are shaved either with a gauge-hook or an ordinary shave-hook and straight-edge, the part at C being shaved with the hook-compasses. Great care should be taken, when shaving the straight parts, not to dig in the point of the hook so as to unnecessarily reduce the substance of the lead, and neither should the shaving be too wide, in which case the lead would not have the parts which are thinned by the hook strengthened by the solder. Some men think that by shaving wide angles it makes them appear to be stronger, whereas the reverse is the case, as it is very rarely that

LINING SINKS AND CISTERNS.

the solder is left full up to the edge of the shaving, as was explained in one of the chapters on "Joints."

Figure 109 shows a section across the soldered angle of a cistern, showing the shaved parts filled up with solder right up to

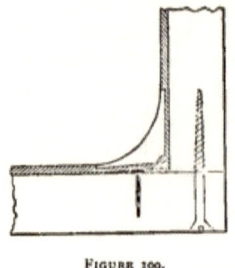

FIGURE 109.

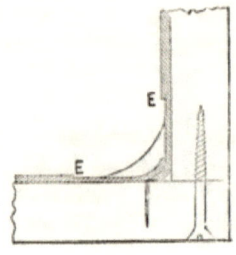

FIGURE 110.

the edge, and Figure 110 the common way and the evils complained of at the parts marked E, E.

Supposing the piece of lead to be prepared, soiled, and shaved, as described; the piece to form the undercloak at the angles should have its outer edge rasped down to a feather-edge, so that the overcloak part may lie quite smooth over it. If this was not done it would be very difficult to wipe the angle nice and true and leave the solder of a proper substance at the edges. These undercloak pieces should have a sharp-pointed hook run where the edge is to be folded, so that they will fold up without any dressing or knocking about to dirty the shaving or scratch the soiling, which would necessitate its being done over again. The sides should next be folded upward and the angles around the bottom "set in" with the edge of a dresser, but not too sharp, and only sufficient to crease the lead in straight lines. These angles should then be dressed well on the outside; if this is properly done the lead will not have become reduced in substance, whereas an indifferent tradesman would have set in these angles so sharp with his dresser as to nearly cut the lead through.

The bottom should now be bellied or hollowed from the underside; the flat of the hand is quite sufficient for this, and the use of tools should be avoided as far as possible. The sides should be bellied inward in the same manner. The result of this bellying is to make the lead, when turned up, rather smaller than the wood case it is going into, so that it will drop into its position without any dressing; if the lead should be rather tight, by lifting up one end of the cistern a few inches off the floor or bench and allowing it to drop, and then the other end in a similar way, the lead will generally shake down to the bottom. The sides with the undercloaks attached to them should have their edges pressed into the angles of the wooden case, and then the bellied part pressed back with a smooth piece of board; the other two sides should be treated in the same manner, and the result will be that the lead is home in the angles without any dressing or working of any kind. A piece of board should be laid in the bottom, and the plumber should stand upon it to press down the bellying of that part and so force the angles of the lead back into their proper position tight up to the woodwork. An iron punch or a blunt chisel can be used with advantage about every 3 or 4 inches in the angles of the parts to be soldered, so as to punch in a series of little spurs to prevent the angle opening by the expansion of the lead when the solder is poured on. About 2 or 3 inches from the top end of these soldered angles, a copper nail with a tinned head should be driven in, but only one to each angle, so as to fasten down the lead and keep it from rising off the bottom when the top edge is worked over on the woodwork. Some men will run a knife or the point of a shave-hook on the back side of the lead, and on a level with h top edge of the wood cistern, so that the lead will fold over and leave a straight and sharp arris without much dressing, and although when this is done the work looks smart and free from tool marks, still it should be condemned, as it will not be so well able to support the sides of the cistern, which, to a certain extent, are suspended from the top. It is a common practice to have the top edge of the cistern dished or hollowed at the angles, so that the lead can be dressed into it, and then where it meets in a niche it can be soldered "flush."

The cistern should now be placed on its side or end, ready for wiping, so that the undercloak piece stands upright, for if this lies flat, it sometimes happens that if the lead does not fit tightly in the angle, solder will get underneath and so get wasted, in addition to making the surface of the lead so uneven that the wiping looks irregular and patchy, for a plumber can no more wipe an angle straight if the sides are not flat than a plasterer could run a straight cornice or moulding if his running rules were not straight.

It is always best *not* to "touch" or grease the angles of a cistern until after it is lined, as there are always dust and bits of rubbish about that stick to it, and when it is "touched," some should be rubbed on the soiled portions as well, as the solder does not stick to it so much, and the metal works freer. Good thick cloths should be used for wiping cisterns; a thin one is sometimes useful to get away from the start, as a thick one would leave a great lump in the corner; but a good thick one is best for straight-away work—so thick that only one finger is necessary, and so that pressure can be applied to it. A thin cloth gets so hot as to burn the hand; it also requires two, and sometimes three, fingers behind it, so as to press on the edges of the soldering and so wipe them clean and avoid the necessity of trimming them afterward with a pocket-knife at the risk of cutting half through the lead as well; and, speaking from experience, when using a thin cloth, sometimes more pressure is applied by one finger than by another, and the soldering is not of equal substance throughout, or on one edge the lead shows and on the other an extra thickness of solder is left. When three fingers are used with a thin cloth it is not at all uncommon for the little finger of the wiping hand to keep touching the spare solder at the side of the wiping, and so get burnt, and sometimes a piece will get under the finger nail, causing great pain. Care should be taken to keep the lead *pressed* tight back to the woodwork when soldering, for as soon as the hot solder is poured, or splashed, on, the lead expands and bulges out, and this must be pushed, not dressed, back before commencing wiping, as if done afterward, when the metal is partly set and in a brittle condition, it frequently cracks, and so fine as to be invisible. Some men will have a small piece of wood, about 8 or 10 inches long, to press back the sides, but others push it with their cloths, and as the lead is very soft from being heated, it generally goes back very easily. The process of wiping is very difficult to describe; five minutes' practice would be worth more than pages of writing about it.

The points to be attended to are to properly prepare everything as described, taking care that the shaving is perfect, and to have a good heat, but not hot enough to burn holes in the lead for the solder to run through. The irons should be hot, but not so as to burn the solder and convert it into dross. Pour on sufficient solder to go a few feet, but not too much, as that which is behind the worker gets cold, and so has to be heated up with the iron. As soon as a heat is got up, begin wiping while the solder is in a semi-fluid condition, and don't play with it until the heat is almost lost, or until the metal is half cold, so that it pulls apart or perhaps away from the lead. A good plumber will keep his iron at work heating up the solder at one part while he is wiping the portion already heated. A good many men, in trying to do this, have their attention so riveted on their wiping hand that they forget to use the iron, and the result is that their angles look patchy. Others, again, will keep their iron buried in the solder to keep it heated, but forget to keep moving it about, and the result is that it often burns a hole through the lead in such a way that perhaps he has to stop soldering until he has shaved around the hole so as to solder it over.

A new beginner at cistern soldering generally wastes so much time at patting and heating his solder, that he gets part of it so hot that it runs away from the angle and perhaps beyond the soiling, so that it tins on the unprotected lead, while other portions are almost set, or at all events too cold for wiping. The result is that he only wipes a short portion at a time. While he is playing with the solder, the heated portion of the lead is extending more and more, and as fast as he pushes back the bulged parts they come out again, and at last it perhaps takes him longer to keep the lead back than it does to wipe the angle. It need scarcely be added that the work, when done, is very dirty and has a very ragged appearance, whereas a good plumber quickly gets up his heat and wipes, perhaps, 1 foot or 1 foot 6 inches at one stroke, and so keeps on, for the quicker it is done the cleaner it looks, and the less the sides bulge inward.

We will now proceed to describe how to line a cistern with the bottom and sides in one piece, and the two ends soldered in afterward.

LINING SINKS AND CISTERNS.

Cut out the lead to the proper size and allow for turning over on the top edge, also the angle-pieces on the corners at A, A, Figure 111, and the ½-inch for the undercloak. Soil, shave, and drop it into the wooden case, as previously described. Put a few clout-nails through the undercloak to hold it in its place, but drive them well in so that the heads do not stand up beyond the surface of the lead, and so make bumps which will show through the next piece of lead which laps over them.

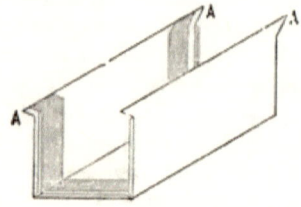

FIGURE 111.

Next cut out the pieces of lead for the ends, not forgetting the top angle-pieces. This piece of lead should be ¼-inch wider than the place it is to fit, so that it pinches tightly into its place, the centre part being bellied upward to insure this, as shown in Figure 112. It is a good plan to slightly bevel the edges by rasping, so that if it should fit too tight they will give a little, and so avoid the necessity of *dressing* the bellied part in to get rid of the superfluous lead. When cutting out lead for these cisterns it is always best to get a wooden lath and cut it to the length to fit in the cistern, and mark the lead from that, so as to insure it being the size required; for if the lead is too small, the solder will run through, and if too large, the dresser has to be used at the risk of spoiling the shaving and soiling, as before mentioned.

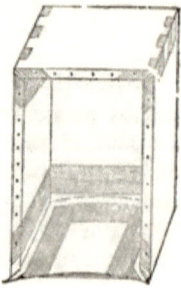

FIGURE 112.

For larger-sized cisterns the lead is generally cut out so that an end and side are in one piece, and the bottom is put in by itself. According to principles that have been laid down, the bottom should be put in first and the edges turned up all around, so that the solder cannot by any means run through, but in practice this is very rarely done, as the workmen's feet, and tools laying about, scratch the lead, and, in addition, it is not so easy to get the sides in afterward. One reason is that lead will not slide on lead, and another is that the bottom edge of the piece forming the side and end wants stiffening by turning a small portion, for it must be remembered that the whole weight of the piece has to be supported by the bottom edge, and if this is not looked to, perhaps one end will drop a little, and so throw the whole piece out of square with the cistern-case, with the result that the lead angle will not be tight home.

To line a cistern in this manner, cut out the lead, leaving corner pieces on the top as before described, and 1 inch extra in length for each angle, and also enough for laying on the bottom and turning on the top edge. These surplus pieces for the ends and bottom should be turned up and dressed perfectly straight, by placing a piece of timber inside to insure this; the piece of lead should then be folded to fit the upright angle, and "set in" lightly, and then well dressed on the outside to a moderately sharp arris. The piece of lead should then be folded two or three times for convenience of removal, care being taken not to spoil the part which is prepared to fit the angle, and the inch margin on the bottom edge which is turned up should be doubled inward, not outward. The piece should then be carried and placed in the cistern-case and partly unfolded, so that the angle part can be placed in its proper position. Next proceed to unfold the sides and to belly them the reverse way, so that the ends can be forced back into their respective angles. Drive a clout-nail in temporarily—*i.e.*, only half way in—to hold one angle until the other one is treated in the same manner. Force back the bellied parts a little, and then look to see that the angle is tight home; if not, pinch the ends of the lead toward it with a chisel, and then drive in a part of the small return pieces. Re-nail, temporarily, the ends, and force back the remainder of the bellied parts, and at the same time dress down the buckles on the bottom edge. Thin the outer edge of the bottom return with a shave-hook, and drive in clout-nails about a foot apart, to keep it back. Serve the ends in the same manner, and then use a leaden flapper to take out all bulgings or irregularities; work over the top edge and nail it. Prepare the other side and end in the same way, and fix it in its position as described for the other; but in this case the return pieces on the ends should be cut off, and only leave about $\frac{1}{8}$-inch, which can be driven in afterward, so as not to project far enough to show through the soldering if it should happen to be wiped too bare by using a thin cloth. Put a nail in temporarily, near the top of each soldered angle, to keep it from rising when the top edge is being worked over, after which it should be taken out again, so that the lead can be properly shaved.

When nails are used for keeping the lead tight in the angle when soldering, they should be warmed first, as, if damp, they *blow* and leave a hole through the solder. It is scarcely necessary to add, the angles should be shaved before the nailing is done. The writer never uses nails for holding the lead back, excepting for the undercloak. After the sides have been placed in their position, the bottom piece of lead should be cut a little larger than the cistern, and the edges curled up, so that the lead will drop down into its position; or, if the piece is too large to move about when open, it can be folded up, and the edges which are curled up buckled inward, and afterward unfolded in its position; the edges must be worked tight against the sides—in fact a sharp-edged dresser and chase-wedge can be used so as to almost cut the lead through; the surplus should then be trimmed off. When all is ready, and the whole of the lead *flapped* nice and smooth, measure the requisite distances at various points, and then use a chalk line to mark the space for soiling; this is generally done with a pair of pointed compasses, with the result that the lead is scratched and sometimes partly cut. When the soil is dry, shave one upright angle and about a foot each way on the bottom; punch the corners at intervals instead of using nails, and touch and solder this angle before preparing any more, as the sooner the soldering is done after shaving the better.

It is a good plan to keep a shave-hook especially for cistern angles, one just the right size, so that it will take off a shaving the proper width, without the trouble of measuring and gauging or using a straight-edge. This shave-hook should not have a very sharp point, but one slightly rounded, as Figure 56. For getting into the corners where the ordinary straight hook will not reach, some men will

FIGURE 113.

use a bent hook, as shown in Figure 113, while others prefer what is called a spoon-hook, as illustrated in Figure 114. When soldering the bottom, great care should be taken not to move about more than can be helped, especially in a small-sized cistern, as great risk is run of breaking the soldered parts before they are thoroughly set. This sometimes happens, and the crack is so fine as to escape observation. When the bottom is being wiped there is sometimes a little trouble when passing the soldering of the upright angles. Some plumbers will paste a piece of brown paper over them, while others will only smear them with chalk to prevent the bottom solder from tinning too high up; others can get away very well without doing either, by simply being careful not to splash the solder too high up the angles, and not getting too much of a body so as to be unmanageable and beyond control.

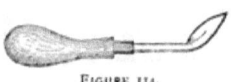

FIGURE 114.

CHAPTER XIII.

LINING SINKS.

SINKS of all kinds are made in such quantities and of such a variety of materials, that one would think the plumber would never be called upon to make one. Some of these sinks are made of galvanized wrought-iron, but these are not much liked, as dirt accumulates around the rivet-heads, and so they are difficult to keep clean. Enamelled-iron ones are very good until the enamel gets chipped, but they are unfit for wash-up purposes, as, being very hard, they are destructive to crockeryware and any other frail articles that may be cleaned in them. Vitrified stoneware and potteryware sinks are easily kept clean, but are very liable to get chipped, or perhaps broken, and are quite as ruinous to glass or chinaware as the other kinds of sinks described. These sinks are also very difficult to get true in shape, as they warp and twist very much in drying and burning. The sinks above described can only be had in what are called stock sizes, and if any other is wanted the buyer has to pay for the moulds and patterns, and also has to wait for several weeks until they are made. The manufacturer generally makes two, although only one may be ordered, as there is always a probability of one of them being unfit for use, and in which case, if that precaution was not taken, an interval of several more weeks would elapse before another one could be made, and attended with the same risks as for the first. Wooden sinks soon wear out, although they are the least destructive as wash-up sinks. Sinks made of slate slabs break by expansion when hot water is used in them. After looking all around, the plumber need not have any fear of these rivals, for it is almost impossible to have anything better than a good lead-lined wooden sink. Any plumber can make them, and to any size and shape; they are not destructive to what may be washed in them, can be made at once, and easily repaired when required, or the lead cut out and new put in when the old is past repair. The old material is a valuable set-off against the cost of the new. The only disadvantage they labour under is that hot water soon causes the lead to rise in buckles and eventually to crack and break, but this can be avoided by taking precautions which will be described. Sinks, as a rule, are lined in the same manner as cisterns, and generally about the same substance of lead is used. This is mistake number one, and is a false economy. If anyone would take the trouble to calculate the quantity of lead required for a sink, they would see at once that a few shillings or a dollar more would be in some cases sufficient to perhaps pay for lead double the substance, so that it would wear longer. As a rule it will be found that the ends and back side of the sink need not be quite so thick as the bottom and front side, but at the same time, as an economy, the four sides and bottom are sometimes made out of one piece of lead, so that the four corners only need be soldered, thus saving that material. This way of lining a sink has been described when speaking of cisterns. Another mistake often made is when the wooden case has its sides and ends at right angles to the bottom. When this is done, the lead lining is so fixed that it cannot expand when filled with hot water, so that the bottom bulges upward, and eventually a ridge is formed. As this ridge stands up a little it gets worn very much, and gradually gets worse and worse until it breaks. When this occurs, it is generally soldered over, leaving an unsightly-looking patch, very few plumbers taking the trouble to pull up the lead and dish out the woodwork beneath, and then dress the lead into it, so that it can be soldered flush with the surrounding parts. Sinks should always be repaired in this manner. It is advisable, before dressing down the lead into the dishing, to place two or three thicknesses of paper, so that when soldering the place the heat may not convert the water in the wood-work into steam, which would keep "blowing" through the solder. The plumber could also make a neater and quicker wipe, as the metal would not chill so quickly by the wet and cold beneath.

If sinks were made with sloping sides, as shown in illustration,

Figure 115, the lead would not be held so tightly, and could expand and contract without such a risk of breaking, and would consequently last much longer. Of course it is very wrong to put a cistern where it is liable to be affected by frost; but if cisterns so exposed were made as described for the above sink, the expansion of the ice would not force out the sides and disjoint the dovetailed angles, as

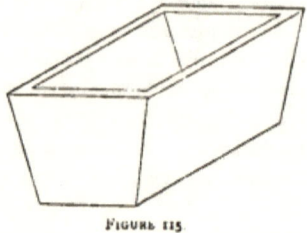

FIGURE 115

is so frequently the case. Another good plan, when lining sinks, is to have a hollowed wooden fillet fixed in each angle, as shown in section, Figure 116. In addition to distributing the effects of expansion, it avoids any sharp angles in which dirt and grease can accumulate beyond the reach of the scrubbing-brush. A great many men are in the habit of not only turning the lead on the top edge, but also down the sides for about 1 or 1½ inch. This is a waste of lead, and it also leaves a dirty black mark on the dress of the person using the sink.

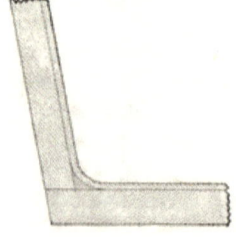

FIGURE 116.

It is generally done on the front side, so as to prevent the lead being knocked and buckled over into the sink, as shown in section at Figure 117, by standing pails, &c., upon the edge; but this can be avoided by putting an oak capping-piece and fixing it with brass screws as shown in section, Figure 118. This capping should be "weathered" inward, so that any water falling upon it will run into the sink. Where the sink is fixed in an enclosure, the capping should extend to the front so as to prevent any water falling down between, and also to hide the joint, which generally gets filled up with grease and other dirt. In hotels, clubs,

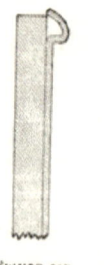

FIGURE 117.

FIGURE 118.

and other large establishments, this capping can be made of galvanized iron, but in private houses the oak looks cleaner and neater.

When soldering brass gratings into sinks, the mistake is generally made of not having the dishing around the waste-hole made deep enough, so that tubs or pails may not bend or injure the grating. This remark also applies to gratings in stone sinks. It is not at all uncommon to see the grating stand higher than the bottom of the sink, so that to get rid of the water it has to be swept into the waste. It may have been all right when first fixed, perhaps, but the bottoms of these sinks generally get torn away by the iron hoops of tubs or pails, or when scouring copper cooking utensils.

When metallic or metal-lined sinks are required for washing-up purposes, it is necessary to have a plug over the waste-pipe instead of a grating, so as to retain the water in the sink. If these plugs, which are ground into what is commonly called the "washer," is not sunk below the level of the surface of the sink bottom, they often get jammed in by pails, tubs, or dishes, so that it is difficult to get them out. When this happens, the usual thing to do is to take hold of the chain and pull until it breaks, or, perhaps, the brass ring by which the chain is connected to the plug. This can be avoided by not only having the washer soldered to the waste low down, as stated, but in addition to have the plug itself hollowed so that the ring can drop down, as shown in section at Figure 119. Another reason why the ground in

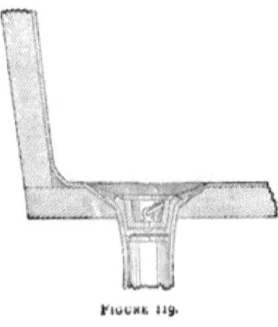

FIGURE 119.

plugs sometimes gets fixed in the washers, is because the *plug when cold* is put into its place when the *washer is hot*, so that it is expanded, and which, on cooling, contracts and grips the plug so tight that in some cases they have had to be unsoldered before they could be got apart. It is always advisable to have wire cross-bars in the washer below the plug, to prevent spoons falling down the pipes and getting lost, or other matters getting into the waste which would choke it up. Some plumbers have gratings instead of the bars, but they generally reduce the water-way, so that the sink takes a long time to empty, or small matters get into the holes and clog them up.

For sink-tubs for washing vegetables, it is a very good plan to have a tinned-copper perforated strainer fixed in the angle to prevent anything getting into the waste pipe or plug-washer; this also prevents the plug from being jammed in.

It is important that all sinks with plug waste pipes should have an *overflow* pipe fixed, and this should be large enough to take away the water as fast as it runs in. Neglect of this precaution has sometimes resulted in serious damage to property, especially when the sink has been fixed in an upper storey.

CHAPTER XIV.

SEWERAGE AND SEWERS.

PUBLIC sewers very rarely have any provision for ventilation excepting those in the streets, commonly called lamp-holes, and the gulley-gratings by the sides of the streets.

In some districts may be seen here and there a small iron pipe, of about 4 inches in diameter, fixed up the side of a building. These pipes are supposed to be quite sufficient to ventilate a sewer of, perhaps, several feet sectional area. The lamp-holes spoken of above, very rarely exceed $\frac{1}{3}$-foot of space for air to pass through, and even that small amount of opening is generally contracted, and in a great many cases choked with street refuse.

The side gulleys are not intended as sewer-ventilators, but anyone, when passing these places, would not require to be told what duties they are performing; in fact, complaints are heard daily about the nauseous smells which escape from them. These

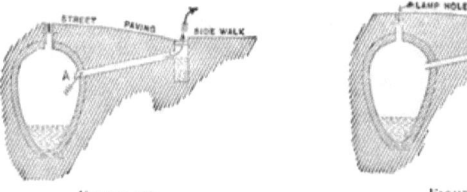

FIGURE 120.   FIGURE 121.

places are constructed as in Figure 120, with a tide-flap fixed at A, but in some districts the silt-box is being taken out, and a gulley-trap, constructed to catch driftings and street washings, is fixed in lieu of same, as shown in Figure 121, on looking at which it will be noticed that as long as the water does not evaporate sufficiently to break the seal, no smells can pass through.

Figure 122 is an enlarged view of this gulley-trap. Figure 123 has a valve-trap in combination with the water-trap. When this is done, a moment's thought will prove that public authorities are themselves aggravating an evil, without providing another source for pent-up air to escape, or stagnant air to be put in circulation. Boxes of charcoal have been tried in ventilators to deodorize the sewer air, but are now discarded as being perfectly useless, for the reason that it is impossible to keep the charcoal dry, it being of no value when in a damp or wet condition; and if packed so closely as to prevent sewer air from escaping without coming under its purifying influence, the free circulation of the air is impeded, or perhaps entirely obstructed.

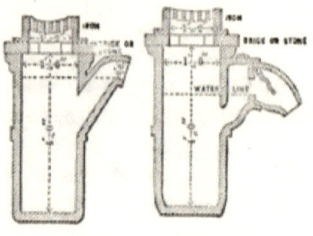

FIGURE 122.   FIGURE 123.

It has been suggested that the street lamp-posts should be adapted so as to act as sewer-ventilators. In some places a stoneware-pipe air-drain has been laid from the sewer and built up in the party-walls of houses, to the no small annoyance of the inmates, who could not tell where the smells came from until the walls were cut away and betrayed the source.

Patents have been taken out for passing sewer air through wire gauze heated with gas jets so as to render it harmless, but, if this is done, it should discharge at some considerable height. As to its being harmless, is open to considerable doubt. It has been proposed to have connections from sewers to any high factory chimney-shaft that may be near, and no doubt this would be a good plan, but these shafts are so far apart that they would be of about as much use as trying to empty the sea with a pail.

This ventilation of public sewers is rather beyond plumbers' work, but at the same time it has a great deal to do with the success of the sanitary arrangements of houses, and in more ways than one. For instance, in London the rainfall is discharged into the same sewer as the sewage proper. The result is, after a heavy fall of rain the sewers are so fully charged as to reduce the air-space, and if no vent is provided for this compressed air, it will force its way back up the tributary drains, which convey sewage from the houses, and possibly gain access to the habitation to the injury of the inmates by breaking through the water-seal of the traps; for it must be remembered there are thousands of houses that have no ventilation pipes to the soil pipes or drains, or any other protection beyond a trap under the water-closet or other fitting, although, in some instances, the pipes which conduct rain-water from the roofs have been connected directly to the drain and so act as vents. In some cases where the sewers are not efficiently ventilated, when the water subsides after a storm, the seal of any of the traps in the house can be broken by the water being syphoned or forced out, by the air rushing through to fill what would otherwise be a partial vacuum in the sewers.

Again, in a badly-ventilated sewer the gases become more concentrated, and putrefaction of the contents takes place more rapidly, and even if these gases do not actually force themselves through the water in the traps (which some people maintain they do) still *the water* becomes so impregnated as to be converted into sewage, and so give off foul emanations inside the house.

These sewage gases, or sewer air as some prefer to call it, seem to play a very peculiar part. Hundreds of people die from the effects of breathing them—so the doctors tell us; but as a fact, the writer has questioned scores of sewer-men, some of whom had worked in sewers for more than twenty years, and not one had ever had any illness or suffered from their effects in any way. On asking a leading question as to what complaint these men suffer from most, the answer is almost invariably "rheumatism." The writer has been down into a great many sewers, but never felt any ill effects, although two or three times he has had slight diarrhœa and a sickly feeling when working near an open drain, or when removing dirty old fittings, &c. Dr. Richardson, in a lecture at the Parkes Museum of Hygiene, made the remark on this question that "he supposed they (the sewer-men) got so used to it that they were not influenced by these emanations, and cited a case where he went near some men who were removing refuse from which sulphuretted hydrogen was escaping, and which made him vomit and gave him diarrhœa. These men only laughed, and said it was *rather warm*, and went on with their work as if it was nothing unusual to them." As doctors make the statement that these odours from sewers do make people ill, and in a great number of cases cause death, it behoves plumbers to take every precaution that human ingenuity can invent for keeping them out of dwellings or places where they may be inhaled. Dr. Shirley Murphy told the writer some time ago that the improvements made in sanitation during the last few years had been the cause of adding quite two years to the average period of life.

Dr. Lyon Playfair made the remark in the House of Commons on the 4th of March, that "in one generation the span of human life had increased by two years." This was on the question of improved dwellings for the poor and the influence it had upon health.

At a meeting of the Executive Committee of the Parkes Museum, in 1880, Sir William Jenner made the statement that "the sewers of this metropolis are always filled with the poison of typhoid fever, with diphtheria, and a series of other terrible maladies."

Coming from so high an authority we cannot do otherwise than accept this as a truism, and do all that can be done to obviate the effects.

The first question to consider is—Where is the starting point for the sanitary plumber? The answer is—If these evils are in the sewers, keep them there in preference to allowing them to pass into the house drains. If they can be got rid of by destruction, or be dispersed in a proper and harmless manner, by all means do so, but keep them away from the dwelling. The proper place to start from is as near the general sewer as possible. There are several contrivances for this purpose, and one of the most in use is what is commonly called the "tide-flap," which is fixed on the extremity of the house drain inside the sewer. This consists of a galvanized-iron plate, suspended on a hinge, or in some cases on two pairs of iron chain links over the end of the pipe, so that it fits as closely as possible. This contrivance is not to be trusted in any way; a match, piece of orange peel, piece of paper, and the thousand and one things that float down the drains, get between the flap and its seating, and so render it useless for keeping back smells, and it is a well-known fact that rats can open them and so pass into the house drains.

There are a great many different kinds of these flaps—Figures 124, 125, and 126—and no doubt some are better than others, but, for the reasons given, it is almost impossible to get

FIGURE 124.

FIGURE 125.

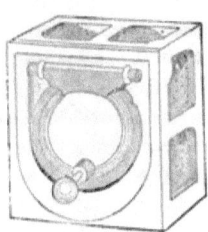

FIGURE 126.

any piece of mechanism that will keep back smells and at the same time allow the sewage to flow freely away.

There are other contrivances specially constructed to prevent the back flow which occurs when the drains discharge from a house at a low level into the sea, a tidal river, or sewer which is liable to flooding. Figure 127 is a sectional elevation of a ball tide-valve, which explains itself, and although there are other kinds in the market, they are mostly the same in principle. For drains discharging into ordinary sewers these tide-valves are of no use for keeping sewer-air from entering the house drains, but are only intended for keeping sewage back. This is a good description

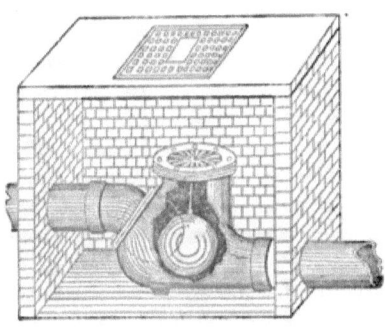

FIGURE 127.

for its purpose, as it will be seen that sewage cannot very well get on the top of the ball so as to clog it and prevent it from fitting over the end of the incoming pipe when any back pressure of sewage takes place.

It has been argued that all house drains should enter the sewer at a level just above the surface of the water, at its ordinary level, in the sewer, so that when an extraordinary quantity of water is discharged into them, the ends of the drains are covered. This will prevent any compressed air in the sewers forcing its way back into the houses. But this is not a good plan by any means, as in rainy weather the ends of the pipes would *always* be covered, and so render the drain, if it is trapped, "air-bound," and thus prevent the free flow of sewage from the house, with the result that discharges from upper parts of a building would escape through the lower fittings, even if the water in the sewer was low enough to allow them to run away if the connection had been made at a higher level. Again, if the connection is made at a low level, rats could much more easily gain access to the house drains, and although most people object to these animals in a house, still it must be admitted that they sometimes act as scavengers, and no doubt get rid of a great deal of garbage in the sewers which would otherwise putrefy and give off foul emanations. Where there are signs of rats in a house, defects in the drains are generally found. It may be noted that when the drains are of iron or vitrified stoneware, it is very rarely that these pests are discovered, although in hotels and such-like public buildings, one or two will often get under the floors and propagate very rapidly, but with pains they can generally be starved out or poisoned. This way of getting rid of them is objectionable, as the dead bodies lay about and putrefy, and, if in a damp situation, the smells arising from them are noticed for several weeks afterward.

Watts & Co., of Bristol, make a very good machine called an "Asphyxiator," which is of great use to plumbers, both for testing overground-pipes for defects, and also for exterminating vermin, as well as for distributing an aerial disinfectant. This machine will be referred to again at a future time, as I find it of great help when making examinations for smells.

CHAPTER XV.

HOUSE DRAINS.

IN TOWNS house drains made of brick are now becoming things of the past, but in some parts of the country, builders stick to them and believe in no other description. Bad as this description of drain is found to be, there are others worse. In a country parsonage, beneath the wine and beer cellars, were found drains large enough for a man to crawl through, the sides and arch built of random rubble—that is, of irregular-shaped stones. The stones in the bottom had been worked to a face—rough, it is true—so much so that solid sewage lay all over the bottom. Liquids glided over the top of the solids in a zigzag course made by themselves, and part escaped into the surrounding earth.

Most of the drain pipes used in London and other large towns are what are commonly called glazed socket pipes. They are made from 2 inches external diameter up to 2 feet, and are 2 to 3 feet long in all sizes. There are several kinds of these pipes, but for the present only two will be mentioned, namely, the *lead-glazed* pipes, which are made of inferior clay and which will not stand the necessary heat for burning them properly, and the *salt-glazed*, which must be made of a clay or composition of clay and other ingredients which will stand a very high temperature, so that the sodium of common salt, used for the purpose of glazing, will combine with the silicate of the clay and the other necessary constituents, to convert the surface of the article into a coating of glass which will withstand the action of sewage gases. Lead-glazed pipes are reported as not being capable of resisting this action. With the laying of the branch drain from the house to the sewer, the sanitarian's trouble commences.

Figure 128 represents in section how a disreputable tradesman was found to be laying a new drain from a house. To save a small fee he was making his own connection, and, in addition, scamping his work by burrowing under the roadway, instead of opening out in a proper manner. It appeared as if he had taken out an old brick drain, and in knocking out the connection with the sewer by means of a crowbar had made a larger hole than was necessary, and then had simply pushed his new pipe to within about 6 inches of the sewer.

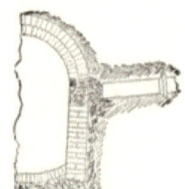

FIGURE 128.

This was done in such a way that a great deal of liquids would not reach the sewer at all. He had then pushed a lot of broken bricks around the opening left, some of which had fallen through so that solid sewage would cling around these pieces. The "shore-man" (sewer-man) said this was very commonly done.

Another experience was where a new drain had been laid in a house, and, although all fees had been paid, the connection with the sewer had not been made good in a proper manner The sewer-man's excuse was that he expected to have two or three more to do in a few days, and was going to do them all at the same time. But before this was done a violent storm took place, and the sewer was so full that water ran out, which, following the excavation made to lay in the pipes, found its way into the basement of the house to a depth of about 14 inches, and which had to be carried out in pails. These drains had been laid for about eight or nine days, had been plugged at the outlet, charged with water and proved to be perfectly water-tight, and this was proof that the sewage was not escaping out of the pipes when the storm took place, but found its way beneath, or by the side of them, and so followed the trench, the earth in which had not sufficient cohesion to resist the hydraulic pressure brought to bear. This is one reason why all pipes laid under the above circumstances should be enveloped in concrete.

The best system to commence with is to dig the trench as neatly as possible, taking care not to loosen the sides more than can be helped, to dig the bottom even, and also to remove all loose earth from the bottom. After carefully levelling from end to end of the trench to see what amount of fall can be had, drive in wooden stakes at intervals of about 8 or 10 feet, and projecting above the bottom 4 to 6 inches, as intended for the thickness of concrete, with which cover the bottom of the trench to the height of the stakes; make the concrete slightly hollow for the pipes to lay in, and have spaces at each socket, so that the man can get his hand and tool beneath to trowel up the face of the cement-joint. This way of preparing for laying the pipes is very important, and for want of it serious results frequently occur. One of these is, that very often a settlement takes place in the earth beneath the pipes. If this is soft earth or a loose sandy soil, sooner or later it is sure to give way, and if there should be water in the soil the circumstances are worse. Or, if one of the joints in the drain should leak, the water trickling through causes the sand or earth to run and leave a hollow.

At a banker's house, at Sunningdale, a drain was stopped. Figure 129 is a sketch of what was found. It appeared that a

FIGURE 129.

leaky joint had caused the sandy soil to run, and at last the drain broke in such a manner that all the sewage from the house ran into the hole and made matters worse. This hole was about 6 feet in diameter and 5 feet 6 inches in depth, measured from the top of the loose stuff which fell in when the stone paving was taken up.

Such was the nature of the soil that no doubt this hole would have become so large that the whole of the paving in the yard would have eventually fallen in, and it is just possible that the wall of the house would have fallen at some future date. Some think if

FIGURE 130.

drains are laid on rocky earth no concrete is wanted. Experience teaches otherwise, because if the bottom of the trench is too hard it will cause the pipes to break by crushing; and in a case of a hard gravelly bottom, drain pipes have been crushed, as shown in section, Figure 130.

This drain was about 7 feet deep. On looking at the fracture it was seen that the pipes were ordinary vitrified stoneware of very fair quality. We all know the strength of an arch, and the resistance a cylinder will present against a crushing force *if it is distributed over the entire surface*, but if applied to particular parts the resistance is soon overcome and the whole collapses. This more readily takes place in drains in towns where heavy traffic in the streets causes the earth to vibrate, and shakes the particles into closer contact, so that anything rigid must either sink with it or break. This is more especially the case in new districts, and no doubt there are hundreds of cases of fractured drain pipes from this cause that are not discovered until something leads to a search being made. We have only to take note of how often a water or gas main requires repairs, especially in newly-made roads, to understand the importance of protecting a frail piece of stoneware, which, in most cases, is no stronger than ordinary crockeryware.

After the trench has been prepared with a concrete bottom, the next thing is to select the pipes, taking care to reject all that have cracks in them, looking particularly to the part where the socket or hub is joined to the barrel. A great many will be found to have holes through them. Pass the hand carefully over the inside surface of the pipe to feel if it is rough; don't be in a hurry over this, or a piece may be cut out of the hand. Next see that the spigot and hub are true, so that the joint can be properly fitted. Any pipes that are very much bent should be laid on one side for curved positions. Very few indeed will be found to be straight; those that are not should be laid so that a perfectly graduated line of fall is at the bottom. Most men are anxious that the drain should look nice and straight in the trench, and they put the bagged part of the pipe downward, as shown at Figure 131, which is intended as a side view. A moment's thought will enable the reader to understand that in the case of a rather

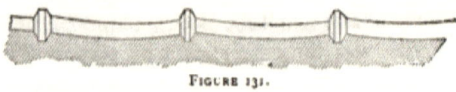

FIGURE 131.

sluggish fall the hollow parts will retain a portion of sewage, which will then decompose and give off some of the smells most people are anxious to keep out of their houses. If these pipes were properly laid, with the bent part at the side instead of the bottom, this would be avoided, and, although the course of the sewage would be rather serpentine, the whole of it would drain away provided the current of water was sufficient to carry away the solid matter.

After selecting the pipes, the next thing is to put them in position, and as many as possible at the same time. The most common way, and which cannot be too strongly condemned, is to put a portion of cement on the bottom of the socket or hub, and then lay in the next pipe. As this pipe is pushed in, it forces a part of the cement forward in a body, which stands up in a heap, as shown at A, in Figure 132. This also prevents the end of the pipe from getting home to the bottom of the socket, so that a space is left for filth to accumulate in. Sometimes the above cement excrescence is wiped or scooped out, but very often it is not, with the result that a stoppage invariably takes place sooner or later, and this is independent of the amount of sewage that is retained by these series of weirs. On looking again at Figure 132 it will be noticed that another evil is shown, and that is the joint is not true all round; the spigot is hard on the bottom of the hub, which in itself forms a check to the free flow of sewage. Some men will bed the spigot-end with clay and then face up the joint with cement. This has the advantage that should the clay project inside the pipe it will get washed away by the water that passes over it. The usual way is to now place more cement carefully on the top of the joint where it can be seen, the sides not always being considered of much importance. Perhaps the drain-layer really does take pains that some of the above evils do not take place, and trowels his joint all around, and takes so much time over it that the cement gets nearly set.

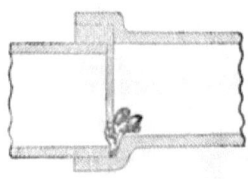

FIGURE 132

The next thing he does is to prepare for and lay in another pipe in the same careful (?) manner, and, in endeavouring to get it tight home, he jars the last pipe that was laid and so breaks the joint; but he could not help it, or was so earnest in making the next joint that he did not notice what mischief he had done.

HOUSE DRAINS.   133

To avoid all the above evils it is best to prepare the concrete bed as described, and, while it is still in a soft condition, to bed all the pipes on it, and take a crowbar or lever of any kind and force all of them home into the sockets. This applies when there is a long straight length. To insure the joints being fair and true inside it is best to give them one strand of gasket or yarn (tarred is best, as it will not rot away so soon) in each socket, and then, as quickly as possible, make all the joints in succession, with Portland cement and a very small quantity of clean sharp sand, and carefully trowel the surface to a smooth face and not be content to smear over a portion with the hands.

Some men use neat Portland for this, but I have had so much trouble with the sockets or hubs bursting, that unless the cement is old and nearly dead I use a small portion of sand. To show the power of Portland cement, it will be enough to state that in one case the central portion of a pipe drain with cement joints was lifted over 2 inches from the bottom of the trench; the centre part was raised like an arch.

At a Bank the smells from the drains were so bad that, although they had not long been taken up and relaid, it was found necessary to have them again examined, with the result that there was scarcely a sound joint found in the entire length; nearly every socket had burst with the swelling of the Portland cement with which the joints had been made.

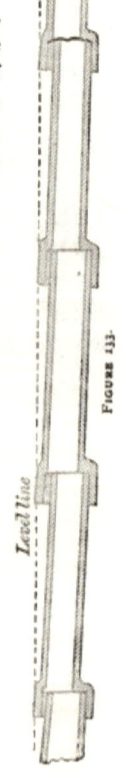

Figure 133.

A new 6-inch pipe drain had been laid, with good concrete foundation, through a house; after an interval of two days it was tested with water, and about half-a-dozen sockets were found to have split open sufficiently for the water to ooze through. When laying this drain, the bricklayer was asked what fall he was giving the pipes. He answered, "⅜-inch to each 2-foot length." Thinking he must have made a mistake of about ¼-inch, I asked how he obtained so much fall. He proved that he was right, but it was in the manner as shown in sketch, Figure 133, which is a longitudinal section. On looking closely into this it will be seen that although each individual pipe had that amount of fall, when levelled from end to end the drain would have little or no fall at all; indeed, it could be laid to fall the wrong way, and still be proved by the spirit-level that each pipe had a fair fall in the proper direction.

If it is required that the direction of the drain shall be changed, and there is not a properly-constructed bend at hand, the usual way is to cut the end of a straight pipe to a bevel, and get around the corner in that way. The joint cannot be made in a proper manner, as it is impossible to fit an oval end—as shown by dotted lines—into a round socket, as in Figure 134. The result of trying to do this is shown in Figure 135, and it is almost

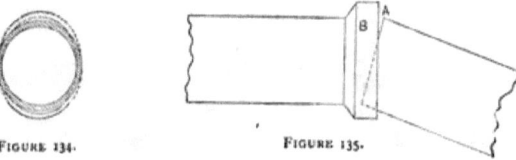

FIGURE 134.   FIGURE 135.

invariably found that there is an opening left, as at A, and a space left inside the socket, at B, for filth to accumulate.

It cannot be too strongly insisted upon that there shall be no parts where sewage can be caught or retained in any conductor.

Anyone, who has any knowledge at all of drains, would be able to form a very clear opinion as to their condition by simply smelling around the top of the ventilation pipe, and if the drains are trapped or disconnected from the public sewer, and yet he finds an abominable stench issuing from the vent pipe, he would know at once that something was causing these foul emanations, and as it could not be the materials of which the drains were constructed, it must be from something they contain and which should not have been there, and would not if they had been properly constructed and plenty of water discharges sent down them.

CHAPTER XVI.

DRAINS AND TRAPS—*continued*.

WHEN it is necessary to have one drain branched into another it is important to have a Y junction pipe, as in Figures 136 and 137, instead of using a square or right-angled one, as in Figure 138. The reasons were fully given when illustrating soil-pipe branches. In the case of drains with horizontal square branches, when sewage is discharged down a branch, as much flows back up the main drain as in the direction of the current, with the result that the liquid portion afterward dribbles away, leaving all solid or semi-solid matter behind. This will lie in the bottom of the pipes until a flush from a higher level of the main drain removes it.

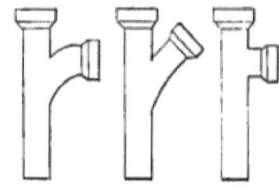

FIGURE 136.   FIGURE 137.   FIGURE 138.

Where a drain lies flat, or has only a slight fall, any sewage flowing down very frequently eddies, and runs a short distance up any branches, especially when they and the mains are of the same size. It would entail a certain amount of expense to have branches especially made so as to offer resistance to this, but the evil can be avoided by raising the branch a little, as shown in section, Figure 139. Where a small pipe discharges into a large one this evil will not happen, unless the main drain is running so full that sewage reaches up to the branches.

FIGURE 139.

Figure 140 represents a branch drain that was continually stopping up, and, having to be frequently forced, it was decided to open it until the cause was found. This proved to be at its junction with the main drain. Figures 140 and 141 are section and plan illustrating the cause of it all. The main drain was 12 inches

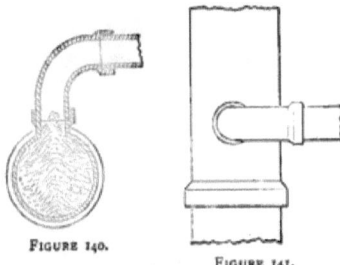

FIGURE 140. FIGURE 141.

and the branch 4 inches in diameter. The 4-inch drain was from a soil pipe, with four water-closets upon it. In the bottom of the 12-inch pipe was a pyramid formed of wet paper, and every time a water-closet was used, the wet paper, falling vertically on the bottom of the pipe, as shown, gradually accumulated until it eventually choked the end of the 4-inch pipe. If any reader were to take a handful of wet paper and let it fall from a height, and then two or three more, one after the other, on the top, and then try to push it over, he would be surprised to find the resistance it would offer. In the above case there was very rarely any water passed down from the upper portion of the main drain—in fact, only when rain was falling.

Most of the best sanitary engineers have inspection-chambers, or manholes, built at all junctions of drains, and also at any change of direction, and have only a half-pipe or half-bend at that point, so that if any obstruction should lodge there it can easily be removed; or, if the obstruction should be in the straight lengths, any ordinary drain-machine (which is similar to a chimney-sweep's) can be pushed through, and so clear it away. A great many people dub themselves "sanitary engineers," and, by dint of copying others, sometimes make a fairly good job, but, for want of experience, very often make serious mistakes. The writer saw a

FIGURE 142.

manhole, arranged as shown at Figure 142, and the drains being used. The branch drain from the water-closet was joined to the channel pipe in such a way that the sewage was discharged up the main drain, and then had to flow back again to pass into the sewer. Excreta lay at A until a current passed by from the upper part of the drain to wash it away. Another and very common mistake of some of these so-called sanitary engineers is to bring a junction in at right angles in a manhole, and also to have the bottom channel too shallow, so that anything coming down the branch washes up and lies on the side ledges formed for a man to stand upon when examining or clearing a drain. Figure 143 is a section showing this, and at B the writer has sometimes seen as much semi-solid sewage as would nearly fill a bucket. Even in some cases where these man-

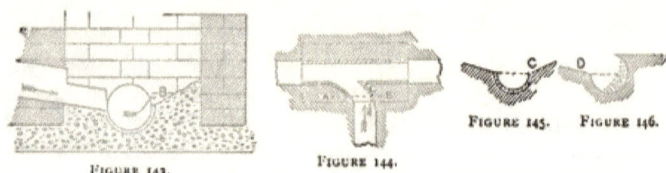

FIGURE 143.   FIGURE 144.   FIGURE 145.   FIGURE 146.

holes have been fairly well arranged, want of thought, and, in some cases, an ignorance of hydraulics, has led to another error, which is shown in Figure 144, Figure 145 being a section on A, B. 'It must not be supposed that because the branch channel is curved that the water will follow it. As a matter of fact, it will wash up, and sometimes leave a deposit at C. To overcome this difficulty it is necessary to curve this bent channel, as shown in section at Figure 146. The dotted lines show the surface of the water as it passes around the bent portion. It is not necessary to raise the side, D, so high as the other, but if it is not done the manhole does not present a smart appearance when finished.

Most manholes have half-pipes fixed for the sewage to pass through, and these channels have the sides built up with brickwork. There is no reason why a proper channel pipe, with junctions, should not be made all in one piece, as in sketch, Figure 147; or bends, as in Figure 148, which could be fixed quite easily, and thus avoid the patchy appearance these places generally present as now done. In some cases it would be necessary to have channel branches instead of pipes, as shown in Figure 147. Before leaving the question of manhole junctions, it would be as well to draw attention

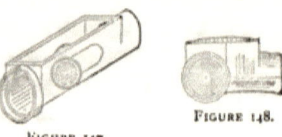

FIGURE 147.   FIGURE 148.

to another important point, and that is, not to allow two branches on opposite sides to be directly facing each other, in which case water or sewage coming down one pipe, with any velocity, would rush up the one opposite. Figure 149 shows this evil, and Figure 150 the precautions to be taken to avoid it. Ordinary manholes are built with common bricks, and where of considerable depth, strong wrought-iron loops are built in the angles, as shown at X, Y, Figure 150, so that a man can get down without the aid of a ladder, which would occupy space, and leave little room for him to work in. Sometimes these manholes have a coating of lime-wash, which makes them look clean and smart. Some sanitarians will have them rendered in cement and sand, worked up to a face.

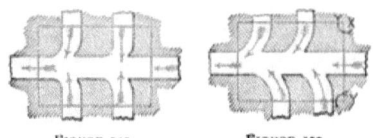

FIGURE 149. FIGURE 150.

Where expense is no objection, it is a good plan to use white glazed bricks for the face-bricks of manholes, as they always look clean and are impervious to moisture, and if they should become dirty, a man with a pail of water and a broom will soon wash them as clean as when first new.

Drain-traps, sometimes called disconnecting-traps, are now considered by all sanitarians to be very necessary to disconnect the house drain from the sewer, as many sewers are in such a condition that people are unwilling to incur the risk of using their soil pipes as sewer-ventilators. Each house should have a separate drain and connection, and each drain should have its trap. By doing this, it is argued that in the case of certain contagious diseases breaking out, any germs which find their way into the sewer shall not pass up the drains into adjoining houses. It has been disputed that water-traps form an effectual barrier, but from the experiments that have been made, and the eminent men who have made these experiments, we cannot do otherwise than accept their ruling that water-traps answer their purpose thoroughly well; and, until something better can be devised, they are the only fitting that can be applied with success. At all events they keep gases of decomposition from pouring in a continuous stream from the sewers, or cesspool, into and through the house drains—that is, when properly constructed. One of the oldest-fashioned traps that has been made for this purpose is what is called a dipstone trap, shaped as in Figure 151. They are simply small cesspools. Some are made 3 feet square and a corresponding depth, with a stone built in the side walls, so that its bottom edge dips into the water which is retained in the bottom of the chamber. I have seen them with the dipstone so low down that the fingers could be passed

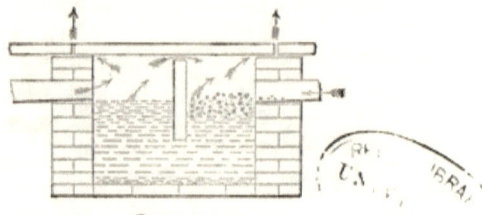

FIGURE 151.

between it and the cover-stone. The cover-stone is very rarely bedded down air-tight, and cases have occurred when it has been removed for access for cleaning out the trap, and then simply replaced without any bedding at all, so that when a smoke test has been applied to the drains it has poured out all around the cover. Another fault of this kind of trap is that when the top is removed for any purpose drain-air can pass freely through during the whole time the man is doing his work Experience has proved that this kind of trap very readily stops up. Paper and fæcal matter float on the surface of the water on the inlet side until quite a heap has accumulated so as to form a dam across the end of the inlet drain. More sewage flowing in will detach and perhaps immerse a portion of this matter below the dipstone, on passing which it floats in a body and chokes up the end of the out-go drain. By making the trap larger these evils are aggravated, but their discovery is postponed for a longer period of time. So much fæcal matter is retained in these traps that an enormous quantity of sulphuretted hydrogen and other noxious gases are given off; in fact, they act as gas generators, and are as bad, or possibly worse, than the evils they are intended to obviate. Sometimes nothing but solid matter is found in these traps, the liquid portion having leaked away into the earth through defects in the brick walls. In the middle of a public school playground one was found with no water in it, having leaked through the walls, and the stench from the drains pouring out in a continuous stream.

The trap most commonly used is shown in Figure 152. Although it has some advantage over the dipstone trap, it retains

Figure 152.

some of the evils, although in a lesser degree. For instance, excremental matter is retained in the middle pipe, and it is very rarely that one is seen without floating matter on the surface of the water on the inlet side; small discharges of water pass below this, and larger discharges do not have sufficient power to overcome its buoyancy, to drive it out of the trap, and float it away into the sewer. Not having any foot or base, in the hands of an incompetent workman these traps are likely to be fixed so as to be useless for their purpose of keeping back smells, by being placed so that the water level is below the throat or dip. This kind of trap is also too large and contains too much water to be thoroughly flushed out with ordinary discharges through them. In choosing what kind of trap to use on house drains, capacity is of as much importance as shape. Take Figure 152 as an example. A 6-inch trap holds, when properly levelled and fixed, something like three gallons of water. In London, and several other towns, we are limited, by means of special apparatus, to not more than two gallons of water at each usage of a water-closet, and by the time this has passed through the drains and pipes it has lost all force or scouring power as it passes through the trap. If a pailful of water—about two or three gallons, is thrown down a sink and has to be strained through a grating, and then down a waste pipe into the drain, it has even less power of cleansing than the water-closet discharges; or if a bath is emptied it is generally through such a small waste pipe as to be useless for removing any floating matter out of the drain-trap.

The writer found on trial that fifty gallons of water, discharged from a bath into the drain, failed to get rid of a piece of crumpled-up paper, no larger than the hand, that was thrown into a similar kind of trap to the one under discussion

Some manufacturers have made traps with the inlet-side a few inches higher than the outlet, the bottom of the outlet end representing the water-line. This is an advantage, as the incoming water falls on the top of any floating matter that may be in the

DRAINS AND TRAPS. 141

trap; but after all this is not effectual, as, if the stream is small and the trap large, eddies are formed, and the trap is not cleansed or the contents entirely displaced. The writer had special instructions to examine a drain that was frequently being stopped up. On seeking for the cause, it was found that originally the drains had not been trapped. It being found necessary that they should be, one had been fixed. This trap, which was supposed to have been of an *improved* description, was similar to the one mentioned. It had been fixed to the old drains, with the result that it had about 7 inches water seal, and it was choked full of floating matter, the incoming water simply draining away, leaving all solid matter behind. A glance at Figure 153 will explain the cause.

There are several patent traps in the market, all more or less good, but there are only a few which fulfil their requirements, or

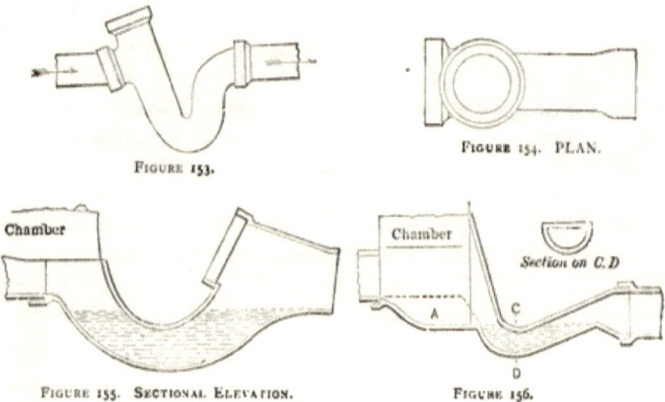

FIGURE 153.

FIGURE 154. PLAN.

FIGURE 155. SECTIONAL ELEVATION.

FIGURE 156.

that can be called self-cleansing. The best the writer knows are shown in Figures 154, 155, and 156, although there may be others he does not know of. On looking at Figure 154, which is a vertical view, it may be noticed that, although the inlet and outlet ends are made to fit 6-inch pipes, the waste or body of the trap is considerably smaller The water-fall is 6 inches into this trap. Figure 155 has a water-fall of about 3 inches. Figure 156 shows a sectional elevation, and also a cross-section of the body of another patent trap. The flat top, as shown in small section, is a new departure, but when we come to think that this represents the actual form the water would take in a 6-inch pipe only half filled—ordinary discharges rarely fill a 6-inch pipe so full as this—the inventor can argue that he has reason on his side. And although at first sight one might question the use of the part A, on further consideration it may be found to be a necessity. Supposing that discharges come down the drain to fill it full, or nearly full bore, the water would head up in this part so as to get a greater pressure, and so accelerate its speed through the trap. It may be suggested that the water would splash about this chamber when only small discharges pass down, and that it would be better for the incoming water to fall vertically into the trap as shown by dotted lines.

Figure 157 is a sectional elevation of a very good drain-syphon, which is patented; the projecting lip at E causing the water to fall vertically on to any floating matter that may be in the trap, instead of trickling down the sides.

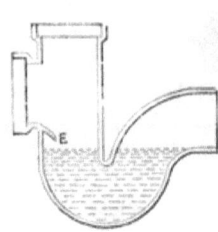

FIGURE 157.

There are several other traps which nearly approach those illustrated, but a great many have the fault of being too large in the body.

It is becoming a common plan with sanitary advisers, when a drain has been fairly well laid, but not trapped, to have one inserted, but much smaller in size than the drain. For example, there are several houses with 6-inch or 9-inch drains, where a 4-inch or 6-inch would have been quite large enough. In these cases the smaller-sized common-shaped traps are fixed and connected to the pipes by

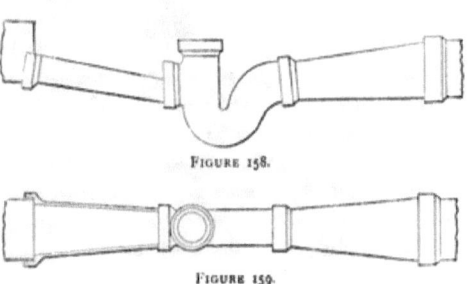

FIGURE 158.

FIGURE 159.

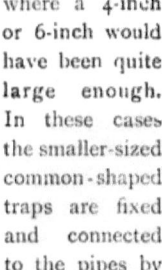

DRAINS AND TRAPS. 143

taper pieces. The incoming pipe is generally fixed as a tapering channel, with a fall of 2, 3, or 4 inches in 2 feet, so that the water runs down with greater speed into the trap.

Figures 158 and 159 are respectively elevation and plan, showing how this is arranged. In one case a 4-inch trap had been fixed in the course of a 9-inch drain. The drain had a very sluggish fall, and was difficult to make a good job of because of the tapering pipes being improperly made. This can be explained better by a sketch, to which I beg to refer the reader—Figure 160—which is a side elevation, and at A, A the hollow parts,

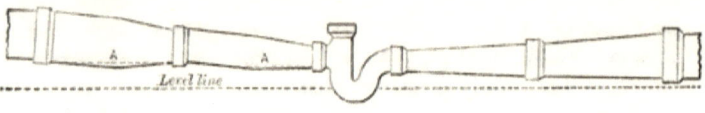

FIGURE 160.

which always retain a quantity of sewage. These pipes, to be properly made, should only be tapered on the one side, and the other should be perfectly straight, as shown in Figure 161, the straight side being placed downward—as mentioned in a previous chapter when writing on bent pipes, and how they should be laid, when, of necessity, they must be used. At the risk of wearying the

FIGURE 161.

reader, I must repeat that it is these little pools of decomposing sewage which is the principal cause of the abominable stench generally to be found issuing from ventilation pipes. I have been down a great many public sewers, and I generally find that those which have a continuous stream of flowing sewage are not nearly so offensive as those which have intermittent discharges into them. The result of these kinds of discharges is that very often the solid matters are retained by friction over hollow places, with a corresponding rise beyond them, the liquids simply dribbling away, so that certain matters are only moved by stages, instead of being washed clear away. Sanitarians are becoming more and more aware of the defects of stoneware-pipe drains, and are slowly but surely discarding them in favour of iron ones.

CHAPTER XVII.

IRON DRAINS.

ANY reader who has noticed my remarks on stoneware-pipe drains will most likely be inclined to think (or he may probably have found out for himself) that although they may be very much better than the old brick drains, still they are very far from being perfection; and his experiences most likely are similar to my own—viz., that they are very costly when done as they should be; that, in spite of all pains that may be taken with them, they very rarely will stand an hydraulic test of only two or three pounds per square inch without leaking somewhere or other. For my own part I should be very sorry to guarantee any stoneware-pipe drains to last any length of time, and, although they may have been left in as nearly a perfect condition as possible, after a time they sometimes are found to be leaky. It is just possible that expansion and contraction of the pipes may have something to do with this, as sometimes hot and cold water alternately passes through, or it may be the traffic in the streets, or, in some instances, railway trains, shaking the earth. At all events, sanitary engineers are now falling back on cast-iron drain pipes as being more trustworthy, especially when fixed inside of buildings. When iron drains are fixed, smaller sizes, in proportion, are used; for instance, where a 6-inch stoneware would have been only 5-inch iron is used, and very often 3-inch and 2-inch iron, where in the ordinary way a 4-inch stoneware pipe would have been laid.

The iron pipes used vary very much in substance and treatment. Some people argue that plain iron does not rust to any serious extent, and answers the purpose very well, provided it is strong enough. A friend of the writer's told him that he had been making an alteration to an iron drain that he laid about twelve years ago. The metal had not been protected in any way to keep it from rusting, and yet he was much surprised to find that scarcely any oxidation had taken place inside the pipe, although it was in constant use. It must be admitted this is in direct variance with ordinary experience. A coating of fresh slacked lime, made into a wash, applied to the inside of the pipes, has been found to prevent oxidation, and to last for a very long time. The Bower-Barff process of protecting iron from rusting has been much spoken and written about, but the writer has never seen nor fixed any iron drains treated by this means, nor yet been able to hear of anyone who has. He has seen a few instances where ordinary galvanized-iron rain-water pipe has been used, but very little faith can be placed in the zinc coating, which is soon eaten away, thus leaving the iron unprotected. Pipes coated with Dr. Angus Smith's preparation (which consists of immersing them in a mixture of pitch, coal tar, and a small proportion of linseed oil, with, sometimes, a little resin, melted and heated to about 300° Fahr.) are mostly used in England for drain-work, and in some instances for soil pipes.

Some plumbers are using, for drains, pipes very little stronger than ordinary rain-water pipe. Others use it of the same substance as generally fixed for hot-water heating pipes. A great deal of what is called "light underground pipe" is used, while the best engineers have the heaviest underground pipe. Figures 162, 163, 164, and 165, are drawn to scale, and show the relative thickness

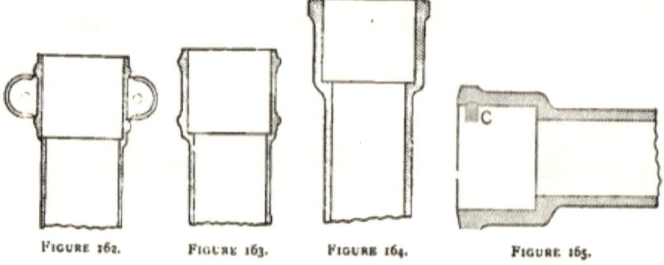

FIGURE 162.   FIGURE 163.   FIGURE 164.   FIGURE 165.

of the iron pipes as used by various people. Figures 162 and 163 should never be used when the joints are made with metallic lead, the reason being that the socket, or hub, is not strong enough to resist the lead being "set up." If this is not done the joint will sometimes leak owing to the lead shrinking as it cools. Figure 165 is the sort generally used for the best kind of work, and also for water-mains, which have to withstand a very great water-pressure, also the enormous strain of traffic over them, in some cases dead weight, and in others shocks or sudden strains, when fixed in streets. These pipes are usually cast in 9-feet lengths, with sockets 4 inches deep. This length is a great advantage for drainwork, so many joints not being required as in stoneware pipes. It is true that skilled labour is required to lay iron pipes, but this is an advantage. For laying stoneware-pipe drains an ordinary navvy is very often considered to be quite good enough for the work, but my experience is very different, and I have found that the best tradesmen very often fail to make sound work of them.

Iron pipes, where the fall is sluggish, have another advantage, as they are much straighter than stoneware, are smoother, and consequently more water can be discharged through them in a given time, other conditions being equal. It has already been stated that skilled labour is required to lay these pipes—*i.e.*, men who know how to do their work properly. Neglect of this may lead to serious trouble. As an illustration, a foreman in the same employ as myself took on a stranger to lay some iron drains; things were going on very nicely until one day it was found that he took about twenty-eight pounds of lead to make one joint on a 4-inch pipe. It need not be explained that the lead had run through the joint and lain on the bottom of the inside of the pipe.

It may be well to dwell on this question of joint-making for a short time. All pipes, when cast, have a bead on the spigot-end, as shown in section, Figure 166. This helps to make the joint "fair" inside—that is, causes the end of the pipe to lay in the socket so that the inside faces are in a straight line, and there is an equal annular space all around the inside of the socket. This bead does not fit quite tightly in the socket, so that it is possible to force the yarn or gasket past it when the pipe is not tight home in the socket. The result of this is, that there is nothing to prevent the

FIGURE 166.

molten lead, used for making the joint, from running through. When it is necessary to cut a pipe to a required length, an unskilful tradesman will do it in such a bungling way that the end is left very ragged and uneven. When this is so, although the spigot-end may be rammed home in the socket, the yarn will very often fall through and so leave no protection for preventing the lead following. Figure 167 explains this. Some men, to prevent this, or if they happen to cut the pipe irregular, will cast a lead ring and push it on the end of the entering pipe, and jam it tight to the bottom of the socket; then another, which can be set up or caused to expand until it fits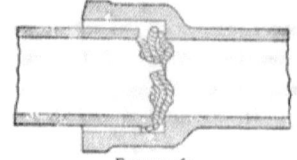

FIGURE 167.

quite tight, and so on until the socket is quite full, when the last one can be worked quite flush with the outside face of the socket. This takes about three or four times as long to make the joint as the ordinary way.

There is a little tact required in yarning a joint. Some men will use the yarn just as it is made, no matter whether the socket fits tight or loose; others will take off a strand for a tight joint, or will add a strand—or, if occasion requires, twist two yarns together—for a loose joint, so as to insure the socket being yarned an equal thickness all around. Black or tarred yarn is the best to use, as if any of it comes in contact with water it will not rot so soon. White yarn is best for water-mains, as it does not impart any taste to the water, which the tarred does.

After yarning the joint, a roll of clay is placed around the outside in such a way that when the lead is poured in it projects

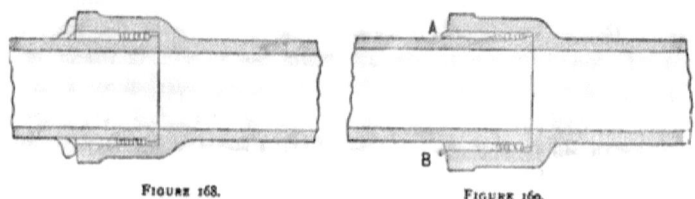

FIGURE 168.   FIGURE 169.

slightly beyond the face of the iron socket—Figure 168. Sometimes the projecting lead is cut off with a chisel, but this is a mistake. The proper way is to take a hand-chisel and hammer to slightly separate the lead from the pipe, as shown at A, Figure 169, and then a thin calking tool, work it all around keeping it close to the pipe, and then use a thicker tool in the same way. This causes the superfluous lead to bulge outward, as shown at B. A good tradesman will cut this off with his calking tool by its coming in contact with the edge of the socket, and never use a chisel for that purpose, and at the same time will leave the surface of the lead so smooth that the tool-marks can scarcely be seen.

If the joint is small, so that it does not require much lead, it should be poured very hot, or it will not run right around, but if the joint is large and requires a good thickness of lead, it should not be made very hot for the reason that the hotter the lead is made the more it expands, and, consequently, the more it shrinks as it cools. Another reason for not having the lead too hot is that it burns the yarn and causes it to smoke, which, bubbling through the lead before it sets, leaves a lot of little cellules in it, through which the water can afterward find a way of escape.

The way to test an iron pipe as to its freedom from cracks or splits is generally to "ring" it with a hammer; if a jarring sound is emitted, it is a certain sign that there is some flaw in it, but if it "rings like a bell" it may be taken for granted that the pipe is in a sound condition. Sometimes a pipelayer will "set up" his joint too much, so as to split the socket or hub, and this is more likely to occur when hard lead is used. Or he will perhaps drive his tool in between the pipe and socket so as to burst it, but a man who is used to the work will at once detect this by the sound. When a socket is burst or cracked it should always be taken out, it never being safe to leave it, and although the crack may have something applied to it so that the iron will rust so that no leakage may occur, still it is always liable to break open. Iron clamps and plates can be bolted on, but no matter how well done, it is only a botch, and the cost is almost equal to that of a new length of pipe. Another test generally insisted upon by engineers is that all pipes for best work shall be not less than the following weights:

3-inch pipes, 1 cwt. 0 qrs. 14 lbs. the 9 foot 4-inch length.

4	,,	1	,,	2	,,	0	,,	,,	,,	,,
5	,,	2	,,	0	,,	0	,,	,,	,,	,,
6	,,	2	,,	2	,,	0	,,	,,	,,	,,
7	,,	3	,,	0	,,	0	,,	,,	,,	,,

It is an open question if these weights should not be modified, as, although the substance may be about equal in all sizes, still the larger pipes should be made thicker in proportion than the smaller ones, as the perimeter of the larger pipes has to resist a greater hydraulic pressure. Another test sometimes insisted upon is that these pipes should withstand a pressure of 400 feet head of water. I was carrying out some iron pipework some few years ago which was tested to this pressure, and there was not one defective pipe found. Two or three joints had to be reset, and that was all in a £4,000 job.

One firm of London plumbers use cast-iron pipes of the following weights:

3-inch pipes, 45 lbs. per 6-foot length.

4	,,	60	,,	,,
4½	,,	70	,,	,,
5	,,	82	,,	,,

These weights may do very well for overground work, but I think the heavier pipes are better for underground. The heavy pipes have the further advantage of being made in longer lengths, as well as having stronger sockets to resist the necessary setting up of the joints.

Some makers have a groove inside the socket, as shown at C, Figure 165, but this is not of much advantage, as the socket is generally rough enough inside for the lead to key to.

When cast-iron pipes, coated with Dr. Angus Smith's solution, are used for drainage purposes, they may be buried in the ground without much risk of deterioration by rusting, but it is advisable, in some kinds of soil, to have a concrete bed for them to lie upon. One eminent firm of sanitary engineers makes a concrete bottom to the trench, and then supports the iron drain pipes on small brick piers, and then builds a wall on each side, laying flat stones over, so as to leave a space all around the pipes. This space is continued to the external faces of the house-walls where air-gratings are fixed. By this arrangement a current of air may circulate through from end to end, and all around the outsides of the pipes. Figure 170 is a sectional view showing this arrangement. Another

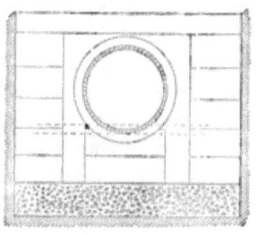

FIGURE 170.

advantage claimed for this system is that the jarring of the floor cannot disturb the pipe drain in any way, and if a light be held at one grating, a person looking through the other can see if any leakage is taking place. It is a further advantage for the iron pipes to lie on horizontal iron bars, as shown by dotted lines, so as to leave the space beneath perfectly clear.

In one case iron pipes were fixed in a building which had been erected on virgin ground. These pipes were broken soon afterward by the settlement of the earth outside of the walls, the pipes being built in and held so firmly that they could not move. Figure 171 explains what actually occurred. To avoid this a small relieving arch should be built over all drains (whether iron

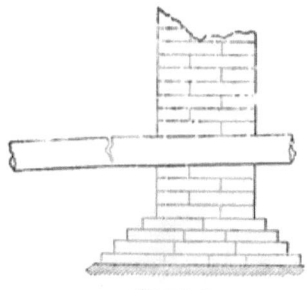

FIGURE 171.

FIGURE 172.

or vitrified stoneware) when they have to pass through a wall, in the manner shown in Figure 172, so that if the house-walls, or the earth in which the drains are laid, should sink, no strain would be brought to bear on the pipes. In some cases—and they are very few—it is possible to fix the pipes on the house-walls instead of burying them under the floors. No doubt this is a good plan, but in nearly all town houses the basement floor is used as servants' offices, and as these rooms require sinks and other fittings with waste-pipes to them, the drains must of necessity be fixed at a lower level. A speaker at the late Plumbers' Congress, held at the Health Exhibition, advocated a tunnel to be built under houses ranged in rows, through which the drains should be carried. These tunnels were to be large enough for a man to pass through, so as to be able to examine the condition of the pipes. It is doubtful if this would be a good plan. Setting aside the question of cost, in a great many cases these tunnels would be lower than the foundations of the house, and would thus interfere with the stability of the building. Again, where the soil is charged with water from land-springs or other causes, it would ooze through the walls and probably half fill the tunnel.

CHAPTER XVIII.

IRON DRAINS—*continued*.

IF drains are properly and carefully laid with an inspection-chamber or manhole at each end, it is not difficult to plug up the lowest end and fill the whole length with water, and so test its freedom from leakages, without actually examining every inch of its length. Another good way for testing drains is to hermetically seal the ends, and pump air into the pipes. An ordinary gas-pressure gauge can be attached, and by watching the column of water it can be seen at once if any leakage is taking place; or the ventilation pipe may be plugged air-tight, and the whole of the drains, soil and other pipes tested at the same time and by the same means. A greater air-pressure than ten or twelve-tenths need not be applied; in fact, a greater pressure will be found to break through the water-seals of the various traps.

It may be interesting to some readers to know how these pressure-gauges are made. Anyone can make them. I generally buy a piece of $\frac{3}{8}$ or $\frac{1}{2}$-inch glass tube; smaller sizes are not so good, as capillary attraction causes the water to appear to stand higher in the tube than it actually does. By holding the centre of this tube, which should be about 12 or 14 inches long, in a common gas-flame until it is red-hot, it can be bent to a U-shape; or the tube can be had about 6 inches longer, and bent to ∞-shape. This bent tube can then be fixed to a small piece of board, on which should be pasted a strip of paper divided by horizontal lines into tenths of inches, the centre of the scale being zero. The tube should then have water poured into the U until it stands at zero on both sides of the leg. Figure 173 represents this gauge ready for use.

FIGURE 173.

A piece of indiarubber tube can be attached to the leg, A, and the other end to a pipe securely connected to the drain or pipe to be tested. It is important that the end, B, should be left open, or the water in the gauge would not rise by the pressure applied to the other end, or, more properly speaking, would only rise in proportion to the extent the air in this end of the tube is compressed. As $\frac{15}{10}$ represents $1\frac{1}{2}$ inches, a pressure very slightly in excess of this would burst through any water-traps which had only $1\frac{1}{2}$-inch "dip" or "seal," or what North country plumbers would call "drown." I have very often used this gauge to test gas pipes inside of a house when seeking for cause of smells, and have suspected them to proceed from some leakage in the gas pipes or fittings. In these cases it is only necessary to attach the other end of the indiarubber tube to one of the burners, and then turn the cock, when from $\frac{10}{10}$ to $\frac{20}{10}$ will be registered on the gauge, according to the pressure from the Company's main. Next turn off the main cock, which is generally near the meter, and then watch if the water in the gauge falls back to zero on the scale. A little drop of almost any kind of dye in the water in the gauge renders it easier to see.

There is some peculiarity about this pneumatic testing of pipes that I do not clearly understand, and that is, I never yet saw a pressure-gauge stand steady at any point above zero. I have tested pipes with water which betrayed no leakage, and yet, on applying the air-test, the gauge has fallen to zero in a very short time, varying from ten seconds to three minutes. It is just possible that the indiarubber tubing may have been porous, but still not sufficiently so for water to ooze through. When testing drains and soil pipes, &c., the pressure can be applied with a small air-pump. In the absence of this instrument a small tube, with a cock in it, can be attached, and a person blowing through this will find that he has sufficient power of lungs for the purpose. When testing a considerable length of pipes, it will be found necessary to take several inspirations and subsequent blowing through the tube before the necessary result is attained, the small cock being alternately opened and shut to allow the air to pass through and then retained in the pipes.

To return to our iron drains. The several precautions mentioned in earlier chapters on lead soil pipes and stoneware pipe drains have to be taken with regard to iron drains. This is in reference to the necessity of their being perfectly true and straight. On no account allow what is commonly called "square junctions" to be used. Even when a vertical is branched into a horizontal pipe, a stoppage is always liable to occur, from reasons already given, and also from foreign matters, such as small bottles and sticks of firewood, which would pass around an ordinary bend if at the end of the pipe or Y-junction if at any intermediary position. These articles were discovered in a square junction in a drain, they having previously passed through several bends in a 4-inch lead soil pipe. Although it is not usual to make rubbish-shoots of water-closets, still it sometimes happens that when bedroom slops are emptied down these places other things get thrown down with them. Sometimes these things, when not too large, will get washed away and no harm is done, but it is always advisable to have means of access for removing anything that may lodge in certain parts of the drain or other pipes. This remark applies more especially to iron drains, as they cannot be cut into and made good with a cement patch in the same manner as stoneware pipes, or cut open and soldered over again, as can be done with lead pipes. A piece cannot be cut out of a cast-iron pipe without drilling a series of holes around the part intended to be removed. This is the neatest way of doing so, when the necessity arises, as afterward the hole can have its edges filed smooth and an iron plate made to fit over it, with the piece that was cut out riveted to it so as to fit neatly and not leave any serious obstruction inside the pipe to catch passing objects. A series of holes can be drilled near the outer edge of the capping-piece, and others to correspond, on the pipe. These last holes should be a little smaller and tapped with a screw-cutting tap. Brass or gun-metal screws should be used for fixing the cover-plate, as they could be easily removed again should occasion require, whereas iron screws would rust in so as not to be readily taken out again. Care should be taken that the screws for fixing the cover are not too long so as to project inside the pipe. If it is not probable that this plate would want removing at a future time, red-lead cement, or what is better still, rust cement, could be used for bedding it on to the pipe, but if frequent removal is necessary an indiarubber (vulcanized) ring could be used. There are other things that also make a good packing, such as prepared asbestos and a kind of felt which is specially prepared for this kind of work.

It is always best to make provision for removing obstructions in iron drains when they are first laid, so as to avoid the necessity of disturbing them afterward. If they are laid in perfectly straight lines, with an access-chamber built over each bed or junction, or at the extremities of the drains, channel, or half-pipes could be used at those points, as described when speaking of traps and manholes; but as these places necessitate air-tight flap covers over them, and are sometimes very near the surface, it is an advantage to have fittings specially made, with an access or movable cover. Some makers have an extra socket cast on, in which a brass screw-cap can be bedded, but, no matter how well done, this always looks patchy, and, if hot and cold water is discharged through the pipes, it is open to the objection that the various metals used would expand and contract unequally, so that after a time they may become not air-tight. Again, these screw-caps are not large enough for removing some of the objects which find their way into the drains, nor for a man to pass his arm through to grasp them. These kind of places should be large enough for this, and also to pass drain-rods through to force any obstacle which is out of reach through and clear of the drains, and so shaped that the drain-rods could be sent up as well as down the drain from the same inspection hole. The shape of the hole should be similar to that described above as being drilled out, and should be cast at the same time as the pipe, with a flat flange around the hole if the pipe is to lie horizontal, on which could be bolted a flat cast-iron plate. If this cover is for an upright pipe a

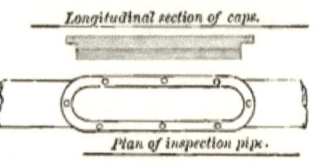

FIGURE 174.

piece should be cast on the back side to fit into the opening, so that the inside of the pipe shall be perfectly cylindrical, as shown at Figure 174.

Makers of iron pipes and other fittings supply junctions cast from the same moulds as those used for water-mains. Now, water-mains can be laid perfectly level, or even up hill, but drains must always fall in the direction of the current that passes through them. Square junctions for iron pipes are made as in Figure 175. What are known as branches are made as in Figure 176, and have a spigot-end at A instead of a socket, and no doubt that is why Figure 175 is used. This is all right for water-mains, but if they are used for drain-work the result will be that all the joints on the branch will be made imperfect, as the lead will have to flow upward to entirely fill the socket. The reader is referred to Figure 177, which illustrates what would occur under these circumstances. On looking at this, which is drawn in section, it will be seen that the joint is not solid, as there is no way of escape for the pent-up air at B, whereas, if the inclination of the pipe had been the other way, the air would be displaced as the lead was poured in. This would occur at every joint in the branch drain. It could be overcome by using a double socket or collar, and making the joint on the branch before laying it in

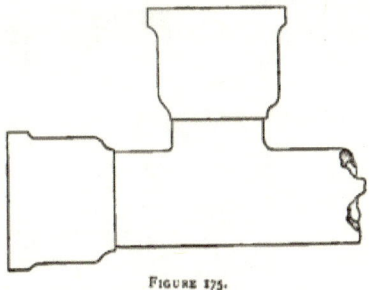

FIGURE 175.

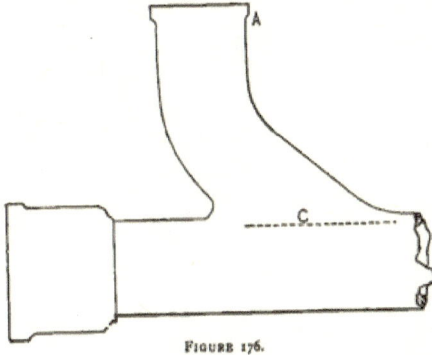

FIGURE 176.

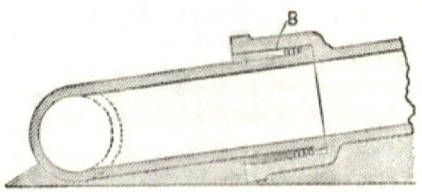

FIGURE 177.

position. Figure 178 is a section of the ordinary kind of collar used for connecting two spigot-ends of pipes, and it need scarcely be added that two lead joints are required to be made to it. A good job can be made in this way, but it is much better to have a socket cast on the branch, and so avoid the extra labour, lead, &c. The ordinary warehouses keep only those which are made for water-mains, but will always make castings to the purchaser's order and directions.

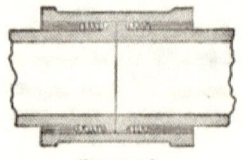

FIGURE 178.

Where the end of a drain turns upward to receive the end of a stack of pipe, it is a good plan to fix what is commonly called a "duck's foot" bend. This is a bend with a foot cast with it, as shown at Figure 179, and which helps to support the weight of the vertical pipe. A bed of concrete should be made for this to stand upon; indeed too much care cannot be taken to avoid settlement of any part of the drain. It is almost an every-day experience to find the end of the soil pipe not made good to the drain. This no doubt has arisen in a great many cases when—the walls of the building having good foundations and the soil pipes being firmly fixed—the earth containing the drain has subsided and the joints have become broken. At the risk of being thought wearisome I must repeat that concrete foundations for drains, under almost all conditions, are of the utmost importance. For want of concrete, a drain was found to have subsided and become so defective that it had to be taken out, and the earth beneath was found so charged with sewage that it had to be dug out to a depth of *five feet* before it could be considered to be free from offensive matter. This was a private residence, in which eleven children and two servants had all been ill. This and other

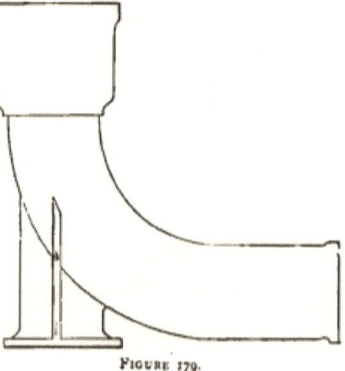

FIGURE 179.

experiences also prove that the drains cannot be too carefully laid, &c., and that they should be of the very best materials. If properly done the first cost would be the last cost, and would be much cheaper in the end, although some think the cost excessive if they have iron drains rather than others which are considered cheaper. These latter entail a permanent tax on the householder for repairs or renewals, to which may be added the annoyance and inconvenience of continually having the house upset by workmen.

There is one little matter in connection with junctions for iron drains that should not be lost sight of. They should always be inspected inside, at the point of junction, as it not infrequently happens that where the cores meet a serrated ridge of iron is left, around which passing objects can cling. The position of this ridge is shown by dotted lines at C, Figure 176. This (in some

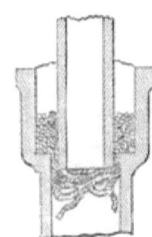

FIGURE 180.

cases they are only little spurs) can be knocked off with a long chisel and hammer without doing any injury to the fitting.

When it is required to reduce the diameter of a drain, it is always advisable to use a proper taper pipe, and not to attempt to get over it by making a large lead joint, in the same manner as shown in section, Figure 180.

All the tools necessary for iron pipe laying can be bought at the usual tool shops, but I much prefer to make my own yarning and staving-irons, although I daresay an ordinary smith would make them in one-half the time.

FIGURE 181.

The yarning-iron, as it is called, is generally made of steel, but I think a good tough piece of iron is better. As it is not necessary to use a hammer with this tool, I like the handle-part rather heavy, so that it does not jar the wrist so much when working the yarn tight home, and if it is slightly roughed it does not make the fingers ache so much by gripping the tool. A good set should be given to this tool, so that the fingers do not get pinched between it and the pipe. It is not at all uncommon to have the skin chafed off the little finger by rubbing against the pipe when yarning. Figure 181 is an illustration of this tool. The staving, calking, or setting-up

IRON DRAINS.   159

tools should be made of steel, tempered rather soft. Octagonal-sectioned steel is best, as it is better for holding. The hands, after using the clay, get so smooth inside that it makes them ache when having to grip anything smooth. These tools should be as short as convenient, so that when working in a trench they would leave more room for swinging the hammer than if the tools were longer. Another reason is, that with short tools the hands do not get hit so often with the hammer, especially when setting up the under side of the joint. When doing this, the man generally stands astride the pipe, the tool in one hand passed round one side of the pipe, and the hammer in the other passed round the other side, so that he cannot see what he is doing to the under side of the joint, but a good pipelayer can tell by the touch if it is properly made. For trenchwork these tools should be made with a set, as Figure 182, but for pipes fixed in upright chases in walls they should be rather straighter, as Figure 183, as there is generally objection to having much brickwork cut away for a joint hole. In some cases, where it is not allowed to cut away

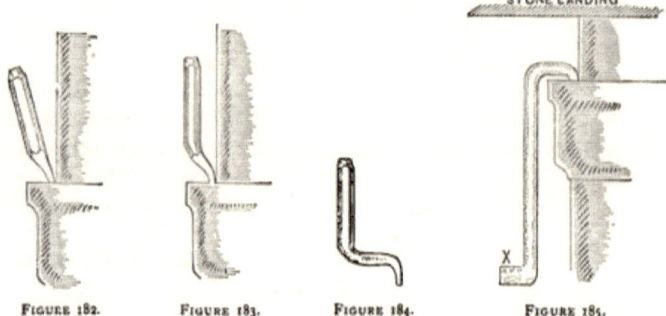

FIGURE 182.   FIGURE 183.   FIGURE 184.   FIGURE 185.

for access for setting up a joint, it is necessary to have specially-made tools. Figure 184 I have used under these circumstances, but a pair of them is required when it is necessary to work from both right and left hands. Figure 185 is another tool I had to have especially made for setting up a joint beneath a stone landing, and which could not be done any other way, X being the part struck with the hammer. The hammer used for jointing has usually a short club head and a short handle, the blows given with it being short and quick. The hammer should be held lightly; if firmly grasped it makes the fingers ache by long usage, and at the same time the constant jarring causes a pain in the wrist, and, moreover, the hammer falls with more force if not grasped too tightly, as in the act of striking the face adjusts itself to the head of the tool better.

CHAPTER XIX.

IRON DRAINS—*continued.*

MOST of the soil pipes in first-class London houses are of lead. It is important that these lead pipes should be securely made good to the drains. This is sometimes done as shown in Figure 186, and it would scarcely be credited the numbers that are found to be made in this bungling way. Even when the socket of the drain is as it should be, it is often found that the joint is badly made, as shown in Figure 187, and a deposit of

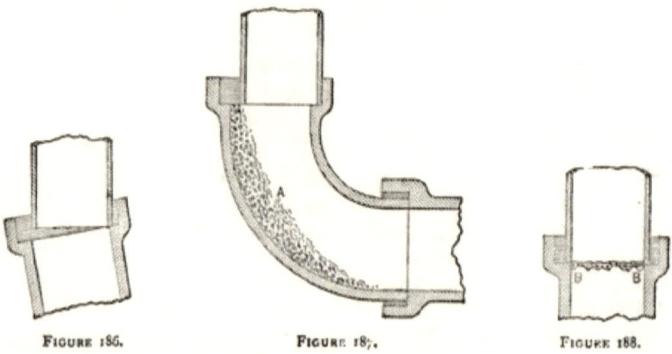

FIGURE 186.   FIGURE 187.   FIGURE 188.

cement which has fallen through and set in a hard lump at A. An improved way is shown in section, Figure 188, where the lead pipe has a flange worked on its extremity, so as to fit tightly into the drain socket, thus insuring the bore of the pipes being perfectly straight. Sometimes putty or red-lead cement is used to bed the lead flange in the earthenware socket. This prevents the Portland cement, with which these joints are usually made, from running through into the drain, but it is not a good plan, as a part of the red-lead gets squeezed out, and projecting, as shown at B, causes an obstruction, and this evil is aggravated if the work has to stand some weeks before any discharges pass down to break it away before it sets so hard as to resist this action. The best way to make the connection is shown in section, Figure 189. In this case the soil pipe enters the drain pipe about

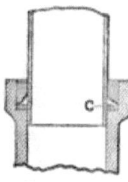

FIGURE 189.

an inch. A lead flange is soldered on the soil pipe to prevent its slipping down, as shown at C. A bed of common putty can be placed under this flange, and the part of the pipe telescoped into the drain will prevent this being squeezed out to form an obstruction, and the socket can then be filled in with Portland cement. Joints made in this way will withstand any hydraulic pressure not exceeding that to which the rest of the work may be subjected. This is a good way for connecting lead soil pipes with iron drains, but a lead joint is better than one made of Portland cement. Sometimes a lead pipe is joined to an iron pipe, as shown in

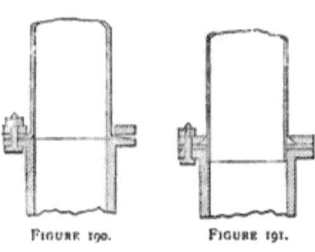

FIGURE 190.   FIGURE 191.

Figure 190. Although this can be made perfectly tight, it is not so good as the next illustration, Figure 191. In Figure 190 the lead is flanged out and very much weakened, while in Figure 191 a flange is soldered on, and about an inch of lead telescoped into the iron pipe. The remarks on connecting lead and stoneware pipes also apply in this last case.

A great improvement has been made this last few years by soldering a harder metal, such as copper or brass, on to the lead pipe, so that an ordinary lead joint can be made to the iron. Brass is not good for this, as the air in the drain causes a corrosive action. The same may be said of copper, but this is an open question. Copper pipes which have been fixed several years have been found to be in perfect condition, and although submitted to a severe test they showed no signs of any defects. But where this metal has been used as part of a water-closet apparatus, and subjected to the action of urine on both sides, it has not been found to last long.

IRON DRAINS. 163

The usual way of making the connection of lead soil pipe with iron drains is shown in section, Figure 192. The copper thimble *is passed over the end of the lead pipe*, which is then flanged over it at the end, and they are then soldered at D, as shown. The joint is then yarned and run with lead in the same manner as those on the iron pipe. By doing it in this way neither the copper thimble nor the lead joint are exposed to the action of either sewage or its gases. Sometimes the soldered joint is omitted, and instead of flanging the end of the lead pipe, it is entered into the iron in the same manner as shown in Figure 191, but as illustrated in Figure 192 is much better, as by no possibility can sewer-air pass out at D. There is less liability of a voltaic action being set up between the various metals used, as the lead only is exposed, the iron being protected by the coating applied to it.

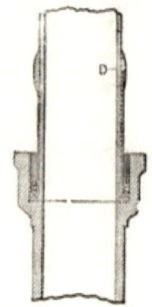

FIGURE 192.

In the late International Health Exhibition a specimen joint was shown for connecting lead with earthenware pipe. Figure 193 is a section. It was called an improved patent collar-joint. A is an earthenware collar, B is red-lead cement, C is a ½-inch wrought-iron nut and bolt, D is the flange of the lead pipe, E is Portland cement. The earthenware pipe, F, has a flange with corrugated surface, and is left unglazed for the cement to key to. There is no doubt that it is simple and easy of application, and could be made at a moderate cost, but it has several defects, which may be enumerated as follows: The lead flange is shown of the same thickness as the lead pipe, but in practice it will be found to get very much reduced in working or tafting it over. In the exhibited drawing the pipe was shown to be quite square at G, which is a source of weakness, and if the reader refers to what has been said further back, and then looks at Figure 193, he will see that there is nothing to prevent the red-lead cement squeezing out and

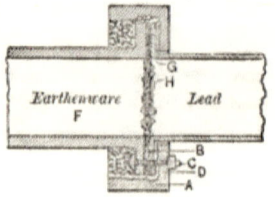

FIGURE 193.

M 2

projecting inside the pipes where they meet at H. Again, earthenware is too frail to stand the strain of being screwed together, even if the bolt-holes were made perfectly true so as to come opposite each other. When made the joint looks large and clumsy, and after all is not so strong and reliable as the one shown at Figure 189. Again, it can only be used in a horizontal position.

While writing, the idea has suggested itself that a gland-joint would be very good for connecting a lead pipe to an iron drain, something as shown in section, Figure 194. By using this

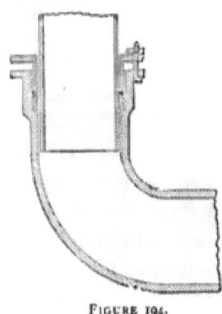

FIGURE 194.

kind, any hot water passing down the vertical pipe causing it to expand, it would do so without hindrance, and in cooling would again shrink back to its original dimensions. Another reason could be given,—that if the building or wall, to which the pipe is fixed, were to settle downward, it would not affect the joint, or, if the ground in which the iron drain is laid was to sink a little, the joint would still be air and water-tight, provided the lead pipe was telescoped into the iron pipe a few inches. Asbestos would be a good packing to use for this kind of joint, as it would not be so liable to perish as yarn, and the nature of this material is such that no lubricant would be required. This is an important point, as part of the object aimed at would be lost if the pipes were so securely connected that they would not freely slide, one inside the other, and would run the risk of breaking the weakest—the lead.

Soil pipes used to be generally fixed inside the house or building, but it is becoming more frequent nowadays with sanitary engineers to have them fixed on the external face of the wall, but there are numbers of cases where there is no available external wall-face, so that it is imperative that the soil pipes should be fixed inside. A great many people think that when the soil pipes and drains are fixed outside so much care need not be taken when making the joints, but no greater mistake could be made. Take the drains; if they leak the earth at first may act as a deodorizer, but a continual leaking will soon so saturate it with sewage that the house may be said to stand in a sewage-bog, and there is the further liability of this sewage soaking through the house-walls or passing beneath the footings, so as to render the earth beneath the basement floors in the same condition as that outside. A case occurred where the occupants of a house were very much troubled with smells. The whole of the drains were taken up and relaid, but before the people had been back more than a week or two they found that the smells were as bad as ever. The builder had to submit to a severe reprimand, and was threatened with an action at law. He pleaded that there was no defect in his work, and proved it to be sound by charging the drains with water and letting them stand several hours without showing any subsistence at the sight-holes. The idea that a cesspool might be buried somewhere next suggested itself, and the only way of finding this out was to excavate the whole of the basement. The result was that sewage was found to be oozing through the wall which separated the next house. On inquiring next door if they were troubled with smells, an indignant "No" was the answer, but the necessary pressure being brought to bear, the drains were examined and found to be completely choked up, and the whole of the earth beneath the basement floor was found to be so charged with sewage (in an abominable state of putrefaction) that it was passing through the party-wall into the next house. The writer has met with men who argued that in certain soils it is a good plan not to make the underside of a drain-joint *so that any water in the soil would pass through the joint and run away down the drain*, and who then gravely and deliberately proceed to put cement around the top and sides of the joints—for what reason the reader must find out for himself, as the writer cannot give any. An exception might be made when the drain passes away down a field to a cesspool or irrigation system, but in towns the rule should be laid down that all sewage-conduits should be perfectly water-tight.

CHAPTER XX.

DRAIN-VENTILATION.

THE question of drain-ventilation does not appear to be so clearly understood, nor the thought given to it that its importance demands. It is only the *bonâ fide* sanitary engineer or sanitary plumber who attaches any value to it. There are hundreds, or it may be said thousands, of houses that have the drains connected directly with the public sewer. Other houses have a trap placed in the course of the drain to disconnect it from the sewer. This trap may keep back sewage-gases, it is true, but for want of ventilation the house drains become a retort for generating these gases on the house side of the trap. The more experience the writer has the more convinced he feels that the best materials for house drains are heavy cast-iron pipes, properly protected from rusting, laid on a firm foundation, and the joints made with metallic lead. The next important item is thorough flushing with water, and they should also be ventilated, or air-flushed. There is a great diversity of opinion on this question as to the best way of arriving at success. Some sanitarians argue, and practice, that the soil pipe should also act as the drain-ventilator, and that it is necessary to have another pipe near the trap, as shown at A, Figure 195, carried up to the roof, to let fresh air into the drains, which would then pass in the direction shown by the arrows and up the soil pipe. Others maintain that the air-current should always be in the same direction as the water or sewage-current, so that any discharges from the water-closets, or other fittings, will accelerate the ventilation by driving the air forward in the drain and up the ventilation pipe A, fresh air being drawn in at the top of the soil pipe to supply what would in some cases be a partial vacuum. If the arrows were reversed it would explain what is meant. It requires very little argument to upset this theory. In the first place the discharges down the soil pipe and drain are intermittent, so that the current of air is only carried in the same direction at intervals. Between these discharges the air-current will be found to be reversed, and pass up the soil pipe.

On testing the air in a drain it will generally be found to be slightly warmer than that outside. Anyone with even a slight knowledge of aerokinetics will understand that even one degree rise of temperature is sufficient to rarefy the air in the drains, and which would then be forced upwards by the pressure of the

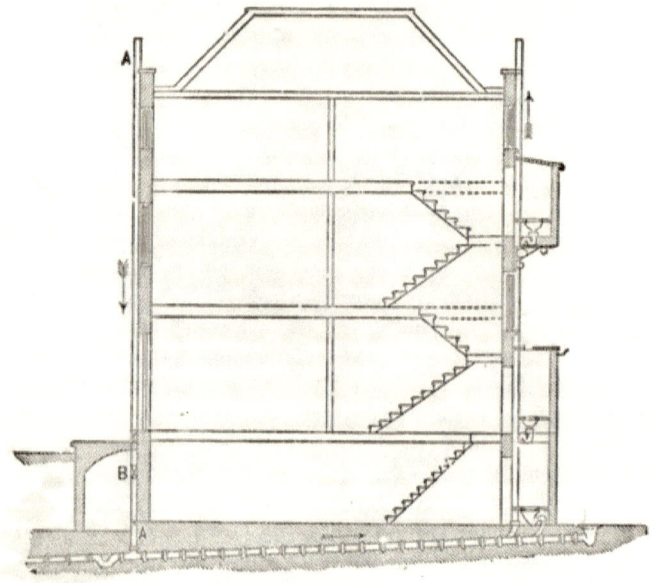

FIGURE 195.

cold air without entering at the lowest end. A very simple experiment will prove this. Take a glass tube, say 1 or 2 inches in diameter, lay it horizontally, with a slight inclination, and hold a piece of smouldering smoke-paper near the highest end, the smoke will be found NOT to enter. If hot water is poured on the tube, or a spirit-lamp held beneath, it will be found that the air inside will become rarefied and pass through with a current strong enough to carry the smoke away; but if the smoke had been applied at the lowest end, after an interval of a few seconds, it would be found to enter the tube and pass through it. By a careful study of some of these properties, which are generally considered too trivial to be noticed, results can often be achieved by natural laws, instead of resorting to mechanical or other means, which is now becoming so common.

Some people maintain that if the ventilating pipe, A, was carried to a position a few feet higher than the top of the soil pipe, it would act in a way similar to the long leg of a syphon when applied to hydraulics, and in its inverted position would syphon the air down the soil pipe and through the drain. This theory no doubt looks very well on paper, but the facts are found to be quite different when applied in practice to drainage-work. An explanation may be given by pointing out that the horizontal pipe or drain is generally in what may be considered a warm situation, and discharges of hot water through it not only raise the degree of temperature of the air, but also transmit heat to the surroundings, which again acts upon the air passing in through the inlet, expanding that, and so really assisting the ventilation. The steam, also, from hot water discharged from baths, sinks, &c., will always rise to the highest point, and it is absurd to think that warm vapour of any kind would pass downward through a drain perhaps 50 or 100 feet long, so as to then ascend a lofty ventilation pipe, simply because the pipe was placed there for that purpose. Why people should persist in saying that the principles of ventilating downward are the only right ones is a mystery. If they simply stated that they had provided two stacks of ventilating pipes, one at each extremity of the drain, and did not care which was the inlet or which the outlet, so long as the drains were properly air-flushed, no person of experience would find much fault. Some people place too much dependence on the fact that carbonic acid (one of the constituents of sewage-gas) being so much heavier than air, would gravitate to the lowest point in a drain; so that it is necessary to ventilate in that direction to remove it. A man who understood his business would not advance this doctrine, knowing that, in the first place, this gas would, if generated, become diffused with the air in the drain; secondly, that it is almost always accompanied by sulphuretted hydrogen, a lighter gas, but heavier than air; and thirdly, that if the drains are properly laid, flushed, and ventilated, these gases cannot be generated, and so it is useless and a waste of time and money to provide for something which ought not to occur.

There are some few houses where this system of drain-ventilation (which is sometimes called the syphon system) is carried out, and where specially constructed cowls are fixed on the top ends of the soil and ventilation pipes to insure a current of air passing through. It is usual in these cases to use on one pipe a cowl so made that when the wind blows past it part is caught and induced to pass down and into the drain, the cowl on the top end of the other pipe being so constructed as to act as an air-extractor, and there is no doubt they answer admirably *when the wind blows*. But it often happens that there is little or no wind, and when this is the case the question may be asked, Are not these cowls an obstruction to the free passage of the air through them? and if so, would not open ended pipes answer just as well? If anyone had seen the number of mechanical (in distinction to fixed) cowls that the writer has, lying about on the roofs of houses, the owners in happy ignorance, and the system of drain-ventilation still going on in a satisfactory manner, they would come to the same conclusion that he has: that cowls are worthless, excepting under very peculiar circumstances.

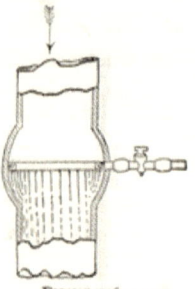

FIGURE 196.

Another principle for inducing an air-current through drains has been practised in a few cases. Figure 196 is a fractional section illustrating this. The air-inlet has a perforated pipe inside, and water is laid on to this and regulated by a stop-cock, as shown. The principle is simple and acts very well, the falling water drawing in a current of air. There are objections to this, however. In the first place, water must be plentiful; secondly, during a severe frost this place would become choked with ice, and so rendered useless, but this might be overcome by fixing the sparge pipe underground, beyond the reach of the frost. But it is an open question if the results gained would be worth the cost. Another firm of engineers drive air through the drains by means of either a centrifugal or an Archimedean fan, propelled by water-power. This is done by causing water under pressure to impinge against the perimeter of a horizontal wheel, which is connected by the necessary gearing, &c., to the fan.

It is doubtful if any mechanical means will come into general use for ventilating drains, as, in addition to the first cost, there would be that of attendance, wear and tear, and future renewals, the provision of motive power, and, after all, the liability of a breakdown through neglect and inattention.

There are a considerable number of houses which have been built during the last few years, in which the syphon system of ventilating drains has been carried out, and an air-current caused to pass upward through the pipe A, Figure 195. Near the position marked B the pipe has a cast-iron chamber inserted. Inside this chamber is arranged a specially constructed gas-burner, which has to be kept continually burning. In addition to creating a strong current upward, the inventor claims that what he calls "sewer-gas" is rendered harmless, and that disease-germs are destroyed by the burning gas. Some years ago there was a long paper war on this principle, in which I am not sure that the inventor did not get the worst of it. But as this work is intended to be on principles and not inventions, the reader is simply referred to what has been said about other means for inducing air-currents.

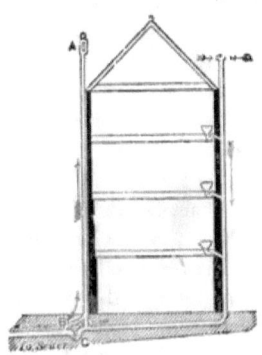

FIGURE 197.

Another illustration is given—Figure 197. In this case, so much faith was placed in the cowl at A that it was considered powerful enough to pull a current of air through the whole of the soil pipe at the back of the house, down the drains, and up the vent pipe fixed on the front of the house, and, in addition, *act as a ventilator to the public sewer*. This was arranged, as shown in above sketch, by fixing the pipe, B. It would be an economy, and simplify matters very much, to remove the trap, C, as it was rendered useless by the by-pass, B. The practice usually followed by most first-class sanitary engineers, is to make a provision for air to enter the drains, at a low level, near the trap placed to disconnect them from the public sewers, or, in country mansions, from the cesspool, when there is one. In this latter case, Figure 198 represents a section of a common way of arranging this air-inlet. A is the drain from the house or mansion; B is the disconnecting trap. The top part of this trap is continued to ground level, where a stone, with a grating let in, is placed for air to pass through, as denoted by the arrows. This grating is generally made strong enough to resist being broken by cattle walking over. It is not a good plan to fix the grating level with the ground—if fixed in a cultivated field, earth can fall through and choke up the trap. When fixed in a grass field, the gratings have been found to be entirely covered with long rank-growing grass, which has become so matted together as to entirely prevent any air from passing through, and so they have become perfectly useless.

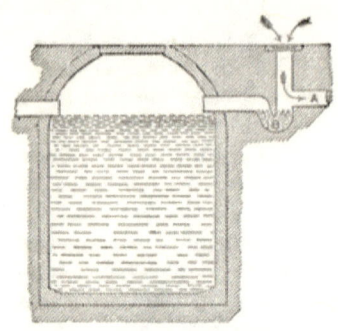

Figure 198.

The best plan, when the trap is near the surface, is to build a small hollow brick pier near the trap (not over it, as it would be necessary to pull it down for access to the trap), with a stone over the top, and a grating let in the side about 1 foot or 1 foot 6 inches above the ground level. This prevents grass or weeds growing over the grating, and is high enough to be above an ordinary snowfall. Another advantage claimed for this is that leaves or other matter cannot drift and lay on the grating in the same manner as when it is laid flat. Straight-barred gratings are the best to use, and the bars should be placed upright so that nothing can lodge on them. Perforated air-bricks are not so good as the grating, as the holes are liable to become choked. In one case earth had drifted into the holes, in which grass seeds had taken root, with the result that there was scarcely any air-passage. Figure 199 represents the raised chamber with the grating let in, as described above.

FIGURE 199.

In a long length of drainage one of these chambers could be placed at intervals of 50 or 100 yards, so that should any offensive air become generated in the drains, it could pass out of them instead of being conveyed up to the house and distributed, by means of the ventilating pipes, in the air surrounding the house. But these chambers can only be applied under favourable circumstances. In some cases they are objected to as being unsightly. They can be hid by planting shrubs around them. In other cases old tree-stumps and roots and ferns can be planted so as to disguise them. A great many people object to any kind of drain-ventilation being seen, remarking that they are suggestive of a stink, but if the drains are well flushed and ventilated there cannot be much to complain of in the way of smell.

At a nobleman's house in the country, during an examination which took three days to make, it was found that the whole of the drainage from the house, offices, and stabling (for about thirty horses) discharged into a cesspool 30 feet long by 10 feet wide, and in which was 6 feet of sewage in a horrible state of decomposition. None of the drains were trapped, or flushed except by the discharges from sinks and water-closets, and which kept continually stirring up the contents of the cesspool, thus setting free enormous quantities of sewage-gases. The cesspool was situated partly beneath the milk dairy, in the floor of which was a bell-trap. The dairymaid said she had to fill this trap with water every day to keep down the smells. *A branch drain from the cesspool was carried to the dry area round the house for the purpose of taking away any water that might get in there,* AND THIS DRAIN HAD NO TRAP IN IT. On raising a stone for access to this area, the stench that escaped was so bad that the men had to beat a retreat for some time until it had cleared away. The cesspool was emptied, and the drains from the stables were separated from those from the house. New ones were laid to convey the sewage about a quarter of a mile, where it was distributed by intermittent irrigation. Entirely new drains were laid round the house, with the necessary trapping and ventilation, and at the highest point of each main drain was fixed a forty-gallon automatic flushing tank. In the park was fixed air-gratings, similar to Figure 199, and, after an interval of two years the whole system is working to the owner's satisfaction. All solid sewage is retained in a specially-constructed chamber, so as not to choke the sub-irrigation pipes.

To return to air-inlets to drains, Figure 199 has been fixed to several London suburban houses, where there has been a small shrubbery or garden, but in these cases the drains have been entirely relaid; where old drains have been left in, other precautions have been taken to prevent bad air issuing out of the grating during a reversal of the air-current by discharges down the drain. This is when a path or walk has been near.

Figure 200 is a sectional elevation of a manhole found at a house. A large iron grating was placed over it, and as this place was surrounded by trees and shrubs, the manhole had a considerable quantity of rotten leaves lying in the

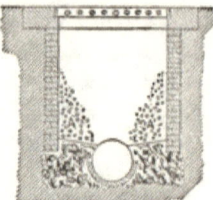

FIGURE 200.

bottom, as shown at A, Figure 200. Rain fell on the leaves, causing the place to smell offensively. On moving the top leaves a great deal of fungoid matter was found to be growing. The drains were well laid and flushed so that whatever fell into the channel was washed away, and nothing offensive could be found excepting these rotten leaves.

In some cases a branch drain has been laid from the manhole to a distance away, with an upright pipe and domical wire grating

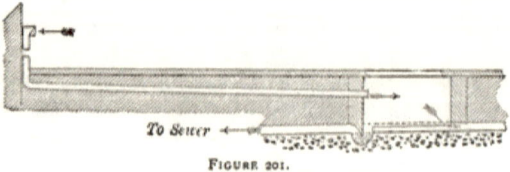

FIGURE 201.

on the end to allow air to pass into the drains, as shown in section, Figure 201. The domical wire grating on the end of the upright pipe has been objected to, as rain can fall through. If the pipe and horizontal air drain should retain any dust which may have had a vegetable or animal origin, and which would have been carried in by the air-current, any moisture would cause it to further decompose and give off offensive smells, and which would be discharged through the ventilation pipe at the head of the drain. It might be said in that case there would be no ill effects from smells, but it must be much better to avoid their cause than to do that which would assist their generation. If sufficient rain fell through to scour away anything that lay in the horizontal pipe it would be an advantage, but as only just sufficient falls through to do harm, it is better to put a conical cap over the grating (or use a louvred cap instead of the grating) to prevent this. The same objection could be raised against some kinds of blow-down cowls, which would catch a great deal of water in case of a driving rain. These arguments only apply when any part of the air drain is horizontal.

CHAPTER XXI.

DRAIN-VENTILATION—*continued*.

THERE is such a variety of houses, and they are so arranged, that the drains nearly always have to be treated differently in the way they are ventilated. The greatest trouble arises when an old house has to be redrained; with new ones the necessary provision can be made when building.

Figure 202 is a plan of the front part of a common description of terrace house, and Figure 203 a section on A, B, drawn to a

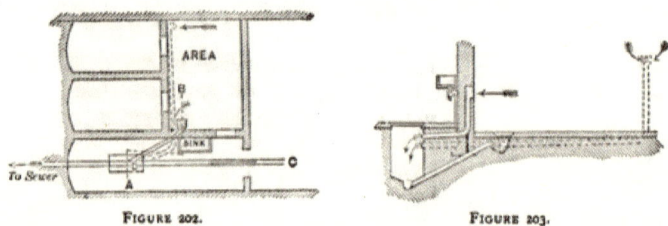

FIGURE 202.   FIGURE 203.

slightly larger scale. In this case the drain, C, from back part of house has to pass beneath the kitchen floor and through the scullery (which is partly situated beneath the sidewalk of the street). The manhole, D, must of necessity be inside the house beneath the scullery floor, and as these places must be perfectly air-tight when so situated, and yet easy of access, a removable cover is placed over the top. Sometimes this cover is of stone bedded in mortar. Some architects specify it to be bedded in putty and white lead mixed together, but these are both bad plans, as the joint cannot be guaranteed air-tight, and if disturbed for any purpose proper care is not always bestowed when refixing. It is much better to use an air-tight iron cover. There are several kinds of these in the market; those are best that are hinged in an iron frame and which can be locked. Those which are not hinged and locked are untrustworthy. Curiosity sometimes leads people to see what is below, and the cover does not always get properly replaced. In one instance three empty wine bottles were found hidden away in the manhole, and that leads to the conclusion that these places could be made use of for other improper purposes unless properly secured.

In Figures 202 and 203, the fresh-air inlet is shown as a branch drain leading through the outer wall where an upright straight-barred grating is placed for air to enter. In this case the grating should be not less than 9 or 12 inches above the area paving, so that dirty water cannot run into it off the surface, or, when the paving is being washed down, cannot be thrown or swept into it. The reasons for this were given in an earlier chapter. This air drain should have a slight inclination, so that should any vapour condense in it the water would run back into the manhole. Too much fall should not be given to this air drain or it would come too low down in the manhole. It is better to keep it as near to the top as convenient, so that should any of the lighter sewage-gases accumulate at that point, they would become diffused with the incoming air and so get carried away up the drain and ventilation upcast pipes.

Referring again to Figures 202 and 203, it should be stated that the drains must be well laid and flushed, or, at times, smells will issue outward through the grating when any discharges are sent down the drains. When iron drains have been laid I never knew anything offensive to pass outward through an open grating, and I have sometimes seen and used them to stoneware-pipe drains. In the case of drains badly flushed or improperly laid, the air drains have to be continued further away, as shown by dotted lines, and a pipe carried up 6 feet above the paving of area, and on the top fix a balanced mica valve in a metal box. Figure 204 is a sketch of this valve and box. The valve is shown by dotted lines, and partly open, as when air is entering.

FIGURE 204.

At Figure 203 the air-inlet is shown by a thick line connected to the drain from the surface-gulley in the area. This is a good plan, as the accumulation of foul air in the branch drain is prevented. This drain can be connected to the manhole, as shown in section, Figure 205. By this arrangement the dirty water from the area gulley-trap is conveyed down to the bottom of the inspection-chamber so as to avoid the walls being splashed, and at the same time the incoming current of fresh air would be divided, part escaping out of the top of the T-pipe, as shown by the arrow, and the other falling to the bottom by reason of its heavier weight,

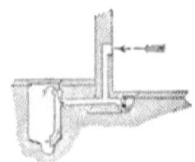

FIGURE 205.

and also to take the place of that passing along the drain and up the ventilation pipe.

To give the details of the whole of the various means for ingress of fresh air to drains would occupy too much space and time. We will now refer to a few places where the application of those described has been difficult. Figure 206 is a small section showing the application to a row of new houses near Belgravia. In these cases the servants' water-closets were fixed in a front vault under the public sidewalk. An air-grating was let into a shaft carried up to the paving level so as to ventilate the water-closet, and a 4-inch pipe was carried up from the manhole to the under side of the grating, with one of the mica valves, pre-described,

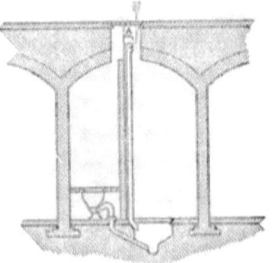

FIGURE 206.

on the top end, so that no unpleasant odour should escape to the annoyance of pedestrians. These soon became covered with street driftings, and it was found necessary to make alterations and shield them, as shown at A. These answer very well when proper attention has been given to them by removing the matters which accumulate over the inlets.

Another principle is shown in section, Figure 207. Here the front area is covered in with thick glass and all the front vaults under the sidewalks are used for storing coal. The coal-shoots could not be used for the purpose, so the only available way was to bring back a pipe from the manhole to the main wall of the building, and then up above the covered-in area to a position a few inches above the sidewalk. A cast-iron grating was placed flush with the face of the wall to protect the valve from injury, and behind the grating louvred baffle-plates to intercept dust and road refuse as much as possible from drifting into the mica valve. In the above case the manhole was covered in with a cast-iron air-tight cover, fixed a few inches below the paving. Over this was laid a strong stone slab bedded on dry sand, so that coals falling on it cannot injure the cast-iron cover.

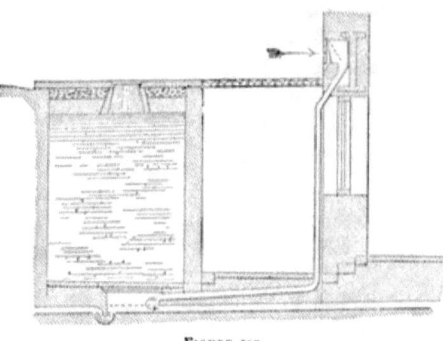

FIGURE 207.

Figure 208 is a basement plan showing the drains of a house. Being a corner house, there is only a very small yard at the rear of the building. All the water-closets and other upstairs sanitary fittings have their pipes brought down and connected with the drain in this yard, the drain from this point being laid in a straight line to the public sewer. A disconnecting trap inserted and a manhole built over it, as shown, with an air-inlet arranged at the manhole, and the two soil pipes continued full size to the roof, the system would have worked first-class; but, on referring to the plan, it will be seen that by far the greatest length of drainage is the branch continued to take the waste from the scullery sink and a continuation to the side area for surface water off the paving. It was proposed, in the first instance, to take a ventilation pipe from the head of this drain to the roof, but the owner objected to a pipe (which he said would be unsightly) being fixed up the front of the house. The best suggestion that next offered itself was to fix an air-inlet valve at this point. This was done, but it was found to be always closed, excepting when discharges were sent down the scullery sink, when it would open slightly for a few seconds and then close again, doubtless by the rarefied air in the drain pressing against the back side of the valve. After several experiments it was found that when the air-

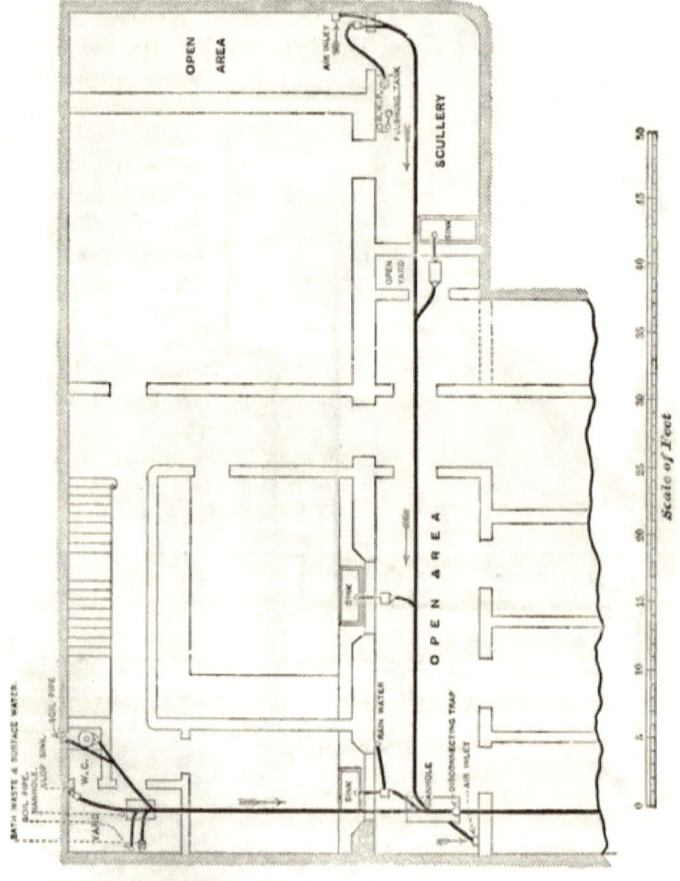

FIGURE 208.

inlet at the manhole was reduced to one-third of its area, which was about 13 square inches, the one in the side area remained open, showing that a current of air was passing through the whole of the system. As this branch drain had nothing but small discharges of water sent down, and which were not sufficient to scour it clean, it was deemed advisable to fix a "Field's" flushing-tank in the scullery (where the frost would not affect it), and a connection was made with the surface water-trap in the side area. This would scour the drain, recharge the trap with water should any evaporate, and also change the water in the trap should it become offensive. The "Field's" flushing-tank was supplied with clean water, as it has been found that when dirty water has been used for this purpose the tank has become very offensive by reason of its foul condition inside, necessitating its being frequently cleaned out. The waste-pipe from the scullery-sink was made to discharge into a grease-intercepting trap.

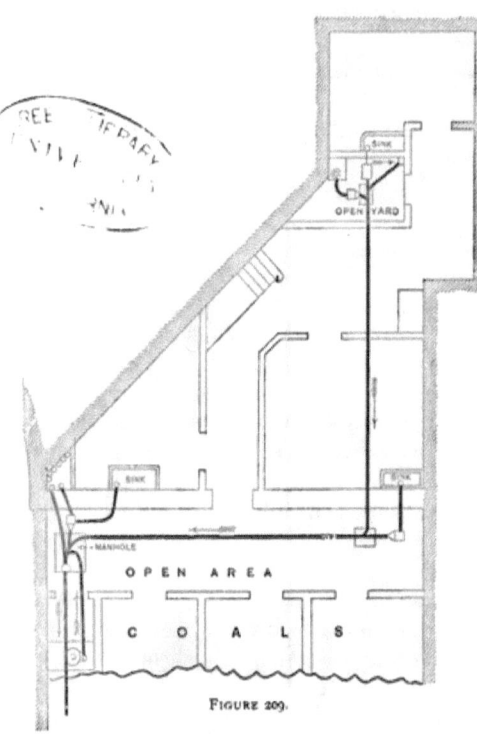

FIGURE 209.

At another house it was found necessary to ventilate the drains in direct opposition to the principles laid down above. Figure 209 is a plan of this house. Being situated at the corner of a street and crescent of houses, it was badly designed, the party-wall forming an acute angle with the front wall of the house. The back basement was used as a kitchen, and the only uncovered surface was a small yard about 5 feet square. It was impossible to ventilate the drains at the highest point without fixing the pipes on other people's property, so it was decided to fix the fresh-air inlet at the highest point of the drain, and use the soil pipe, which was fixed inside the sharp angle at the front of the house, as the upcast ventilator. A flushing-tank was fixed in the small back yard, as it was feared that the drain, having a sluggish fall and only dirty water discharge sent down it, would become foul inside. This flushing-tank can be regulated so that it will discharge itself at intervals of time varying from a few minutes to several hours.

Some engineers say that a fifty or sixty-gallon discharge once or twice in twenty-four hours is the right system to adopt, but I much prefer to have, say, a ten-gallon discharge every two hours. For large houses with long lengths of drainage the quantity might be increased; but with drains well laid, even if in long lengths, ten gallons of water discharged into them through a 4-inch pipe in a few seconds will be found to have a very good effect.

The soil pipe of the house, Figure 209, and which acted as the drain upcast ventilator, was fixed in a recess inside the house, and the hot-water circulation pipes being close by, the heat rarefied the air inside the soil pipe and so accelerated the ventilation of the drains, &c.—in fact, overcame the resistance offered by the rarefied air in the horizontal drain, and which would have a tendency to rise to the highest point where the fresh-air inlet was fixed.

Figure 210 is a plan of a very common description of a terrace house, having two frontages, one on the street and the other on a common garden or lawn, the sides being hemmed in by other houses. One of this kind had to be rearranged after it had been built about nine or ten years. On examination being made it was found that the whole of the soil beneath the basement flooring was saturated with sewage. About fifty loads of earth had to be dug out and carted away. A so-called sanitary engineer had made a few alterations about two years before the writer was sent to see what was wrong, and one of this gentleman's introductions was a cesspool—misnamed a trap, about 3 feet square and 5 feet deep, situated at the point marked A on the plan, Figure 210. This was not sealed down air-tight, and the floor over being of wood, it may be judged that this was a very unhealthy corner. A staircase was immediately over this cesspool, up which the smells used to pass and into the various bedroom windows which looked on to the staircase.

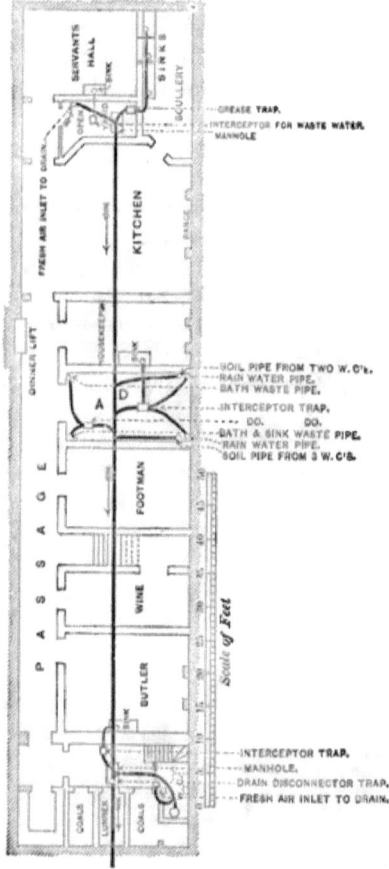

FIGURE 210.

A new 5-inch cast-iron drain was laid through the house in a perfectly straight line, a brick manhole or access-chamber being built at each end, in the front and back areas, so that if a candle was held at one end it could be seen at the other.* There were two stacks of soil pipes, as shown on the plan, Figure 210, fixed in the angles of the staircase, and both were defective and had to be renewed. One was of light iron, and was also used as a rain-water pipe; all the joints were defective.

It was impossible to remove the water-closets so that the soil pipes could be fixed outside the house, or the water-closet chambers ventilated through the walls, unless some of the best rooms had a portion partitioned off. So to make them as inoffensive as possible ventilation-shafts were fixed from each

---

* If an examination is made during the day-time a glass hand-mirror is as good as a candle to reflect light into the interior of the drain pipes.

water-closet chamber to a position clear of the skylight over the staircase—see Figure 360.

It was suggested that the drains should have a ventilation pipe carried up from the highest point to the roof, but the occupier objected to have this done.

The only resource left was to make use of the soil pipes to ventilate the drains. Unfortunately, these pipes were in the centre of the house, and, so that the whole of the drains should have an air-current pass through, an inlet had to be arranged at each extremity. An open grating was left in the front area at the exit end of the drain. A similar grating was left in the back area, but it was found that sometimes when hot water was discharged down the scullery-sinks steam escaped through the grating. When this water was that in which vegetables or fish had been cooked, the smell was so unpleasant that a mica non-return valve had to be fixed to prevent any escape.

This mica valve had to be so adjusted that it would open only a short distance, and the open grating in the front area had to be contracted, so that, as nearly as possible, an equal amount of air would pass in at each end of the drain.

The part D was only connected in a temporary manner until the ventilation had been tested, as it was thought probable that it would be necessary to insert a trap, to insure both sections of the drains being thoroughly ventilated, but this was not found to be required, so the joints were made secure.

CHAPTER XXII.

DRAIN-TRAPS FOR SURFACE-WATER AND WASTE PIPES.

To WRITE a history of traps is quite beyond the author's purpose, but the subject is so important that it cannot be passed over in silence. To describe all that are in the market would only weary the reader. A great many makers would be delighted to see their goods mentioned, while others would be vexed and disappointed at theirs being omitted. Neither of these considerations affect the writer, who is neither a patentee nor manufacturer.

The various traps used for the purpose of receiving surface-water and waste-water from sinks, &c., may be divided into the following heads: The cesspool-trap, as shown by Figure 211; mechanical traps, or those with a flap-valve either in the water, in the body of the trap, or on the outlet or discharging end; traps with floating balls, or those with heavy balls, which, under certain conditions, close the trap from back pressure from the drains or sewers; and traps which are constructed so as to retain water in such a way that under ordinary conditions no drain-air can escape through them. This last description can be further divided into two classes: Those (mis)called self-cleansing, and those specially constructed to retain certain matters which would do harm if allowed to pass into the drains. A further class may be added—those

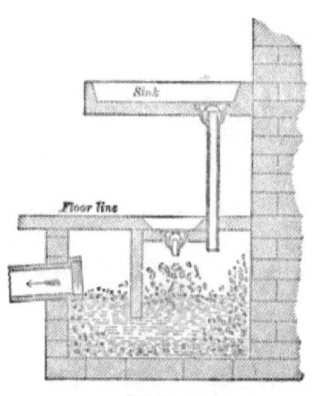

FIGURE 211.

DRAIN-TRAPS FOR SURFACE-WATER AND WASTE PIPES. 185

which have a water-seal, but are so very peculiar in their construction that they may be said to be almost as offensive as the evils they are supposed to remedy. Anyone who wishes to make a study of the various kinds of traps in the market is referred to the advertisement pages of almost any paper devoted to sanitary or building matters, where he will find plenty to interest him.

Figure 211 is a section of a cesspool-trap as usually fixed under a scullery-sink, and it would not be an exaggeration to say that more than half of the houses in the most aristocratic squares and streets in London are as shown. In fact, in some of the larger houses these traps are made bigger in proportion—no doubt so as to be in keeping with the house. No matter how large this trap may be made, sooner or later it will require to be cleaned out. The accumulated fat, which will be found to form a crust on the top, and a quantity of other matter, such as sand that gets washed down when the copper cooking utensils are scoured, mud from vegetables, small pieces of meat, bones, &c., from the dinner-plates, and very often portions of garbage and cleanings from game and other kinds of food—all pass into this trap. If any difficulty arises in this, the bell can be lifted off the trap in the sink and the poker will do the rest. When these places are opened for cleaning out, the stench that escapes is so *strong* that words can scarcely be found *strong* enough to describe it. More often than not the bell is left off the trap in the sink, so that the trap below may be said to breathe into the house, as the air is rarefied and cooled by alternate discharges of hot and cold water down the waste-pipe. The bell-trap, which is generally placed in the floor to keep the smell of the *trap* below from escaping, is very often found uncovered or without water. In some cases, where the servants have been particularly clean, and prided themselves on the clean and white appearance of the stone floor, the grating has been so covered with hearthstone that its presence could not be discovered. Everything about this arrangement of the scullery-sink is wrong. There are very few houses indeed that have not a back yard of some kind or other, where the trap could be placed to receive the scullery waste-water, so there is no (or very rarely any) excuse for fixing them indoors. The bell-trap in the sink is a continual source of annoyance and trouble. The water-seal is so little that a slight puff of air will blow through it; in a few hours the water will evaporate out of it. The water-way through it is so reduced in area that it is difficult to get water to run away without lifting the bell, this very rarely being replaced in a proper manner. The bell-trap in the floor is perfectly useless. It is generally placed there to receive surface-water when the floor is washed down with water and broom. As a matter of fact this is never done, the maids usually doing this part of the work on their knees with pail and flannel. Neither is it a good plan to swill down the floors inside a house. It may clean them, but it makes them and the walls for some inches up so wet that it may take several days to make them dry again; or, if any woodwork gets wetted, rot is engendered. So this floor-trap should be entirely removed.

Scullery-sink traps require to be specially made so as to intercept and retain all grease, or sand, &c., as these matters frequently cause stoppages in the drains. This is the reason, it may be presumed, that the old cesspool-trap has still a great many admirers. An improved way of constructing these places is shown in Figure 212. This was designed by a master plumber under whom the writer was working some years ago, and since then a great many have been made on the same lines. The advantages are that this trap can be easily cleaned out without allowing sewer-air to pass through during the time it is open. The incoming water is discharged below the floating matter, so that if this water is hot it does not remelt the grease, or break it up. The outlet is so arranged that grease cannot float into the drain, and this is the great object sought. The trap itself is a small brick chamber, built water-tight, the inside face rendered with cement trowelled up to a smooth surface. The incoming pipe is the lead waste pipe from the scullery-sink continued to within a few inches of the bottom. At A is a brass screw-cap soldered in. On removing this a cane can be passed

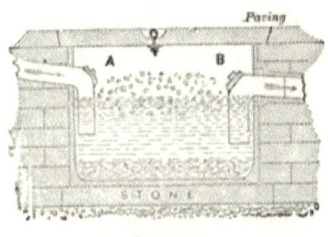

FIGURE 212.

DRAIN-TRAPS FOR SURFACE-WATER AND WASTE PIPES.

through the pipe on each side. The outlet has a lead pipe bend fixed to the drain with a lead collar soldered on to fit into the drain pipe socket, so that an air-tight joint can be made with cement. On this bend, as shown at B, a large brass screw-cap is soldered, so that drain-rods, with a hoe or brush on the end, can be passed into the drain to remove any matter that may accumulate there to form an obstruction. The cover for this trap should be air-tight.

There are three or four patent traps in the London market, made nearly similar to the last one, but have the advantage that they are made in one piece of vitrified stoneware, or highly-glazed potteryware. Figure 213 is one of these that is spoken well of.

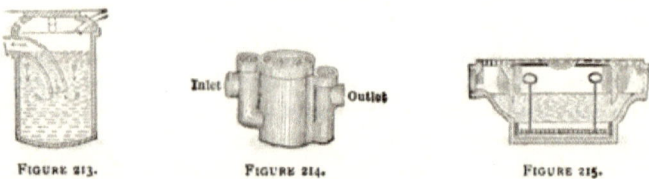

FIGURE 213.   FIGURE 214.   FIGURE 215.

The inlet pipes of this trap discharge on to any matter that may be floating in the trap, but the end of the outlet being immersed beneath this matter it cannot float into the drain. Figure 214 is another trap specially designed for the same purpose. In this case the inlet and outlet ends of the pipes are both below the floating grease, but being placed outside the body of the trap the inside is open and without any objectionable corners difficult to be cleaned.

Figure 215 is another grease-intercepting trap, differing from the last in that it is oblong with round corners on plan. Figure 215 has also a galvanized-iron perforated pan, made to fit into the bottom, with a handle for lifting. This is a great convenience, as at one lift can be taken out the whole of the solid matters that may have accumulated there. These traps all have the advantage that a man with a pail of hot water and soda can wash them as clean as when new.

It is usual to fix these traps beneath the yard-paving, but this is only necessary when they are placed in a position where the contents would become frozen. There is no reason, beyond frost, why they should not be fixed above the paving; indeed, they would be better so, thus making them easier to be cleaned out. It will generally be found that all places requiring periodical attention or cleaning get done in a better manner if no difficulties are in the way.

It is important that all kinds of grease-intercepting tanks should have the covers quite air-tight, and a small air pipe fixed to them and continued to a position where any smells escaping from it would not be offensive. There are evils, and evils, and the grease-tank is a necessary evil; but a trap of this kind is out of place when fixed where there is no grease to intercept. A gentleman amateur sanitarian had five of these traps fixed in various positions round his country mansion to receive the waste-pipes from sinks, baths, &c. He thought these traps " were the best, as they were the most expensive." He also had a "great objection to a complication of traps," and "objected to any being placed in the waste pipes near the fittings." It was very difficult to make him believe that he had made a mistake, but he could not combat the evidence of an anemometer placed over the waste-pipe of the wash-hand basin in his dressing-room, which registered 380 feet lineal as having passed through in five minutes, and this air had to pass over the contents of one of these traps, which were anything but pleasant to the olfactory nerves.

For a large building the consulting engineer designed a grease-intercepting tank similar to Figure 212, but it was made about 8 feet long. During the greater part of the day the contents of the tank are nearly at boiling point, the grease floating about like oil. There need be no importance attached to this, excepting that proper attention must be paid to it, and the liquid grease skimmed off before it accumulates to such an extent as to get through the syphon pipe into the drain, but this place is very offensive-smelling, and, if neglected, the grease floats down the drains.

At another large building, similar to the last one, the drains were opened for examination and found to be in very good condition and free from grease. It was naturally presumed that an elaborately-arranged grease-interceptor would be found in the scullery, but on examination nothing of the sort was discovered; the only trap was a 4-inch P-trap and a 4-inch drain from it. As the whole affair was working satisfactorily it would have been ridiculous to propose any alterations, so nothing was done beyond ordinary repairs to fittings. This last case shows that, under certain conditions, grease will retain its liquid state until clear of the house drains, but the drains themselves must be at a temperature above that at which grease congeals.

Nothing has been said about the sizes of grease-tanks for ordinary dwelling-houses, and, indeed, little can be. It may be said that the tank itself should hold sufficient cold water to congeal the grease, but, as a matter of fact, this is almost impossible without making an enormously large affair. It would be very improper to do this; the smallest traps are very offensive, and there is not the least doubt the evil would grow with the size. Then, again, in some houses more cooking (hence more grease) is done than in others; and even the kitchenmaid is an important item to be considered—one will scrape all grease off utensils into a garbage-tub, and another one will put all into the dish wash-up tub, and which eventually finds its way into the drain, or grease-tank, if one.

A firm of London sanitary engineers, with which the writer is well acquainted, has been in the habit of fixing a small-size grease-interceptor, with an outer jacket, so arranged that cold water can pass between them and so keep the contents cool. Figure 216 is a sectional elevation showing the arrangement. The outer chamber, A, A, is made of stout tinned-copper, with an outlet, C, for connecting to the drain. There are two brass unions, D and E, for connecting a cold-water supply and overflow pipe. The inner chamber is also of tinned-copper, with a pair of handles for lifting out, and with an air-tight cover. There is also another union for connecting a vent pipe to

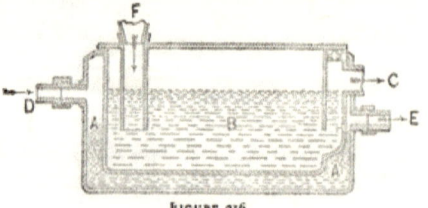

FIGURE 216.

the inner chamber. By a simple and ingenious arrangement the waste pipe, F, from sink can be disconnected, also the part connected to the outlet, C, so that the inner chamber and its contents can be lifted out, emptied, cleaned, and replaced with very little trouble; or the cover can be removed and the contents scooped out without taking away the receptacle. Disinfectants should always be used, and should be applied some time before disturbing the contents of these fittings. When fixed beneath they should be kept as close as possible to the sink, in which case only a very short waste pipe would be required to connect them; but when fixed some little distance away, or out of doors, it is important that a trap should be placed in the waste pipe as close to the sink as possible, to prevent any unpleasant odours from passing through the pipe. Even fresh air passing through one of these dirty pipes would be rendered unfit for respiration. There can be little doubt that to this may be laid the cause of so many cooks and kitchenmaids being laid up with sore throats.

The modern way of arranging the scullery-sink is to fix it near an outer wall, with a trap beneath the sink, and the waste pipe continued to the grease-tank fixed outside in an area or yard. This is shown in section, Figure 217. It is a good plan to fix the sink on cantilevers, so that the whole of the space beneath is in view and easily cleaned. I find the more light there is in these places the less liability there is of their being used as stores for rotting vege-

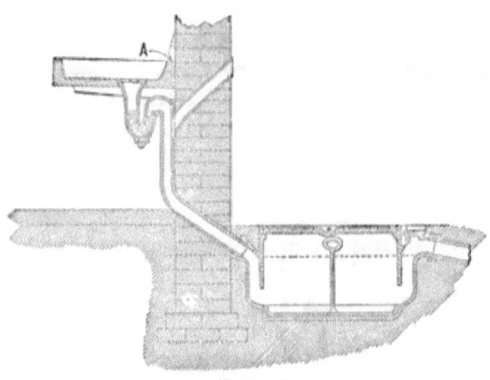

FIGURE 217.

tables, garbage, uncleaned pots, old shoes, worn-out house-flannels, and other things that in some mysterious way find a last resting-place at this spot. In Figure 217 the waste pipe is shown on the face of the wall; it is better to let it flush into the wall, so that the pails or a passing broom should not bruise it and so contract its water-way. An air pipe should always be fixed as shown in the sketch. This is not entirely for the purpose of preventing syphonage of the water out of the trap, as this is not an important item under these conditions. It will be found that a large flat-bottomed sink always retains enough water on the bottom to afterwards dribble into and recharge the trap. But if there were no vent pipe the air between the two traps could not escape, so that the waste pipe would be "air-bound," and the waste water would not pass readily away.

A quibble might be raised as to the position of the vent pipe, but I think this the best, and always fix them so. If they are fixed on the top of the crown of the trap, grease would sometimes wash into it, and, in time, choke it up and so render it useless. Neither could the trap be fixed so close to the fittings if the vent was fixed on the crown. Again, it is an advantage that the air pipe should not be too close to the water in the trap. A current of warm air would assist to evaporate the water out of the trap, and in frosty weather the water would be more likely to freeze by contact with cold air.

In some of the higher-class houses so much grease has accumulated in the branch drains from the butler's pantry and the still-room sinks, that it has been found necessary to provide the needful conveniences near those places for retaining it in the same manner as described for sculleries.

When stoneware scullery-sinks are used, they should not have any corners where dirt can accumulate beyond the reach of the scrubbing-brush. For instance the reader is referred again to Figure 217. A is a ledge that can be avoided either by letting the sink into the wall, or a fillet of cement made with the face sloping into the sink, or a wooden fillet can be laid in and a lead flashing fixed over it; but the cement is the best, not being liable to rot in the same manner as the wood. It is still better to line the walls all around the sink with glazed tiles bedded in cement. This prevents the walls becoming saturated with offensive-smelling matter from frequent splashings of dirty water. Where tiles are not handy, or the cost (which is not much) is an objection, the walls should be rendered to above the splashing point with cement and sand worked up to a smooth face. Stone sinks, when of a porous nature, should be avoided as much as possible, as they generally smell offensive by reason of the amount of matter retained in the pores.

Figure 218 is a plan showing the arrangement of scullery-sinks at a nobleman's country mansion. A is a grease-and-sand-intercepting tank to receive the waste pipes from B, C, and C.

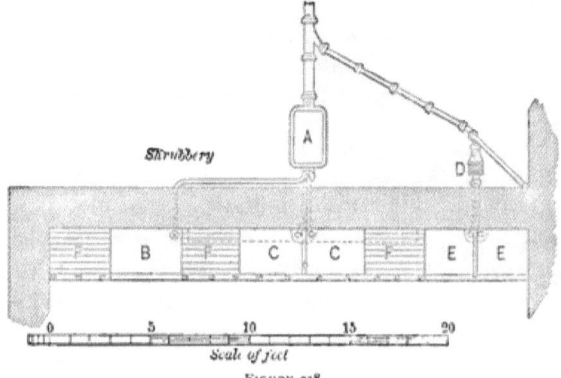

FIGURE 218.

D is an ordinary drain-interceptor trap to receive the waste pipes from sinks, E, E. B is a large-size stoneware sink, white glazed inside, used for tubs or scouring coppers. C, C, are two wooden sinks lined with strong tinned-copper, one to be used for washing, and the other for rinsing, plates and dishes. Over these sinks along the back, shown by dotted lines, is a wooden rack in which to place the plates, &c., for draining. E, E, are the two sinks made of slate slabs bolted together, to be used especially for washing vegetables, salads, &c. F, F, F, are draining-boards between the sinks. These are covered with corrugated tinned-copper, so that the water off cups, basins, &c., will run away from them back into the sinks. Figure 219 is a fractional section

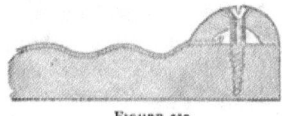

FIGURE 219.

across this drainage-board, showing also a tinned-copper half-round capping-piece screwed along the front of the drainage-boards and continued on the front edge of the copper sinks, to prevent that being bruised by tubs or pails, &c.

Waste pipes from scullery-sinks should always be fixed separately, so that, should one become stopped or choked up, the others will not be affected or thrown out of use.

CHAPTER XXIII.

TRAPS AND WASTE PIPES.

In nearly all town houses the servants' offices are in what is called the basement or lower story. This is situated below the level of the street—in some cases only 3 or 4 feet, and in others as much as 14 or 16. Between the house and the street is generally an open space, called the area, and this is usually a few inches lower than the basement floor. In thoroughfares or streets with shops on each side these areas are mostly covered over, sometimes with gratings, in other cases with thick plate-glass, or iron frames into which are let glass prisms of such a shape that light is reflected by them into what would otherwise be dark corners. In private houses these areas are open, and anyone walking along the street can look down into them. To get rid of rain-water from these places, the most common convenience used is a bell-trap—Figure 220. The most common size is what is called a 6-inch—that is, measures that distance across the top. The water-way or size of the pipe of a trap of this description is about $1\frac{1}{2}$ inch, and it does not matter how heavy a storm may be, this is supposed to be large enough to take away the rain-water as fast as it falls. The part A is

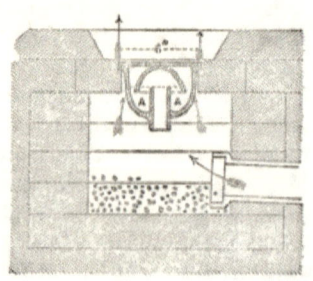

Figure 220.

generally found to be choked with mud (or dust in fine weather), and the result is that, after a shower of rain, a number of servants will be found dodging about on their heels in a pool of water, with a long-handled broom, each vainly endeavouring to find out the position of the trap so as to remove the obstacle which prevents the water running away into the drain. So that this shall not occur again, it is usual to take off the bell-grating and lay it in some corner of the area. The result is that a current of air can freely pass out of the drains and through any open door or window into the house. An apologist for the bell-trap was once heard to pass an opinion that in some instances it was *very useful for admitting fresh air into the drain*, in which case, it may be argued, a plain grating would be much better, provided that one could be sure of the air-current being in the proper direction. In spite of all sanitarians condemning this trap, they are still sold and fixed in thousands, and, in some cases, in what is considered to be first-class houses. A small brick chamber is usually built for this trap to discharge into, as shown in Figure 220; and so that the drain may easily be got at in case of stoppage, the trap is fitted loosely into a stone, with the result that drain-air can pass out all round, as denoted by the arrows.

Some people prefer a trap shown in section at Figure 221. This trap is superior to the last-described, in that, on removing the grating, the water-seal is not broken, and there is a larger water-way through it. This trap also necessitates a brick chamber beneath it. Another mistake frequently made is to have the stone in which these traps are fixed some 2 or 3 inches lower than the surrounding paving, as shown at Figure 220. When treated in this way it will be found that all the pieces of paper, leaves, straw, and other constituents of London dust, drift about until they find a resting-place in this hollow. If the rain-water pipes from the roof are fixed so as to discharge on to these area-pavings, the reader can readily imagine the results. The writer has frequently had to inspect empty houses for prospective tenants, and found several inches in depth of water in front and back areas after a rainfall, and more than once this water has been found to be so deep as to reach up to the gratings in the house-walls placed for ventilating any spaces beneath the floors in the basement, and pass through and lay in a pool until it soaked away into the earth beneath the floors. Further comment on this is unnecessary, as water under these conditions is nearly as bad as sewage.

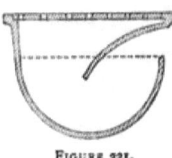

FIGURE 221.

TRAPS AND WASTE PIPES.   195

To give a list of all the traps that are made, and which are suitable for the foregoing circumstances, is beyond the writer's intentions. Only one, shown in Figure 222, is selected, because of its shape, and also by reason of its having from 2½ to 3 inches water-seal. This is an important item, as during a dry season a trap with a small water-seal is liable to have it broken by evaporation. The inlet portion is made large so that the grating may have as much water-way through it as the body of the trap, that being of the same size as the drain from it. The inlet-arm

FIGURE 222.

is for the purpose of attaching the waste pipe from a sink, bath, or wash-hand basin. By doing this the trap is always kept charged with water (that is, when the house is occupied), and the water in its trap is being continually changed, so that it may not become offensive by stagnation.

A sink is generally placed in the window of the butler's pantry, in which case the trap to receive the waste pipe is placed outside in the area. To avoid an unnecessary number of traps the area-paving should be laid so that any water falling upon it should run into the above-mentioned trap, and so save using a separate one for the surface water.

Figure 223 is a sketch section showing the usual way of arranging the waste pipe from sink and the trap under outside paving. Several sanitary engineers are in the habit of fixing the waste pipe

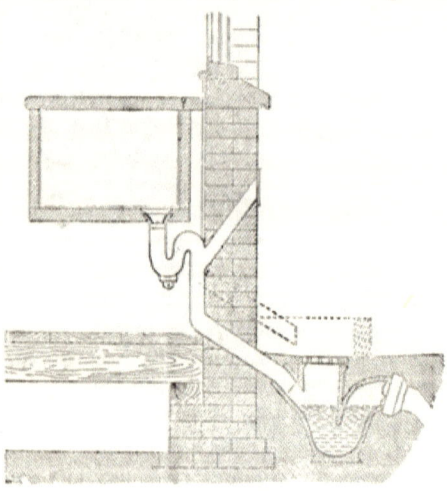

FIGURE 223.

so as to discharge over the grating of the interceptor-trap, as

O 2

shown by dotted lines, Figure 223, with the result that most of the water coming out of the waste pipe is splashed over the surrounding pavement, rendering that in a continually sloppy, dirty, and unpleasant condition. Sinking the outside trap below the level of the paving, as shown in Figure 223, does not get rid of this. In some instances a raised curb is put round these places, as shown by dotted lines, with the result that they require constant attention for removing street driftings, &c., that as surely find their way to this spot as if they had been collected and placed there by someone's hands. When a raised curb is placed around a trap for receiving waste water it is necessary to fix another trap for the surface water.

Some sanitary engineers are not content with having the drains trapped off from the public sewer—a thorough air-flush through the drains and an interceptor-trap fixed to receive the waste pipes—but are so alarmed at the idea of sewer-air escaping and passing up the waste pipes into the sinks or other fittings, that they stipulate for the waste pipes to discharge some distance away from the trap. Taking this extra precaution against an unlikely event leads to the evil of a large surface of space being continually splashed with dirty water. In one case four waste pipes (one was from a cistern) discharged into an open channel leading into a gulley-trap 3 feet away. A few leaves had got over the trap-grating, so that all bath and wash-hand-basin waste water that came down, having no other means of escape, over-flowed the channel until the surrounding gravel and earth was reduced to a sewage-bog, and this was in a back garden close by the conservatory.

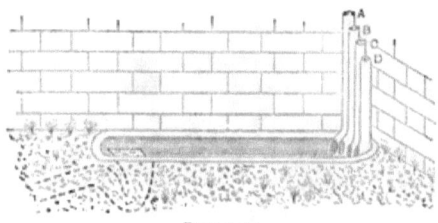

FIGURE 224.

Figure 224 is a sketch showing the arrangement. A is a bath-waste; B, waste from safe under water-closet; C, wash-hand-basin waste; and D, overflow from cistern. Smells were noticed in the bed and bath rooms, which led to the above discovery. In this case no traps were fixed on the waste pipes, so there was nothing to prevent smells passing back into the above rooms. In another case a similar arrangement was found, but the precaution had been taken to place a grating over the channel and trap to keep out leaves, &c., but the channel was offensive-smelling by reason of the sides being covered with a greasy-looking matter, the bottom being quite clean down the centre.

It is an economy and much more sanitary to take all waste pipes into the trap, as shown at Figure 223, but they should be above the water-line, when, to all intents and purposes, they are open to the air, and if ventilated and trapped, as shown there, it may be said that triple precautions are taken to prevent any gases from the sewers passing into the house by means of the waste pipes. It is unnecessary to speak of any other sinks in the basement, as they should be treated in the same manner as the pantry, excepting where a lot of greasy matter passes down the waste pipes. The remarks on scullery-sinks will apply in those cases.

When examining houses (especially those built some years ago) for smells the source is very often found to be in a cellar. The other day, when looking over a noble earl's house, a wine-cellar was found to have two bell-traps, devoid of water, and air from the drains passing through with sufficient power to blow out a candle. Next door was the beer-cellar (both cellars being nearly in the centre of the house), and another bell-trap was found there; this having been improperly used by the men-servants as a urinal no smell was passing from the drain, but the smell of the other matter was so strong as to make the eyes run with water. I have frequently had to complain of cellar-traps being used for this purpose, and for that reason always advise their entire removal. Whenever there is a complaint of a smell in a butler's pantry or footman's cleaning-room, these being also used as sleeping-rooms, it is frequently found that no proper convenience is provided for the men's use, but a sniff inside the sink generally tells how that trouble is got over.

Under no conditions should any trap be fixed in the paving on the inside of a town house, as they are never used.

Passing from sinks in the basement, the next one for consideration in a large house where there are children will be the nursery scullery, this being sometimes on the second or third floor. These sinks do not require such an elaborate provision for intercepting grease as in the case of those in the ordinary scullery, and they may be treated with regard to trapping, ventilation, and waste-disconnection, much in the same manner as described for the butler's pantry-sink.

It does not matter where the sink is situated—whether at the top or the bottom of the house—it should always discharge into a gulley-trap in preference to being connected directly with the drain, or any soil or other foul pipe. Where a long length of waste pipe is fixed, the ventilation pipe from the trap under the sink is very needful to prevent syphonage of water out of the trap. This ventilation pipe should never be less in diameter than the waste pipe itself, and in some cases it has been found necessary to fix it a little larger, so that there is not the least obstruction to air entering as fast as the rush of water down the waste pipe displaces that contained in it. These ventilation pipes need not necessarily be continued to the roof of the building, but may be continued through the nearest external wall, so long as it is a few feet away from any window. The top end should be a few inches above the level of the top edge of the sink, so as to avoid the first rush of air displaced by the water discharge from carrying a few splashes out of it. In some instances a brass grating has been soldered over the end of this air pipe, or the end has been closed and a few small perforations made in it. Both of these form an obstruction to the free passage of air. It is much better to make two saw-cuts across the end of the pipe, which should be slightly enlarged with a turn-pin, at right angles to each other, and two or more pieces of stout copper wire laid

FIGURE 225.

in, as shown at Figure 225, and soldered with a copper-bit. This is to prevent birds placing anything in the pipe to choke it up.

The best kind of sink to use is made of white porcelain or earthenware. This has a clean appearance, or, if dirty, it at once shows the maid's lack of attention. A small bowl should be used for washing up crockeryware so as to avoid risk of breakage as much as possible. It may be said that enclosures to these sinks are a mistake, and on no account should a lid be allowed. It has been found, on opening a lid of one of these places, that it has been half full of dirty crockeryware, plates half full of rejected food, small saucepans of infants' diet, and such-like things, that give off an unpleasant sour odour when neglected. The cocks for supplying hot and cold water should be placed high enough for drawing into a jug. If too low, so that there is a difficulty in getting the jug underneath the nose of the cock, a breakage frequently takes place. Neither is it a good plan to put the cocks at the back side of the sink, unless they are some height above it, but they should rather be at one end, so that hot and cold water can be drawn at the same time into the same bowl, and yet leave as much of the space of the sink for washing-up purposes, without doing any damage, as possible. Sometimes the cocks are fixed clear of the sink to avoid this, but it is not a good plan; and what with leakages and splashings, the surroundings are in a continuous wet condition. These sinks should be moderately deep.

CHAPTER XXIV.

SLOP-SINKS.

Slops from nurseries, and also invalids, should never be thrown into the ordinary sink. It is a daily occurrence to see the housemaid empty a pail of slops down a sink, and then immediately draw water for toilet purposes into a glass bottle from a cock, the nozzle of which may be splashed with slops. This suggestion is not nice for water-drinkers or those in the habit of drinking water out of the toilet bottle. In all well-arranged houses a special provision is made for receiving chamber-slops. Some kinds of slops must go into the water-closet, but these are the exception rather than the rule.

There is no more frequent cause of stoppage in soil pipes and drains than allowing slops to be thrown down water-closets. Nail, tooth, and scrubbing brushes, housemaids' flannels, lumps of soap, the sweepings of the bedrooms, small ornaments, and a host of other matters are frequently recovered from those places. A properly-constructed slop-sink would catch and retain all these matters, and at the same time allow others to pass away. It is unreasonable to expect a person to hold up the handle of a w.c. apparatus with one hand and pitch a pail of slops into it with the other, and yet if this is not done the basin will often overflow on to the floor or into the safe beneath. In some cases a plain basin, as Figure 226, has been fixed for the purpose, but while this is better than the pan

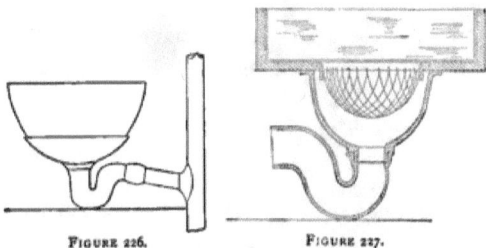

Figure 226.   Figure 227.

SLOP-SINKS.

or valve water-closet, still there is no provision for intercepting improper articles passing through at the risk of causing a stoppage in the pipes. Figure 227 is a sectional elevation of the same kind of slop-sink, with a galvanized-iron wire basket fixed over the top for catching solid articles that would otherwise pass away. Some years ago this was a speciality of one firm of sanitary engineers, but they have now discarded it on the ground that the meshes of the wire basket got clogged with matter, and in time this smells very offensive. Another firm of sanitary engineers have a slate sink perforated in the bottom near one end, and an earthenware basin fixed beneath for receiving slops. Figure 228 is a sketch showing this. In this case no provision is made for intercepting brushes, &c. Figure 229 is an elevation of a basin with

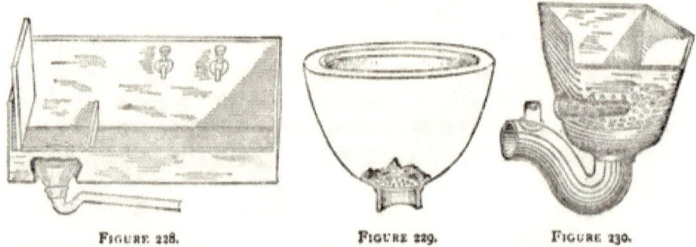

FIGURE 228. FIGURE 229. FIGURE 230.

a piece broken out to show the grating in the bottom. All these slop-sinks are made of earthenware. It not unfrequently happens that the basins get broken by throwing improper things into them; indeed, from the usage they sometimes get, one would think that they had been intended for rubbish-shoots. Figure 230 is a patent one that seems to meet all requirements. It is made of cast-iron and is porcelain-enamelled inside. The back and ends are higher than the front so that nothing can splash over them, and, although rounded in the bottom, the top is square so as to be readily adapted to an enclosure, and, if fixed in an angle, no corner is left to be filled in with woodwork or other material, which would be necessary if it had been round. It is also made large enough to hold rather more than a pail of slops, so that if a housemaid's flannel should cover the perforated grating before the person using this place saw it in time to stop, the sink would not overflow. The faint lines show the position of the strainer in the bottom, but I believe this is now discarded and a metal star-grating bolted to the bottom with a T-bolt and screw so as to be easily removed if required. A slop-sink should not be fixed more than 18 inches high in the front. The higher the pail has to be lifted the less control the person has over it, and the more likelihood of the surroundings being splashed with offensive matter. Receptacles for slops should always be fixed in a well-lighted and well-ventilated position. If in a dark place the slops might be thrown on the floor by mistake.

A rule has been laid down that all sinks should be trapped, and this one is no exception to the rule. The waste pipe should not be less than 2 inches in diameter; a size larger, say 3 inches, in very large establishments or hotels would be better, and the vent pipe should be of the same size as the waste pipe.

It is usual to disconnect slop-sink waste pipes from the drains in the same manner as other sinks, but it may be argued that what is pitched into these places is often so offensive that they ought to be treated in the same manner as water-closets. The greatest reason against this is that very often quantities of hot water are sent through them, and where the soil pipes are of lead it would soon cause them to break. Leaden waste pipes that are used to conduct hot waste water very often break from the same cause, and, speaking from experience, the more firmly and rigidly a pipe of this kind is fixed the more likely it is to break by the alternate expansions and contractions. One firm I have worked for use lead waste pipes 2 inches in diameter, and the substance of the lead from ·2 to ·22 inches thick. The reader may judge of the strength of this pipe by the statement that it requires a pressure of 498 pounds on the square inch to burst it, and yet the writer has had to make good broken places in a bath-waste of this description before it had been used for twelve months.

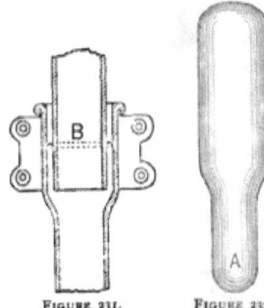

FIGURE 231.   FIGURE 232.

An expansion joint, as shown in section, Figure 231, is better than a soldered joint for hot-water waste pipes. A hard-wood mandrel is first made, as shown at Figure 232. This should be well greased and driven in the end of the lead pipe, causing that to swell out to form the socket. If the end of the pipe is first warmed, it will be found the substance of the lead is not reduced in thickness, and if the part A of the mandrel is made to just fit the pipe it will prevent the pipe buckling sideways or the socket being made more on one side of the pipe than the other. A bead can be worked on the end of the socket, and a pair of very thick lead lugs soldered on the back for fixing to the wall. B is a vulcanized indiarubber ring, sprung on the tip end of the entering pipe. The size of the socket should be such that this ring on the end of the other pipe will fit moderately tight, so that when one pipe is pushed into the other the ring will roll between them and finally remain where shown. The socket should be about 5 inches deep inside, and the entering pipe should only be allowed to enter about 4 inches, so as to avoid its resting on the bottom of the socket, in which case the objects aimed at would be defeated. 2-inch pipes are generally made in 12-feet lengths; if these are cut in halves and each piece treated as shown, the sockets will be about 5 feet 6 inches apart, and that is quite far enough, as the only fixings are at the sockets.

In a newly-built hospital everything was done that could possibly be required for the use of the doctors and nurses, and comfort of the patients. On each floor are baths and other conveniences. All hot and cold-water, waste, and ventilation pipes are fixed inside the building, and supported every 3 feet on iron brackets pinned and cemented to the wall as shown in elevation, Figure 233. The iron band is made in two halves, and after placing the pipes in position these are bolted together as shown, the pipes all being 2 inches clear of the walls, so that they can be seen and accessible all round. A narrow lead flange is soldered on the lead pipe above each bracket to prevent it slipping through, and so avoid the necessity of bolting the band up so tightly as to bruise

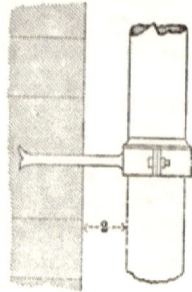

FIGURE 233.

the pipe. Each pipe is painted a distinctive colour, so that there is no likelihood of confusion as to its purpose. For instance, hot-water pipes are painted red; cold, blue; waste, purple; air, white; and so on for other pipes.

To allow for expansion and contraction in the bath waste pipes the joints are made as shown in section, Figure 234. In this case the end of the top pipe is slightly coned, and the bottom one opened to receive it. A vulcanized indiarubber cone-piece, with a bead on top edge, as shown in Figure 235, is sprung on the coned end of the upper pipe, which is then placed in its position and socketed into the bottom one. These joints are so flexible as to allow for all variations in length of piping that are likely to arise.

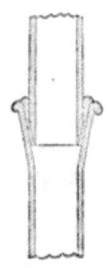

FIGURE 234.

FIGURE 235.

To return to our subject of sinks, these conveniences should be placed in various parts of a house where they would be useful, and so save the labour and toil of having to carry water from a distance and up or down a flight of stairs. At the same time thought should be bestowed on their position. A cupboard or closet in a bedroom is not a good position, or anywhere that injury would be done to the rooms beneath should a pipe break or other leakage arise.

The writer has had to do with several cases where the waste pipes have had to be removed when fixed near an important room, as the rushing noise of water was so unpleasant and suggestive of a water-closet. A slop-sink should not be fixed in a water-closet room, or the same doorway used for the two places. Where gentlemen are about, this would be against the laws of decency. The front portion of the house is not always suitable to fix a sink in. There are cases where a sink so fixed has required a long length of drainage brought to receive the waste water, and a ventilating pipe fixed up to the roof. When sinks are fixed here and there on the bedroom floors of a house, and in such a position that each one requires a separate waste pipe down to the basement, previous remarks will cover all that need be said about them, but when they are situated over each other, additional precautions have to be taken with regard to fixing the waste and ventilation pipes.

The reader is referred to A, Figure 236, where the branch waste pipe is soldered into the main stack above the floor-line. This is the right thing to do, it being an economy of material and there being a short length of branch pipe to keep clean, in addition to having no bends in it. But the one on the floor below should be differently connected to the main stack, as it often occurs that discharges from the upper sink will run up the branch pipe of the lower one, which, if fixed as shown, cannot very well happen. But this arrangement brings up another question—that of syphonage. The upper one is secure against this, provided the branch pipe is not very long and the main stack is continued with an open end to act as a ventilator, but the lower end will have the water drawn out of the trap every time a pail of slops is thrown down it by reason of the outlet pipe acting as a syphon. To prevent this a ventilation pipe should be taken from the branch pipe, as shown by dotted lines at B. This should not be less in diameter than the waste pipe. Another use for this vent is that it prevents the air in the waste pipe being so compressed when discharges are sent from a higher level as to burst through the water-seal of the lower trap. One would suppose that if the main

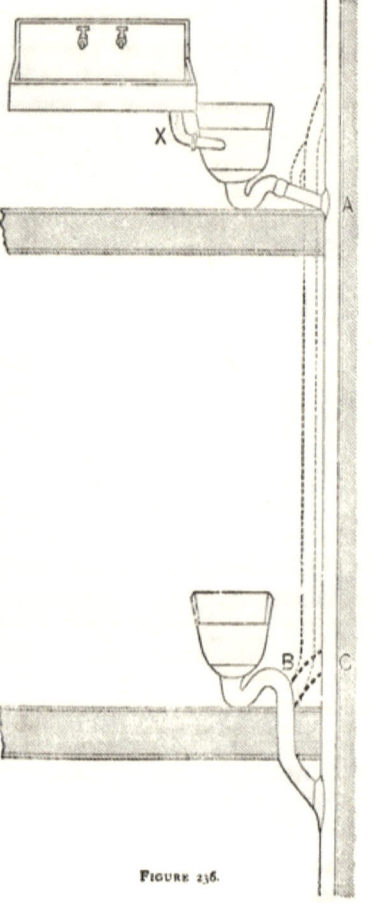

FIGURE 236.

stack of waste was open at the bottom end this could not occur, but it is found in practice that it does sometimes happen, hence the necessity of this precaution being taken. Again, this pipe will prevent an accumulation of bad air at the point B, which would be displaced by discharges from the sink, and driven out at the bottom of the waste pipe, to the annoyance of anyone standing near at the time, or the liability of its entering an open window on the ground or basement floor. Instead of carrying this vent pipe from the trap upward, as has been shown, it has sometimes been branched into the vertical waste pipe at C, as shown by thick dotted lines, but this has not been so successful as the other way described. Some years ago the writer fixed a stack of slop-sink waste pipes, and ventilated each trap through the wall. There were four sinks, fixed vertically over each other, and Figure 237 is an elevation showing one of them. But it was found that whenever a sink was used an unpleasant puff of air was driven out, and this would sometimes bring a few drops of water with it. It was also found that only sections of the system had a continuous current of air passing through, the air in the other sections being stagnant, and this would account for the smells driven out at the bottom end of the waste pipes.

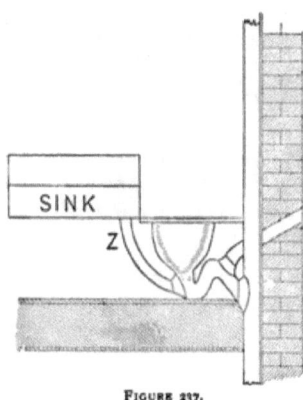

Figure 237.

In addition to the slop convenience it is always necessary to have a sink for drawing water, and this should have a plug over the waste pipe for retaining water in the sink, so that it may be used for washing up any small articles. In the bottom of the plug-washer should be fixed cross-bars to prevent anything getting into the waste pipe to form a stoppage. The waste pipe from this sink does not require any trap. It can be taken into the one under the slop-basin, as shown at Z, Figure 237, or into the arm, as shown at X, Figure 236. The latter is the best plan, as less pipe is required and no stagnant air is displaced when the sink is emptied. The sink also empties quicker, as the air in the pipe does not offer any resistance in the same manner as when the bottom end is sealed with water, as shown in Figure 237. Tinned copper looks cleaner than lead for these sinks. The back side, and, when fixed in an angle, the end as well, should stand at least 18 inches high to protect the walls from splashings and also insure any dribble from a leaking cock falling into the sink. The water service pipes can be brought behind the splash-board, the sink being kept out far enough to allow for that being done, it not being by any means a good plan to bury pipes in the wall. Holes can be made through the splash-board for the bosses to project through, so the cocks can be screwed in afterward. To hide the ragged edge sometimes made when making the holes, or to hide the place should it be made too large, a brass face-plate can be fixed at the same time as the cock. These can be engraved to show what they supply, as shown at Figure 238. At a job recently done there were four cocks engraved "cold," "hot," "filtered," and "soft," fixed to each sink throughout the house.

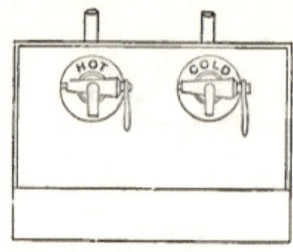

FIGURE 238.

CHAPTER XXV.

BATHS.

It is not often that a plumber has to make a bath. In some old country mansions a wooden tub or casing lined with sheet-lead is found, and generally as shown at Figure 239, the bottom and sides being put in in one piece, and the ends soldered in

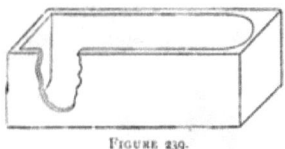

FIGURE 239.

afterward. Baths sold by makers may be enumerated as follows: Sheet-copper, zinc, tinned and galvanized iron, concrete, porcelain, stoneware, cast-iron, marble or slate slabs bolted together, and marble cut out of the solid block. These last are not often seen; they also have the objection that the bath requires to be prepared some time before using, when a warm bath is required, so that heat may be imparted to the material, which would otherwise feel cold to the bather. The stoneware and concrete baths are generally used in hospitals, public baths, and other institutions of a similar kind. The metal baths, and also those made of slate slabs, are usually enamelled inside, either in a plain tint or grained in imitation of marble. When baths are intended to be fixed without an enclosure, a rounded rim is made on the top edge, and they are also enamelled on the outside, and sometimes have ornamental feet to stand upon. For best work the copper baths are preferred, as they last a long time, and when they get discoloured or scratched can be re-enamelled so as to look equal to new. All kinds of sheet-metal baths, when made of light substance, should be well "cradled"—that is, have a skeleton wooden casing fitted to them to prevent the bottom and sides from bulging outward by the pressure of the water and weight of the bather. For cheap work the cast-iron baths are very good, and if the enamel is such that it expands and contracts in the same proportion as the iron they last some considerable time. Complaints have sometimes been made that common iron baths are too narrow in the bottom, so that a stout bather will get wedged in.

A bath is not only a comfort and convenience, but the waste water is a valuable auxiliary for flushing the drains. This is too often lost sight of, and the mistake made of fixing a small waste pipe and discharge-cock, so that the bath is slowly emptied of its contents. The result is, the water dribbles through the drains instead of being sent through with a good scouring force to clear away any matter that may have lodged in them.

Figure 240 is a sketch of a bath as commonly fixed, with ¾ or 1-inch waste and hot and cold-water cocks attached.

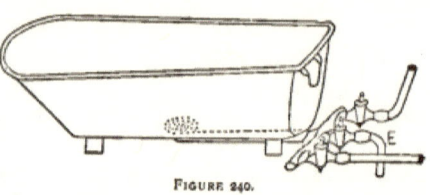

FIGURE 240.

The levers are omitted for clearness. On looking at this it will be seen that the inlet for water is through the waste pipe. This is a very bad plan, as when the bath is emptied the soap-curds which float on the surface of the water, being the last to leave the bath, frequently remain in the waste pipe and get washed back again with the incoming water, making that look dirty and anything but inviting to the bather. In some cases the supply to the bath is by means of standard cocks with nozzles projecting over the bath, as shown in sectional elevation, Figure 241. This is better than the plan last described, and does away with the evil pointed out When these cocks are used they should be fixed at the foot of the bath; if fixed at the head there is danger of the bather knocking his head against them; if fixed at the side the elbows might suffer, and they are also difficult to get at for connecting the unions when fixed near a

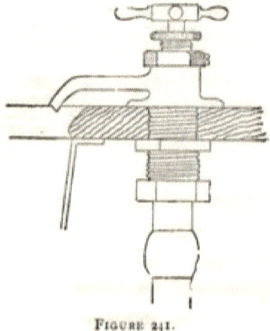

FIGURE 241.

wall or partition. Sometimes, when a bath supplied in this manner is being charged with hot water, the room is filled with volumes of steam, and if the walls are cold this condenses and runs down. Another way of supplying a bath is shown at Figure 242. This is almost similar to the last one described, the only difference being that the valves are combined and discharge through the same nozzle, a porcelain cover with a soap-dish, A, being fitted over the valves, this being more cleanly-looking than the cocks, which soon tarnish unless protected by nickel or silver plating. Figure 243 is an illustration of

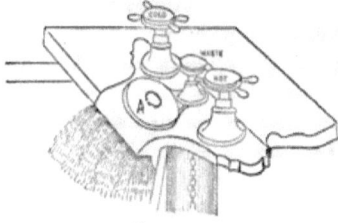

FIGURE 242.

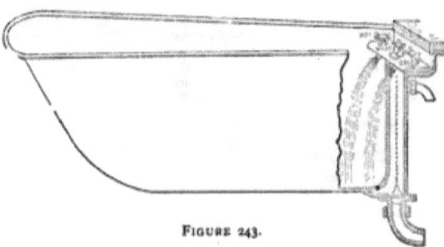

FIGURE 243.

a bath showing the supply and waste apparatus fixed inside the bath at the foot. These baths are also made with the valves at the side near the foot, and are either right or left-handed. Figure 244 is an illustration of a copper bath with side-inlets for the supply of hot and cold water. If these are kept low down the ends of the pipes are soon covered with

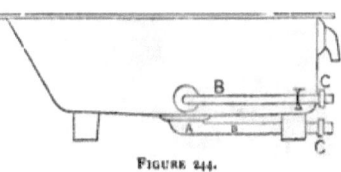

FIGURE 244.

water, and this prevents the free escape of steam into the room. The waste pipe, A, is shown connected near the middle, as done in the usual way, but it is an improvement to fix it as near the end of the bath as possible. This saves a piece of pipe, and as the waste pipe is always more or less foul on the inside, the shorter the pipe the better. Sometimes, on the score of economy, the pipes are left short at the points B, B, and the plumber, when making a soldered joint to them, has allowed the hot solder to touch the sides, and so discoloured the inside enamelling that the bath has had to be sent away to be re-enamelled. Some baths have only one inlet pipe. This is wrong, as, if the hot and cold-water cisterns were at different levels the pressure from the hot-water cistern at the top of the house would prevent the cold water, supplied from the cistern on the same floor as the bath, from entering. Baths of the description under discussion should always have unions attached to the pipes at C, C, Figure 244, so that they can readily be uncoupled should it be necessary to take them out for repairs or renewal of any of the parts.

All baths should have overflow pipes. Indeed, the overflows should be as large as it is possible to make them, for the reason that should the bath be filled too full, even if not to overflowing, when the bather gets in the sudden displacement of water will frequently cause it to run over the sides. In one case a gentleman who liked his bath to be nearly full, had it altered and perforations made all round near the top edge, with a hollow space outside to catch the water, as shown at Figure 245. The holes were ⅜-inch in diameter, and yet this was not found to take the displaced water away quickly enough. This was an extreme case, but shows the necessity of large overflows to baths, and also, when they

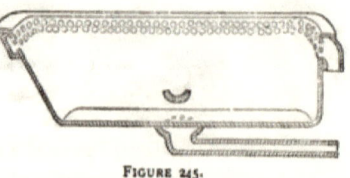

FIGURE 245.

are fixed upstairs, the importance of having safes or other means for catching any overflow of water that may take place, and so avoid doing injury to any rooms beneath.

To describe all the various cocks and valves for supplying baths with water would weary the reader and answer no good purpose. Suffice it to say that nearly all bath makers have either a patent or an adaptation of ordinary cocks, &c., such as are used for other purposes, and are turned, when fixed beneath the inclosure, by means of socket-keys or spindles and levers, or, if lever-valves are used, a connecting chain is fixed and attached to a knob through the bath-top, as shown at Figure 246; the spindle, G, sometimes being triangular in section, or round with a protecting strip on one side so arranged that when the knob is pulled up as far as it will go and a slight turn given to it, it will remain, thus holding the valve open.

For discharging the water out of a bath, very often a common stop-cock, as shown at E, Figure 240, is used. In some cases the size is only ¾-inch, and with the ordinary square way. In others a 2-inch round or clear-way-cock has been fixed, but this is not to be recommended by any means, as it is so difficult to turn them, especially after they have not been used for some time. Figure 246, if made to a good size, is a very good waste-valve, F being connected to the bath waste pipe, but the lever should be as short as possible or it will not open very wide. The same kind of valve has been used without the lever and weight, the spindle being continued as shown by dotted lines; the friction in the stuffing-box of the valve being generally sufficient to keep it open. It is found in practice that the spindle is very liable to get slightly bent or the valve moved sideways, in which cases it is difficult to open or shut this valve. For this reason they are now rarely used.

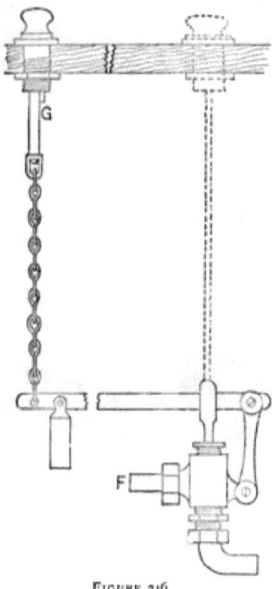

FIGURE 246.

A very simple waste-valve is shown in section, Figure 243, which is an indiarubber ball, with a weight above it, resting upon a turned brass seating bedded near the bottom of a cast-iron tube. This tube is connected with a branch pipe from the bath, and has a union for connecting to the waste pipe. In some cases a smaller tube is used instead of the ball, the bottom end being made to fit into the brass seating, and thus acting as a plug to retain the water in the bath. The annular space between the inner and outer tubes fills with water to the same level as that in the bath. When the bath is filled to a certain height the water begins to overflow down the inner tube, which is made to stand so that the top is level with the highest point at which it is intended the bath shall fill.

When this kind of waste apparatus is used the overflow-connection is generally omitted from the bath. Plumbers sometimes make a bath-waste apparatus with lead pipes, as shown in section, Figure 247. The outer tube is a piece of 4-inch lead soil pipe with the bottom end reduced to 2 inches, and a hole made in the side at the level of the branch pipe, H, from the bath. The inner tube is 2½-inch lead pipe, with a corresponding hole opened in the side, and so fitted that it projects through the one in the outer tube. The top of this inner pipe is cut down so as to act as the overflow pipe, the outer tube having its top level with the rim of the bath. A brass union, H, is soldered in, as shown, for connecting to the bath. K is a small brass ring into which is ground a valve, and this is soldered to the bottom of the inner tube just below the branch, H. L is a lead flange soldered on to support the weight of the apparatus, and also for cementing over the end of the under waste.

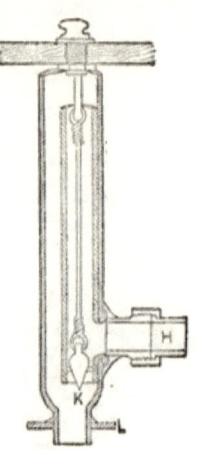

FIGURE 247

CHAPTER XXVI.

BATHS—*continued*.

FIGURE 248 represents in section a waste-valve patented by a leading firm of sanitary engineers. With this valve and a 2-inch waste pipe an ordinary-sized bath can be emptied in two and one-half to three minutes.

Common brass plugs and washers are not much liked for emptying baths, as they generally are made too small; the grating at A, Figure 249, gets choked with small pieces of soap, &c., and

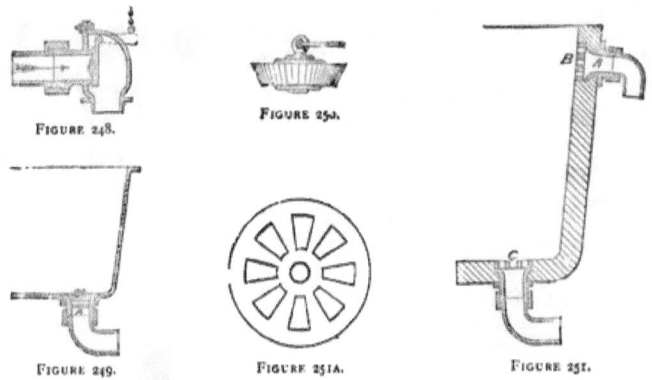

FIGURE 248. FIGURE 250.

FIGURE 249. FIGURE 251A. FIGURE 251.

the links of the chain get filled with matter and always look dirty. The plug, by being allowed to fall about, chips off pieces of enamel. This latter can be prevented by making the plug of vulcanized indiarubber, with a small brass plate on each side for fastening the ring to, as shown by Figure 250.

Figure 251 is a fractional section of a porcelain bath showing the overflow and waste connections. A, A, are brass couplings, with bent unions for joining to waste-valve and overflow pipe. B, C, are porcelain gratings to prevent anything washing into the pipes to choke them up. B should be cemented in to prevent it falling and breaking, but C should lay loosely in the sinking so that it can be lifted out for access to the waste pipe for cleaning it when necessary. Figure 251A is a plan of the porcelain grating.

Some Water Companies insist that all bath supply-valves shall discharge above the water-line of the bath, so that it can be *seen* if the water is running, and thus avoid waste. They also stipulate that the overflow shall not be connected to the waste pipe, but shall discharge out-of-doors in some conspicuous position. When this is done a flap should be fixed on the outlet-end to prevent cold air blowing through. This flap should be light so that it will open with the least pressure behind it. Flaps are generally made of sheet-copper, as shown at Figure 252. Any plumber can make them, and instead of cutting out the centre-piece when making the joint, as generally done, it should be bent as shown at A, when it will prevent the flap opening so far as to fall back and remain open, which now so often occurs when wind is blowing against it. The necessity of a leaden tray or safe on the floor under a bath has already been pointed out. This safe should be larger than the bath or it will not catch any overflow. The wooden floor should be laid so that any water will run down toward the outlet or overflow pipe from the safe. When the bath is fixed care should be taken to fix it level. If this is not done a space will be found at the lowest end between the rim of the bath and the wooden top. Strips of sheet-lead under the feet are the best for blocking up a bath; when wood is used it sometimes gets rotten from the surrounding moisture.

FIGURE 252.

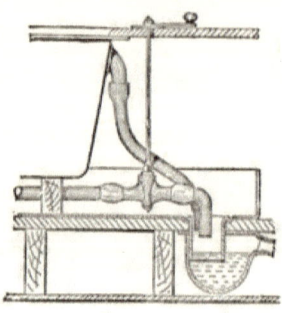

FIGURE 253.

The ordinary way of arranging a bath-waste is shown by sketch section, Figure 253. In this case a large-sized trap is fixed, and the overflow and waste pipes discharge into it as shown. Very often the waste from an adjoining sink or wash-hand basin is also made to discharge into the same trap. This is not by any means a good plan. On looking at a case of this kind it will be found that the discharges will, so to speak, boil over the trap and lie in the bottom of the safe, frequently giving off unpleasant odours. It will be noticed that this trap also receives all that may leak into the safe. So that this boiling over of the trap may not distribute the water over a large area, a sinking may be made in the floor. But this cannot always be done, and sometimes the bath is fixed at a higher level so as to gain the same object. This is an improvement, but it is much better to fix the waste pipe tight over the trap so as to entirely avoid the back-wash, as shown at Figure 254. In this case a small trap (say the same

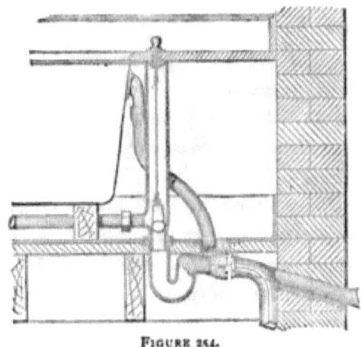

Figure 254.

size as the waste pipe) can be used. An indiarubber ring under the flange will prevent any escape of water into the safe. Where the waste apparatus is arranged in this way a separate overflow pipe must be fixed from the safe through the wall, and a copper hinged flap fixed on the end, as explained for bath-overflow. The overflow pipe from the bath can be made to discharge over that from the safe, as shown in Figure 254.

Figure 255 represents how a bath was connected to a slopsink in such a way that when a pail of water or slops was thrown

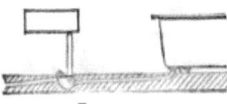

Figure 255.

down the sink it ran back into the bath with force enough to lift the plug out of the waste pipe. On opening the lid of the bath, which had not been used for some weeks, the stench was abominable, arising from matter lying in the bottom, and which had washed up and been left behind, as the water or liquid portion had slowly subsided or passed down the waste pipe.

Figure 256 is a sketch showing a hot-water cistern fixed in a cupboard at the head of a bath. This was at the house of the proprietor of one of the leading London sporting journals, whose wife and servants could not tell how it was that the hot water so often looked cloudy and soapy, and the water in the best bath always looked as if it had been used, although only just drawn. The bath—Figure 256—was on the nursery floor, and the hot-water cistern being so close—" why, it will only cost a few shillings, guv'nor, to lay on hot water to that bath," was the remark of the expert who conceived and carried out the brilliant idea. Only a few words of explan-

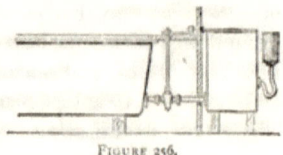

FIGURE 256.

ation are necessary. Bath and hot-water cistern being on the same level the cock was often left open, being unnoticed, as the hot water ceased running when the levels between that in the cistern and bath were equal. During the time the children were bathing, any hot water drawn at a lower level would be syphoned out of this bath; the ball-cock in the feed-cistern, being only ⅜-inch, not supplying the cistern as fast as it was drawn out, with the results given.

Another case came under the writer's notice where dirty water was drawn out of a bath at the tap over a sink on the floor below. In this case the cold-water cistern was small and had an intermittent supply. When preparing the bath, the cistern was emptied and the cock to the bath left open. During the time the bath was being used water was drawn at the sink, but as the cistern was empty it came from the bath. If cocks similar to those shown as Figures 241 and 242 in last chapter had been used this could not have happened.

Bath-rooms should always be well ventilated, and in such a way that there can be no unpleasant draught. Decency forbids open windows being used, even when the climate is such as to admit of this (unless precautions are taken to insure privacy). An unventilated bath-room always has an unpleasant odour in it, arising from a variety of causes. If a sponge is used by the bather the cellules will get full of soapy matter, and a few hours after using this will smell very offensive. These remarks apply to body-brushes and rubbers. The bath itself may be left in a dirty condition by allowing an accumulation of soapy curds to adhere to the sides. The bath may, perhaps, have been allowed to overflow, so that a pool of dirty water lay in the safe until it became offensive, or the cocks may have been in a leaky condition, and where the bath is inclosed with woodwork this is unnoticed. In addition to these remarks there is a probability of smells arising from a defective arrangement of the waste pipe and trapping.

In some cases bath-rooms are hung with paper. In some first-class baths in private houses the walls have been covered with upholsterer's hangings. In these cases there is always an unpleasant mouldy odour arising from moisture being retained in the wall-coverings. When the walls are painted, or covered with glazed tiles, steam will condense and run down. All this points to the necessity of ventilation. In some instances air-bricks are fixed in the walls near the ceiling for this purpose. In others a perforated centre-piece has been fixed in the ceiling, and a tube leading to the outer air to carry away any vapour. One eminent firm fixes a hollow and perforated cornice all round the bath-room, with tubes through the house-walls and with the ends open to the air. Whatever system of ventilation is adopted, care should always be taken to guard against any unpleasant draughts. A fire is generally looked upon as an additional comfort in a bath-room, but a hot-water coil is much preferable. With a very little extra trouble a hot-water coil can generally be made to work off the hot supply to the bath. If made of $1\frac{1}{2}$-inch copper tubes, say about 4 feet 6 inches long and to stand about 3 feet 6 inches high, with about six or seven longitudinal pipes, it will be found a convenient size for airing or warming bath towels. If this coil is nickel-plated it presents a smart appearance. This way of warming a bath-room has the advantage that no attention is required in the same manner as when a fire is used, and as the action is continuous the painted or tiled walls do not get cold enough for steam to condense upon, or if paper or hangings (which should on no account be recommended) should be used to cover the walls, and become damp, the heat would soon dry them again, and prevent mouldy or other smells being generated. Towels, &c., could be hung up to dry, instead of being left in a wet state. Thorough ventilation is absolutely necessary to carry off all damp vapours, which would otherwise condense again on objects or walls in the bath-room. This would take place in most cases during the night-time, when the boiler-fire would be out and the water cooled down, so as not to impart heat sufficient to keep the moisture suspended in the atmosphere in a state of vapour. The floors of bath-rooms should be waterproof, and more especially when on an upper story, so as to avoid any possible injury to rooms below. The floors should also be impervious to moisture, or, failing this, a tray should be provided for the bather to stand in. Carpets should not be allowed in a bath-room. A large flat piece of cork may be used for the bather to stand upon, as it does not feel so unpleasantly cold to the feet as oil-cloth and some other kinds of floor-covering. The only furniture required in a bath-room is a dressing-table and a *cane*-seated chair, and on no account should chests of drawers, wardrobes, clothes, or linen-chests be permitted to be in such a place where they become damp or musty-smelling. All baths should be fixed so that the bather may face the light.

From a sanitary point of view, baths should be open all around, and no inclosures used which would, perhaps, hide any defects or leakages. It has sometimes been found that fungoid matter has grown on the woodwork around a bath. At a noble earl's residence the writer found the cocks of a bath leaking very badly, and mushrooms growing out of the wooden inclosure. These mushrooms had stems about 10 to 12 inches long, and the tops were about the size of a shilling. They were all quite white and free from colour, and as they all appeared to be leaning in the same direction, search was made for the cause. A crack was found in the woodwork, through which a very fine ray of light could pass, and it was toward this ray of light that the fungoids were leaning, although about 3 feet away.

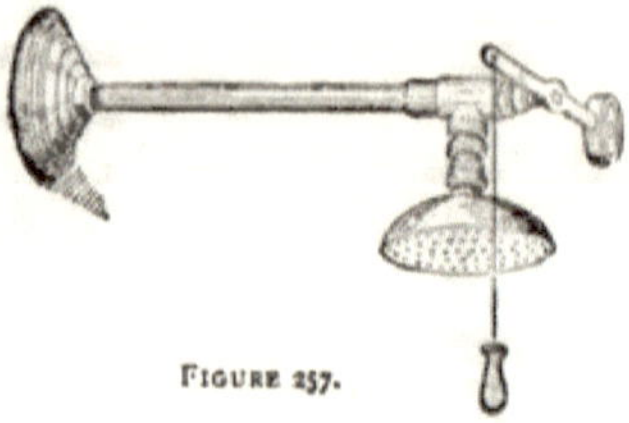

In addition to the baths that have been described there are shower-baths. Figure 257 represents a very simple one that can be screwed on to the wall. A

FIGURE 257.

weighted lever is keyed on to the spindle of a valve at the extremity of the bracket, and on the nozzle of the valve is screwed a copper or other sheet-metal rose or spreader. This can be used for cold water only. It has the advantage that the water runs during the bather's pleasure; immediately the handle is released, the water ceases running.

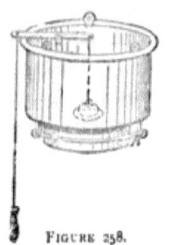

Figure 258.

Figure 258 is another kind of shower-bath, which can be suspended from the ceiling. This can be supplied with a small service pipe, with a ball-valve and ball to keep it full, ready for use. A, A, is a metal ring encircling the bath for attaching a curtain for preventing the water splashing too much. Hot water can be laid on to this bath, but it is dangerous to do so, as the bather has no means of knowing the heat of the water without the aid of a pair of steps and a thermometer, and so runs the risk of scalding himself.

Figure 259 represents a shower arranged by the writer. This consisted of a plain shower-bracket fixed at the necessary height,

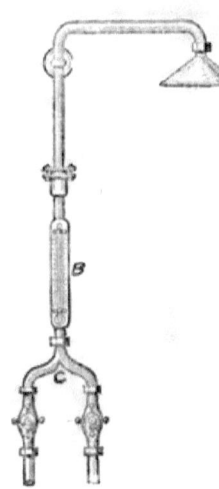

Figure 259.

with hot and cold water attached, the cocks being at a convenient position for the bather to adjust at his pleasure. The two pipes were joined together at C, and at B was fixed a Negretti & Zambra's thermometer, on a level with the bather's eye, so that he could see the temperature of the water, and by simply turning one or other of the cocks, make it hotter or cooler, as required. These baths all require a curtain to prevent the water rebounding off the bather's head and shoulders beyond what he may be standing in—usually a plunge-bath, or else a large tray especially made for the purpose. On the score of economy, a plain canvas or linen curtain is generally used, but unless this is taken down and dried (rarely done), it soon begins to smell unpleasant. Plain vulcanized india-rubber curtains have been objected to because of their unpleasant smell, but after a time this, to a great extent, will pass away, and as they retain no water they are not so unpleasant as the others. In some houses a wooden inclosure has been made for shower-baths; in others, zinc or other sheet-metal, enamelled inside, has been used. This last few years a new kind of bath has been much used. This, commonly called a needle-bath, is shown in elevation, Figure 260. The union, A, is for connecting the service pipe to the shower-bath, and that at B for the needle or spray-bath. In some baths, another union, C, is attached to what is commonly called the "wave." This is a narrow slit inside the bath, from which the water can play on the hips of the bather. In some baths a jet is fixed immediately beneath the shower, or a rose is attached to the jet. This is sometimes called a

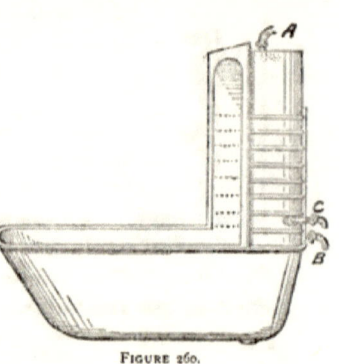

FIGURE 260.

"sitz-bath." These conveniences are only useful for medical purposes, and should not be attached unless really necessary, for the reason that a complication of cocks or valves is required, and these very often get out of repair. Dirty water will get into the rose or sinking of the sitz when the plunge-bath is used. When selecting or making a spray-bath, note should be taken that the tubes outside the hood-part, and the holes inside, are not too large, as, if so, the water will escape so freely out of the bottom holes, or a large quantity will be required to fill the tubes, that the top holes are almost useless. If the hood-part is attached to a plunge-bath of the ordinary shape, it is found that the needle-sprays will splash over the side of the bath, as shown by arrows, Figure 261. This does not happen when the bather stands in the centre of the hood. When a hot bath is required, the water is generally turned on before entering the bath, so as to ascertain the heat, and avoid the risk

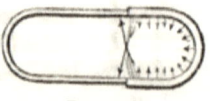

FIGURE 261.

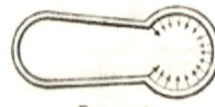

FIGURE 262.

of scalding. Baths made to the shape shown at Figure 262 do not splash over the sides so much as the one last described. All makers have their own system of supplying this kind of bath with water.

Figure 263 illustrates the engraved face-plate of one of the commonest tiers of nobs to the cocks or valves. Figure 264 shows another kind in side elevation. This is more simple than the other, as there is only one hot and one cold-water valve, the others being used to distribute the water to the desired places. This also saves a complication of piping, which is required when cocks are arranged as shown by the face-plates and knobs in Figure 263. For ordinary baths for private house the "douche," "wave," and "sitz," are not required and should not be fixed, for reasons already given. All the baths the writer has seen have the disadvantage that should the bather wish to turn on more hot or cold water, he cannot do so without stepping outside the hood. It would be better if the heads of the cocks or valves were fixed inside, so as to be under the bather's control.

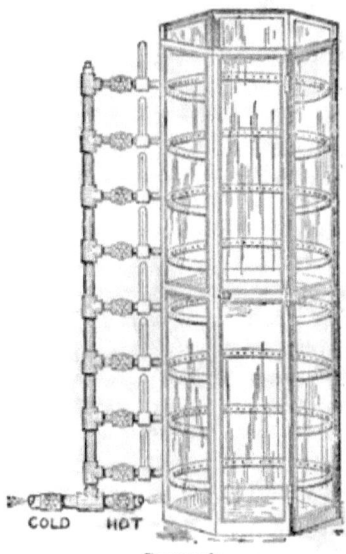

FIGURE 265.

Figure 265 represents a spray-bath as used in some hospitals. Instead of a metal hood with perforations, as shown at Figure 260, perforated copper tubes are used, and inclosed in a framework with glass sides, so that a patient standing or seated in the centre can be watched from the outside. A separate cock is attached to each ring, and a thermometer fixed so that the attendant can regulate the heat of the water to any desired degree, and can also turn it on at any desired point, so as to play on any part of the patient that may be desired. Most medical men have their own ideas with regard to this kind of bath, but it is very little trouble for any ordinary skilled workman to carry them out so as to suit special cases.

In first-class houses a "bidet-pan" is sometimes fitted up in the dressing-room. These can be fitted up as a miniature bath, and have hot, cold, and waste pipes laid on to them. As a rule, the waste pipe from these fittings should be attached to the trap of the bath-waste, and so as to avoid a long length of untrapped waste pipe the pan should be fixed close to the bath. If trapped separately the water will sometimes evaporate out of the trap, and this occurs when this fitting has not been used for some weeks. When fixed near the bath a cover can be made so that it represents a seat. In a house where the bath-room is used by both sexes it is highly improper to fix these pans, and, although fixed by him, the writer never recommends them.

Figure 266 is sometimes useful either for invalids' use or as a child's bath. It is called a "sitz-bath," and can have hot and cold water attached in the same manner as an ordinary bath; also with back-shower, bottom-shower, and wave at the back. All baths can be enamelled

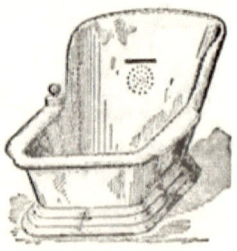

FIGURE 266.

inside and outside, so that no inclosure is necessary. The writer has seen a 5-foot 6-inch plunge-bath nickel-plated inside. It looked very nice, and would not appear so dilapidated as the ordinary enamelled ones do after they have been in use for some time, but the cost of doing them in this way prevents a general adoption of the system.

CHAPTER XXVII.

WASH-HAND BASINS.

WASH-HAND basins are a great convenience in a home, but they, as a rule, have less thought bestowed on their arrangement than any other sanitary fitting. Very few basins are made as they should be. Figure 267 shows an ordinary basin, with the rim drooping, so that when it is used any water that may get into the space between the basin and slab will run outward and rot or injure the woodwork. To obviate this, the under side of the marble, or other material that may be used, should be hollowed,

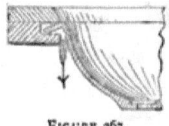
FIGURE 267.
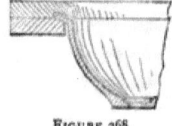
FIGURE 268.
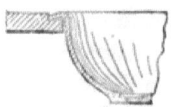
FIGURE 269.

and the rim of the basin made to curl upward, as shown in section, Figure 268, so that any water that may get into the joint will run back into the basin, and the joint should be made as close as possible so as not to have an unsightly and dirty appearance. Figure 267 shows the marble top cut so that any water that may run off the top must fall on the rim of the basin. Figure 268 shows the top slightly oversailing the basin so as to form a kind of dripping-eave for the water to drop into the basin. When marble is used as the top for a wash-hand basin, plaster-of-paris is used for bedding it. This will not stand being wetted, but soon breaks away. Putty, or any oil cement, cannot be used for this purpose, as it discolours the marble. Sometimes the wooden carriage for the basin is not used, but the basin is let into the marble or slate top, as shown in section, Figure 269. This is a bad arrangement, as any water splashed on the top cannot get back into the basin, or, should the rebate be cut deep enough so that the basin does not stand up so high, as shown, a dirty and unsightly joint is seen. When arranged in this way a new basin can easily be fixed in case of breakage, but, for appearance, it is not equal to Figure 268. For fixing near the garden entrance or gun-room of a mansion, or in the kitchen for cook's use, a basin 12 or 14 inches in diameter is quite large enough, but for toilet purposes it should never be less than 18 or 20 inches in diameter. Some people are partial to fancy-painted and coloured basins, but I prefer a plain white or cream tint. A plain gold line is a relief, but adds to the cost of the basin. Anything that tends to disguise dirt should be avoided, so the plain basins are the best. Figure 270 is a view of a bowl and top manufactured in

FIGURE 270.

one piece of earthenware. The bowl is made oval-shaped, and is sometimes called the elbow-room basin. This shape has an advantage over the round one as being more convenient for the user.

Figures 271, 272, and 273, are illustrations of other shapes of cabinet-stand wash-hand basins. Some have skirtings, but when

FIGURE 271.

FIGURE 272.

FIGURE 273.

made to a large size they are very difficult to get true, as they twist and warp very much in the burning; but latterly, makers appear to be improving in this matter and send them out much truer. Another advantage of these cabinet-tops is a raised margin all around the edges, which prevents water from running off on to the floor, or, perhaps, the user's dress and boots. The sinkings for soap and nail-brushes are a convenience, and are much better than the ordinary portable dishes, but the covers very often are found to be in the way and generally get broken or lost. Sometimes a waste pipe is made in the porcelain from the sinkings into the bowl, but a channel, made as shown at A, A, Figure 270, is better, as there cannot be any accumulation of stale soap to give off any effluvia. After usage of a wash-hand basin it always occurs that soap-curds and other matter float on the surface of the water, and as the basin is emptied of its contents these curds adhere to the sides of the basin. If a whirling motion is given to the water as it subsides, the basin takes longer to empty and the floating matter is still left sticking to the sides. Figure 274 is a view of a bowl with perforations

FIGURE 274.

FIGURE 275.

FIGURE 276.

near the top edge with the necessary provisions so that the supply to the basin passes through the holes. Figure 275 is another basin with a flushing-rim. Both these last basins can be rinsed after using by turning on the water for a few seconds. Figure 276 is another patent basin with a brass rose, so arranged that the incoming water is spread around the inside of the bowl so as to rinse it. Figures 274, 275, and 276, are supplied by means of valves fixed beneath, with knobs projecting through the top, the pipes being connected to the basin by means of the arms, A, A.

Figure 277 is a sketch of a wash-hand basin fitted in such a way that the inlets for clean water are connected to the waste-apparatus. At one time these kinds of fittings were thought to be very good, but they are now being condemned by most sanitarians, although a great many are still used. The objection to them is, the incoming water will often bring back any small pieces of matter or soapy curds that may be hanging about the neck, B. The same kind of supply-valves can be used, and the outlets connected to the arms, A, A, of the

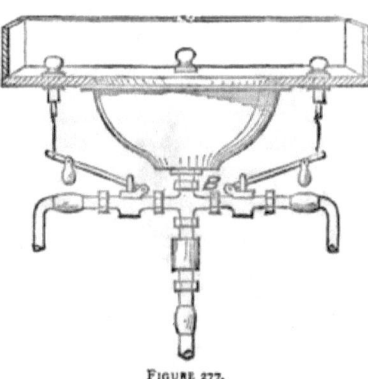
FIGURE 277.

bowls, as shown at Figures 274, 275, and 276. For supplying wash-hand basins nearly all the manufacturers have their own system and their own valves. When the valves are fixed beneath the top, the knobs for opening them are generally fixed as shown at Figure 277. It is better, although more costly, to fix them as shown in fragmentary section, Figure 278. In this case the knob is pulled forward and a slight turn given to it, when the valve will remain open; or the space, C, can be made larger, and screw-down cocks used. The marble shelf, D, can be made to lift off for access to the cocks for repairing. When the knobs are fixed on the top, as shown at Figure 277, the metal-work soon gets tarnished, even if nickel-plated. The guides also sometimes work loose, so that any water splashed on the top will run through. The back nuts are very difficult to screw up, the plumber having to lie on his back and punch them around with his hammer and chisel, unless he has time to get a specially-made spanner to turn them with.

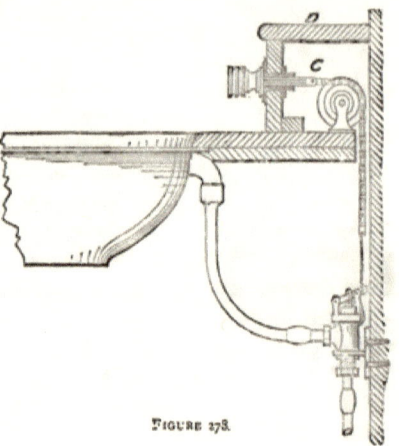

FIGURE 278.

For simplicity and moderate cost, a standard valve as described for baths, Figure 241, is very good. The nozzle should project about ¼-inch inside the basin, so that should it leak it will drip into the basin clear of the marble, so as not to have the water drawn into the joint by capillary attraction. All cocks of this kind (in fact, bibb-cocks or faucets generally) retain a nozzle full of water when shut off, which will sometimes hang up for hours until it has dripped empty, and this occurs more when the nozzle of the cock is bent downward. This has sometimes led to the plumber being sent for to repair the valve, under the impression it was leaking. With a great many people a self-closing supply-valve is preferred, as the water cannot be left running so as to overflow the basin. Figure 279 is a good description. It is known as a "cam-action" valve. By pulling the top, E, forward, the "cam" at the bottom presses down the spindle of the valve and allows the water to pass through. If the valve has a flanged connection at the top, half can be taken off, and any repairs made without taking out the rest of the valve. The nozzle should be bent as shown, so that the incoming water plays into the centre of the basin. If the nozzle is bent downward, as shown by dotted lines, the water will strike on the side of the basin, rush around the bottom and over the opposite side on to the dress of the person using it. When valves of this description are used they should be fixed on the back side of the basin, and some few inches away, so that when stooping to wash the face the head may not come into unpleasant contact with the knobs. When fixed at the sides the elbows of the user sometimes get injured.

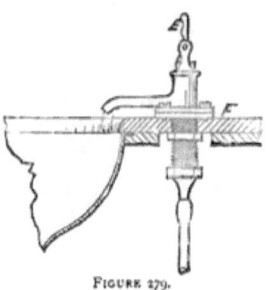

FIGURE 279.

All wash-hand basins should have good large overflows to them so as to take away the water as fast as it can come in. As a matter of fact, very few will do this. Even when they are a good size the mistake is generally made of fixing them immediately beneath the nozzles of the cocks, when the rush of the incoming water creates a depression or vortex, and prevents the overflow from being of any use. When holes in the sides of the basin are used for the overflow, the arm and overflow pipe will act as a syphon and draw the water away very fast, but air is also drawn in through the upper series of holes. A shell made to fit over the overflow orifice will prevent air entering, so that the water is drawn away more rapidly. The overflow pipe should be carried into the trap of the waste pipe, unless a safe is fixed on the floor, when the overflow can be made to discharge into the waste pipe from it. When the overflow is connected to the trap of the wash-hand basin waste, and a shell is fixed over the orifice in the basin, it is necessary to fix a vent pipe. Without this, the air in the overflow pipe would be pent up so as to be "air-bound," thus rendering it perfectly useless. The air pipe should be fixed near the bottom end of the overflow pipe; if fixed near the top end any syphonic action would be prevented, but if fixed too low the water in the trap would seal the bottom end and prevent the air from escaping. The sketch, Figure 280, will explain what is meant. It makes no difference if the overflow pipe is connected to the trap above or below the waterline, as in each case the air cannot escape. Sometimes wash-hand basins are made so that one shell will cover both the outlet and overflow-arms, but when this is done there is generally some difficulty in getting the overflow to take away the water as fast as it should. The reader is referred to what was before said about the overflow being fixed beneath the cock-nozzles for the reason.

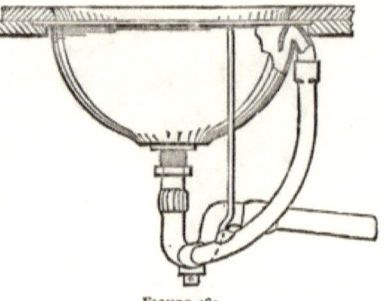

FIGURE 280.

Some Water Companies insist that self-closing supply-cocks should be used, and no overflow fixed; or, if one, that it should be carried through the wall so as to discharge in a conspicuous position.

There are wash-hand basins with the overflow provision made in the earthenware. Figure 281 shows in section how this is arranged, and Figure 282 is the specially-made waste-plug and washer drawn to a larger scale. These basins can only be fitted with a plug-waste. If a valve or apparatus were used for emptying the basin the above overflow provision would be useless.

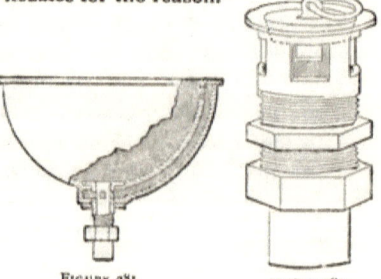

FIGURE 281.   FIGURE 282.

The waste-holes of wash-hand basins are generally made too small, so that a very small plug has to be used, with the result that the basin takes a long time to empty of its contents. Figure 283 represents a wash-hand basin that had to be altered because of the unpleasant smells arising from it. The owner

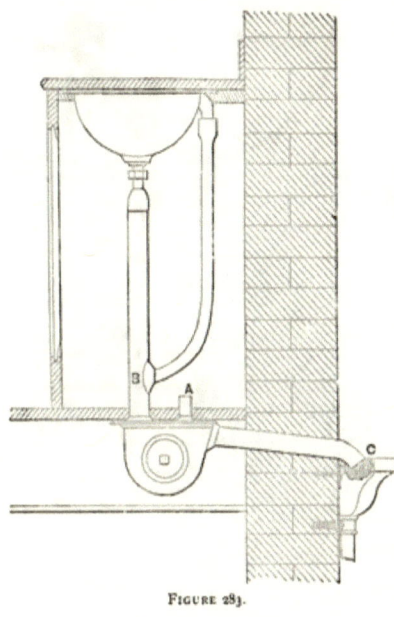

Figure 283.

complained that it had been altered several times, but the evil still remained. The smells could not be accounted for, because the waste pipe was cut off, as shown at C, so that the end of it was open to the air. It appeared as if the overflow pipe had originally been connected to the trap, as shown at A, the stump-end having been left in and the end soldered over. The overflow had been altered and soldered into the waste pipe at B. The trap was a medium-size **D**-trap, about 8 inches deep. The waste pipe was 2-inch lead pipe. The plug was ¾-inch in diameter. The trap being fixed under the floor, there was about 2 feet of the 2-inch pipe very foul inside from the soapy matter, and every time the basin was used the air in this pipe was displaced and driven through the overflow pipe. Here was a host of mistakes made by the man who fixed it. The hole in the basin was of such a size that a small washer and plug had to be used. The waste pipe was too large, the **D**-trap was too large and fixed too far away from the basin. The brass screw-cap was in an inaccessible position. Figure 284 is a sketch showing the new basin and how it was fitted up with trap, &c. The water-way through the washer, trap, and waste pipe, was of the same diameter throughout. The trap vent pipe, E, was continued upward as high as the top of the basin; if this was not done the escaping air would sometimes bring a few drops of water with it. When open heads are used for waste pipes to discharge into, the pipe should be carried down to the bottom, as shown at G, Figure 284. This is to

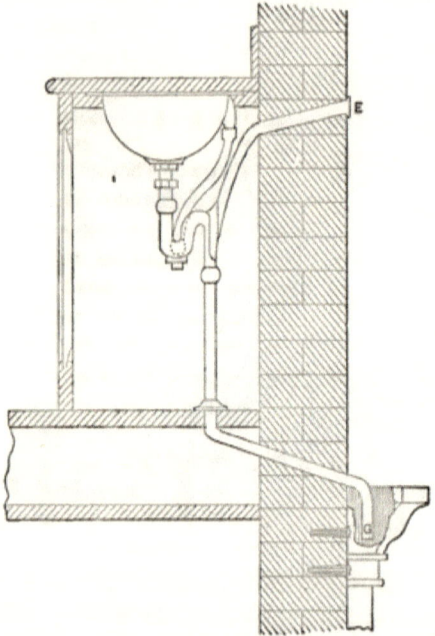

FIGURE 284.

prevent the waste water being splashed over a large surface on the inside of the head, which would soon give off offensive smells. In several cases where these hopper heads have been used for two or three waste pipes to discharge into, the smells have been so objectionable that air-tight coverings have been fixed and ventilation pipes carried up to a higher level.

CHAPTER XXVIII.

WASH-HAND BASINS—*continued.*

REFERRING again to plug-wastes, Figure 285 represents, in section, the ordinary style of making them, and at H is shown how the water-way may be contracted by allowing the lining of the union to enter the washer. Sometimes this lining is butted up to the washer, and a leather grummet placed between. When screwing up the union this grummet will get out of its position, or perhaps get squeezed so that it projects inside, as shown at J, in section, Figure 286. By making the brasswork as shown at

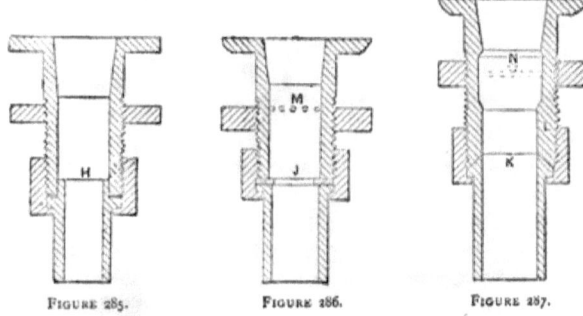

FIGURE 285. FIGURE 286. FIGURE 287.

Figure 287, the water-way can be made of equal diameter throughout, and if the lining is ground into the washer, as shown at K, similar to steam-unions, no packing or grummet is required. To get a shoulder on the lining to fit the cap, the substance of the metal in the brass washer must be a little thicker, as shown at L, than in the other examples given. The flange of the washer should be as large as the sinking in the basin to avoid a space, which always looks dirty. Some makers fix a small brass grating, as shown by dotted lines, M, Figure 286, to prevent finger-rings

WASH-HAND BASINS. 233

or small pieces of soap and other matters from getting into the waste pipe, thus adding to the evils of the already too much contracted water-way. It is much better to drill holes and fix brass cross-bars, as shown by dotted lines at N, Figure 287. These wires need not be very thick, as little or no strength is required, but as they contract the water-way, the brasswork can be turned out slightly larger, as shown in section, to allow for this. When large washers and plugs are used for wash-hand basins, it is advisable to make the plugs of vulcanized india-rubber, as described when writing on baths, as it sometimes happens that a large brass plug will break the basin when allowed to fall into it. Plugs have been objected to, as they require chains to prevent their being lost. The links of the chains get filled up with soapy matter, so that they always look dirty and smell unpleasantly. In addition, when the plug is made too slender in the tapering, it sometimes fits into the washer so tightly that the chain is broken in the attempt to pull it out. Or, perhaps, the plug when cold is placed in the washer when hot. On cooling, the plug will sometimes fit so tightly as to resist all efforts to pull it out, so that the union has to be uncoupled and the plug knocked out from the under side. This only applies to metal plugs.

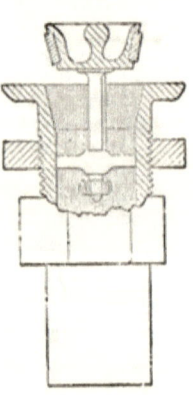

FIGURE 288.

Figure 288 represents a brass plug and washer used by a firm of sanitary engineers. The plug has an indiarubber ring sprung on and a spindle, with a nut on the bottom end, working through a guide-bar. A projecting slip is cast on the spindle, and a corresponding slot in the guide-bar, so that a slight turn given to the plug, when open, will cause it to remain. The knob, O, for lifting the plug, should be flush with the bottom of the basin, to prevent injury to the user's finger-nails, and also to prevent the user unintentionally lifting it so as to empty the basin. If the plug is made a good size, the spindle should be short, so that it cannot be opened too wide, so as to allow anything to fall down the waste pipe. When these plugs are fixed to wash-hand basins each user should discharge his own water, as it is unpleasant for the next comer to have to put his hand into the dirty water to reach the plug to open it.

One maker of sanitary fittings has a "trigger" connection to lift a plug of the kind last described. Figure 289 is a sketch showing the arrangement. P is the trigger, R is an indiarubber joint. The trigger is placed on the front side of the basin, so that the user has only to press his hand on it to open the plug In some cases the trigger is placed at the side of the basin, and a spindle continued through the slate or marble top, and a metal, porcelain, or other kind of knob fixed for pressing down the trigger. This is shown by dotted lines.

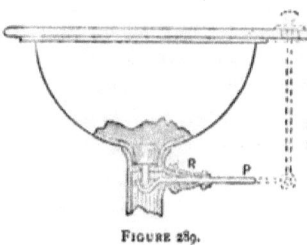

FIGURE 289.

Several makers have "valves" for discharging the water out of the basin. Some of these valves are similar to those described when writing on baths. Figure 290 represents one made by plumbers out of lead pipe. The part S is made of brass, T is a flanged connection, and U is a slip-joint in the pipe which forms the overflow. Ears are soldered on for screwing to the back boards or wall for fixing. When making these quick waste-valves the plumber sometimes makes the valve pipe too long, so that the branch waste pipe from the basin is some 2 or 3 inches above the flanged connection containing the valve. In this case the valve is an obstruction to the free escape of the water. The branch pipe should be as near the valve as possible, so that, when lifted, the valve will be above the branch waste, as shown by dotted lines at V,

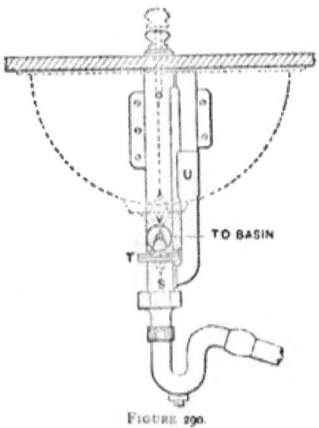

FIGURE 290.

and thus allow the water a free passage. When properly made, this waste fitting will enable a good sized basin to be emptied in three or four seconds; but when badly made, it takes five or six times as long to discharge the contents of the bowl. Very strong wire should be used for connecting the valve to the pull-up knob. If thin wire is used it sometimes stretches by constant usage until it becomes so long as to only partly open the valve. Another reason for paying especial attention to the thickness of the wire is because the whole of the apparatus has to be taken down for access to repair a broken wire, and it is very troublesome to get it the exact length. If too short, the valve would not fit tight in its seating, and if too long, the valve would only partly open. When this waste apparatus is used, it is not necessary to have an overflow-arm to the basin, but when an overflow-arm is used the pipe, U, can be omitted.

When waste-valves are used for emptying the contents of a wash-hand basin, a grated connection should be made to fit the bottom, and should be fixed flush on the inside of the bowl. The grating should be a good size, and the water-way through the perforations should be slightly in excess of that of the waste pipe. That is, should 1½-inch waste pipe be used, the number and size of holes in the grating should, in the aggregate, be equal to, or a little larger than, the pipe. When round basins are used, a vortex will form in the water as it runs away, but this does not take place to such an extent when oval basins are used. Instead of a grated connection, one similar to the section, Figure 291, is sometimes used. This does away with the grating, the holes in which are sometimes thought to be unsightly,

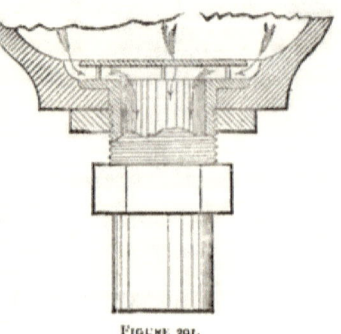

FIGURE 291.

but nothing else is gained. If the cap is raised so that there is a free water-way around it, there is room for a small finger-ring to slip through.

"Tip-up" wash-hand basins are liked by some people, in

spite of the outer receiver very often being offensive by reason of the soapy matter adhering to the inside. Great numbers of manufacturers make these basins, but the original inventor has patented an improvement by means of which the tipping basin can be lifted off the trunnions, and thus gives free access to the receiver for cleaning it. The receiver, also, has been improved by making a back outlet instead of the straight down. This

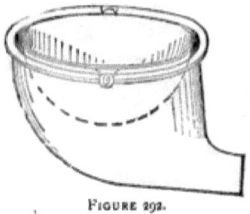
FIGURE 292.

gives more room beneath the wooden enclosure, and the water is not so liable to splash out of the receiver on to the dress of the user when the basin is tipped up. Figure 292 represents the tip-up basin with a back-outlet receiver. Pains should always be taken to make a tight joint between the slab and the rim of the receiver. I have found a piece of indiarubber tubing to make a good packing.

Lavatories should not be fixed in bedrooms if it can be avoided. When of necessity they must be, great care should be taken in the selection of the basin and in every detail of waste pipe and trapping, so as to avoid any possibility of a smell escaping. There is very often a faint odour near an ordinary wash-hand stand, and more especially when highly-scented soap is used. The sponge and nail-brush will retain soapy matter to an unpleasant extent, and an improperly fixed basin, with waste and overflow pipe, and sometimes from the waste pipes of the soap and brush sinkings as well, adds to the evil. The dressing-room is the proper place in which to fix the lavatory.

In first-class establishments the lavatory enclosure is sometimes made to match the rest of the furniture in the room.

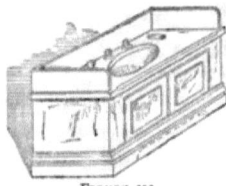
FIGURE 293.

These enclosures are generally made to look like a large piece of furniture, and although, perhaps, nice to look at, are very uncomfortable to use. The mistake is generally made of having the front out as far as the top, as shown by sketch, Figure 293. By cutting out a piece of the skirting or plinth, as shown by dotted lines, room is made for the toes, but the knees lodge against the front. A much better plan is to have a cornice beneath the nosing of the slab and the doors recessed a few inches, as shown by sketch, Figure 294. This gives room for the person to lean over the basin without making the back ache. The above evil is aggravated when the stand is too high, and the basin is placed too far from the front edge. The stand should not be more than 2 feet 6 inches high for persons of ordinary stature. If higher, water will be found to run off the user's elbows and drip on the floor.

FIGURE 294.

When the basin is too far from the front edge, the person has his whole weight thrown on the tips of his toes when washing, and runs the risk of tipping forward on to his face. It would be better if wash-hand basins could be fitted up without any enclosure, but there is often such a nest of pipes beneath that to all (excepting admirers of plumbers' work), they would appear unsightly. In some cases a small brass rail is fixed 1 or 2 inches away from the marble slab, to prevent the user's dress becoming wetted by contact with any water that may be near the front edge. This is shown in Figure 294, from A to B. Or, if the basin is used before dressing, a towel can be hung on this rail to keep the body from touching the cold marble.

Figure 295 is a neat enclosure for an oval basin with porcelain top. The cupboard door, A, being recessed, is a great advantage, and can be made much more cheaply than if it was rounded in the same manner as the front, at B. Some makers insert ornamental tiles in the panels of the doors, or have fancy designs painted on the woodwork.

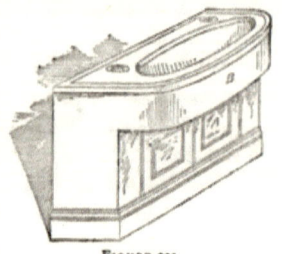

FIGURE 295.

Plumbers vary very much in their ideas as to the size of traps for wash-hand basins. Figure 296 is a view of one shown to the writer by a master plumber, who boasted that it was impossible to break the water-seal by syphonage.

Calculations based upon the measured dimensions show it held about two-and-a-half gallons. The basin was very small, only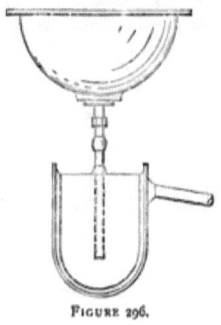
12 inches diameter, and held just enough to stir up the contents of this small cesspool. A wooden stool had to be placed beneath this trap to keep the weight from pulling out the bottom of the basin. The inlet and outlet pipes were 1 inch in diameter. Traps should not be larger in diameter than the waste pipe and plug, but they should always have a good dip or water-seal. If a 1-inch or 1½-inch round pipe trap is used, and made by the plumber, it should have a 4-inch to 6-inch water-seal, as shown by Figure 297. The reason such a deep seal is an advantage is because the trap being fixed above the floor is more likely to have the contents evaporate by

FIGURE 297.   FIGURE 298.   FIGURE 299.

exposure to the atmosphere. If made to the shape shown, and thick pipe used, it could be loaded with sand and bent without crippling the throats of the bends. The shape is also good to prevent the water being carried through the trap by the impetus given when falling from the basin so as not to leave enough to charge the trap. This is found to occur when traps are made as shown by Figure 298. All traps are liable to syphonage, but those sometimes called self-cleansing (that is, the passage of water through them scours the inside so as to prevent fur accumulating) are more liable to this than the box-shaped traps, hence the necessity of ventilation pipes.

Figure 299 represents a trap very often made by plumbers for fixing to wash-bowls. It is sometimes called a soap-trap, and there is a general impression that it will retain soapy matter from passing into and clogging the waste pipes. There is about as much sense in this argument as in that of the old lady living in a house in the north of London, who asked me to fix a grating in the trap of the water-closet to prevent the paper used from passing into the drains, as she was afraid a stoppage would occur.

CHAPTER XXIX.

WASH-HAND BASINS—*continued*.

BESIDES the traps that have been described there are several others very similar in their action, and there are also what are known as mechanical traps. Some of these have a flap-valve fixed inside over the outlet pipe. Others have a floating ball or else a weighted ball; these fit over the end of the inlet pipe in the body of the trap, which has to be made in such a way that when out of action the ball covers the orifice. The advocates for these traps claim that they are proof against back-pressure, but there is no back-pressure when the waste pipes are ventilated. These traps all clog up with soapy matter, and this occurs more frequently in districts where the water used is very hard.

Figures 300, 301, and 302, are illustrations of the principles of the flap-valve, heavy, and floating balls, which speak for them-

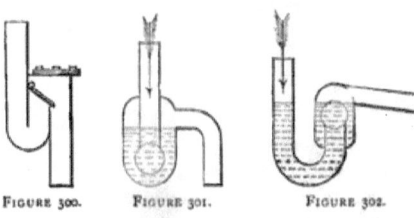

FIGURE 300.   FIGURE 301.   FIGURE 302.

selves. The writer, some time ago, had to take out a mechanical trap, as it was continually becoming choked. After it had lain on a shelf for two or three months, it was taken down to show a friend the condition it was in, when it was found that the matter inside had all dried up, and nothing was left excepting what may be compared to coarse-looking cobwebs. A piece of waste pipe, 1½ inch in diameter, which had become choked with soapy curds, was kept as a sample, and, after an interval of time, the matter had all dried up, and only a thin scale was found adhering to, in some places detached from, the sides, and this stoppage had resisted, when it was fixed in its position, the action of a hand force-pump.

WASH-HAND BASINS.

All wash-hand basin waste pipes should discharge with open ends into an interceptor or gulley trap.

In some town houses, built years ago when little or no thought was bestowed on sanitary questions, and all the fittings are in the centre of the house, it is very difficult to arrange for the waste pipe to discharge into the open air; or the expense of the alteration would be so great that other means have been resorted to. In these cases two traps have been fixed, and a ventilating pipe soldered in the waste pipe between them. This is shown by sketch, Figure 303, at A, and the waste pipe connected to the soil pipe of the adjoining water-closet. When this is done the *air pipe must not be branched into the soil pipe*, but must be continued separately to a suitable position out-of-doors. If another vent pipe is fixed at B, a current of air can pass through. This is an advantage when

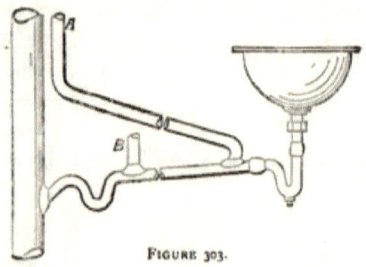

FIGURE 303.

the wash-hand basin is some distance away, thus necessitating a long length of waste pipe. When the soil pipe is of lead, and hot water is used in the wash-hand bowl, it will often occur that the soil pipe will break by the expansion and contraction of the metal caused by the hot-water discharges.

There are several manufacturers who make sets of wash-hand basins as shown by Figure 304, which is a front elevation, and

FIGURE 304.   FIGURE 305.

Figure 305, which is an end view. These are very strong, and suitable for public schools and similar institutions where they are subjected to rough usage, but from a sanitary point of view they are far from being good. As a rule, the trapping of the waste pipe is improperly done. Sometimes no trap at all is fixed, the end of the horizontal waste being continued to discharge over a gulley-trap, or, if fixed upstairs, into the head of a vertical stack of waste pipe fixed outside the house. In these cases a current of air passes from the outside and through the waste pipes. This air may be perfectly pure and sweet when it enters the waste pipe, but when it escapes through the waste or overflow pipes it is rarely in a state fit for breathing. The writer has had to fit traps as shown at A, Figure 304, to several of these sets because of the offensive smells that pass through, but this is little or no remedy for the evil. Where eight or ten basins are fixed in one range, every time one of them is emptied a certain amount of air is expelled through the branch waste or overflow pipes of the other basins. The horizontal waste is generally of iron, 3 inches in diameter, and is fixed perfectly level. On taking these pipes to pieces for cleaning, a quantity of black offensive-smelling matter is generally found inside. Neither is any provision made to prevent the waste pipe from rusting. The branch waste pipes are generally connected at right angles to the main waste. The result of being branched in this way is, that every time a basin is emptied as much water passes toward the stopped end as to the outlet, so that the black matter spoken of is being continually stirred up and moved backward and forward inside the pipe.

Figure 306 is an end view and Figure 307 a front view of one basin out of some ranges designed and fixed by the writer for the young men and women's use at a large drapery establishment. The whole of the horizontal waste pipes and the brackets, &c., were of galvanized-iron. The brackets had projec-

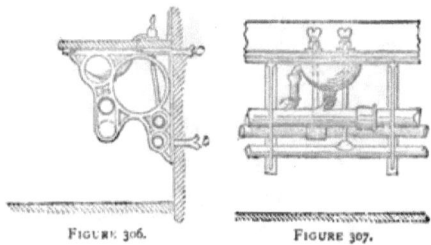

FIGURE 306. FIGURE 307.

tions cast on at C and D, Figure 306, and were firmly fixed to the wall. By doing this the whole of the space beneath was

WASH-HAND BASINS. 243

open, and, as nothing touched the floor, there were no corners in which dirt, &c., could accumulate. The horizontal waste pipes had a fall throughout their length. They were fixed immediately beneath the centres of the basins, so as not to have long branch waste pipes. Instead of round openings in the brackets for the horizontal waste pipes to lie in they were made as shown in Figure 306. The highest end of the waste was in the position as shown, but the lowest end was situated as marked by dotted lines. The branch arms, F, F, Figure 307, to receive the waste and overflow pipes, were all turned in the direction of the current, so as to avoid discharges from the basins running up the main waste. The waste-plugs were locked similar to those illustrated in an earlier chapter, and the unions were socketed into the branch wastes, the joints being made water-tight by the insertion of india-rubber rings. The basins were supported by means of galvanized-iron cross-pieces resting on the brackets, and were so arranged that should a basin get broken, a new one could be substituted without taking away the marble top, or moving it at the risk of breaking it.

"Tip-up" basins are sometimes fixed in ranges, and have a large trough placed beneath so as to receive the contents of the basins. This is far from being a good plan, as the dirty water is splashed over a large area which soon becomes offensive.

In hospitals, and public schools attended by the poorer class of children, it is very probable that certain diseases may be conveyed from one body to another by means of wash-hand basins, and the question arises, would it not be better to dispense with them altogether in those places. I think it was in Paris that I saw something similar to Figure 308, which is a section. Instead of basins, a long V-shaped slate trough, with a channel down the centre, was fixed, and taps at intervals, with a stream of water running all the time a person was washing. By this arrangement the risk of communicating a specific disease is minimised, but the channel should be large enough for the water to pass away without coming up into the trough, and the opening at G should be so narrow that pieces of soap, &c., could not be washed away.

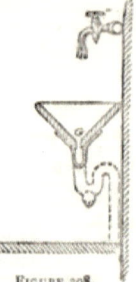

FIGURE 308.

R 2

The tap should be high enough so that the head can be held beneath, but not so high as to cause the water to splash about too much. For children's use the trough should not be more than 2 feet high, but for adults it should be 6 inches higher. With regard to the materials of these troughs, porcelain would be the best, if it were possible to make them large enough, but failing that, slate is the best material, especially if a kind is selected which is not absorbent. The slate could be enamelled, but after a time the enamel peels off, or gets so chipped as to look unsightly.

In some cases long open troughs filled with water have been fixed for the use of the class of people who may be termed casual paupers, or tramps. In these cases it is highly probable that one person, suffering from any skin or other communicable disease, may so contaminate the water that other users may be infected with the same complaint. From this point of view the trough, although simple, and, comparatively speaking, inexpensive, is highly dangerous for using under the above conditions.

A small shower apparatus fixed to a lavatory is a comfortable luxury. This can be made and fixed in a way similar to one illustrated in the chapter on baths, but the crane should be jointed so that it can be turned aside out of the way when not wanted to be used. A looking-glass fixed behind a lavatory is a great improvement, both with regard to appearance and convenience. It also prevents the wall at the back from becoming splashed and thus rendered unsightly. When a back glass is used the above shower-bracket cannot be fixed on the back wall. When the wash-hand basin has a cover, or is otherwise enclosed, the shower-bracket will be found objectionable. The drippings of water from the rose generally continues for some considerable time after using, especially if the perforations are very small, when they retain water by capillary attraction. This is of little consequence when no cover is fixed over the basin, as any drippings would fall on the slab or into the bowl. In some cases hot and cold-water cocks have been connected with a coupling-union, and a piece of flexible indiarubber hose attached. On the loose end of the hose a small rose is fixed, as shown in Figure 309, for spreading the

FIGURE 309.

stream of water. The hose should be wired inside to prevent kinking, which would stop the waterway and prevent the free flow of the water, and if it is covered outside with a silk or worsted plaiting it adds to the appearance. It is unnecessary for the hose to be larger than ½-inch in the bore. The rose should be of ebonite because of its lightness. If a metal one is used and allowed to fall it would, perhaps, break the basin. The rose should not be more than about 2 or 2½ inches across the perforations, H, Figure 309, and the holes should be as small as possible so that very fine sprays of water escape. The perforated face, H, should not be much rounded—in fact it should be nearly flat so as to avoid the water spreading over too large a surface. If the rose is small, the pressure of water not too great, and the hose nice and flexible, the user can hold it in any position he pleases, either over his head, or under, or at the sides of his face. The part I, Figure 309, is better if made rough, so that the hand can hold it more firmly and the user have better control over it. When hot water is used the person should try the heat, by allowing it to run for a few seconds on his hand, before applying it to his head or face; or a thermometer, as described when writing on shower-baths, could be attached.

Figure 310 is a very convenient shower and spray apparatus patented by a leading firm of sanitary engineers. It is supplied with hot and cold water. The crane has ground-in joints so arranged that it can be swung sideways when not wanted. The rose can be turned upside down, as shown by

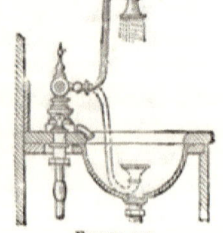

FIGURE 310.

dotted lines, and sideways, or any other desired position. When this fitting is nickel-plated and kept nice and clean it presents a smart appearance. The joints should be made of the very best metals, as constant use will soon cause them to leak, although they do not have to withstand any great water-pressure.

CHAPTER XXX.

URINALS.

THE subject of urinals is one of the utmost importance, and I may premise any further statements by saying that I have never seen one that might be said to be entirely satisfactory, and we have yet to learn the secret of making them sanitary. Take public urinals first. In the older districts of London, and in a great many by-streets, a small recess is made in a wall, generally by the side of a public-house or tavern. Sometimes a screen of slate or cast-iron is put up, or a door is fixed for the sake of decency. Inside no special provision is made for keeping the place clean or for catching the urine, which is ejected against the walls or on the floor, and then has to flow towards a grating, or sometimes only a hole, and so into the drain. Sometimes the walls are rendered, or covered, with mortar or cement, so badly done as to be no protection against their becoming saturated with urine. The floor is sometimes paved with bricks, in other cases with stones. The joints are not always made good, so that liquid matter lies in pools all over the place. In some cases no light or ventilation is provided, and not even a water-supply for washing down the floor. On entering one of these places the person's eyes begin to run with tears, and the pain of the nostrils is similar to that just before a fit of sneezing, so strong are the ammoniacal vapours. Although these places are for the use of the public, a great many of them are private property. When they become so bad as to be a public nuisance, they sometimes get washed down and a coat of lime-white may be laid on the walls. A few days afterwards these white walls are invariably found to be covered with disgusting literature and quack doctors' hand-bills. The sooner the sanitary authorities seek out these places and have them removed the better it will be for those unfortunate people who reside near them, and others who, from sheer necessity, must make use of them In a few cases a stone trough is fixed, but, for want of attention, they soon become coated with yellow matter.

Public urinals should not be made against dwelling-houses; an air-space should always be between them, and if they can be fixed several yards away it is better. In towns, several of these places should be provided to prevent any outrage against decency by the thoughtless, or those driven by necessity.

In busy thoroughfares and streets an independent iron building is sometimes put up. Figure 311 is a sketch showing a very common description. They have from two to perhaps ten or twelve stalls, the whole of the construction being cast-iron. The stalls are continued down to the floor, a channel runs along the floor at the back side, and the foot-stone or slab is laid to fall towards it.

FIGURE 311.

In some cases an attendant visits these places once or twice a week and washes them down with water and broom, or a cock and water pipe are fixed in a convenient position for the attendant to screw on a hose for washing down with. Figures 312 and 313 are plans of two other structures. These have cast-iron floors laid with a fall towards the gratings or traps A. In some districts brick buildings are erected similar in plan to Figure 311, and the walls lined with slate slabs. Sometimes very thick glass is used instead of the slate slabs, but, from a sanitary point of view, there is very little difference between any of those described. They all smell so offensive that users never stay in them longer than they possibly can help. In some instances a perforated pipe, sometimes called a "sparge pipe," is fixed over the part used, and water laid on to run continuously down the back sides of the stalls; or, where it has been found necessary to economize with water, a small tank has been fixed with the necessary apparatus to automatically discharge the contents at regular intervals of time. In these cases the back slab is sometimes quite dry in the intervals between the discharges, so that the salts of urine soon deposit and become so hard that they cannot be scrubbed, but

FIGURE 312.

FIGURE 313.

have to be scraped off. Some people are under the impression that cold water causes the fur to accumulate more quickly, but this is not the fact. Anyone noticing these places will find that the back slab over which the water streams always looks cleaner than the divisions or sides, which get splashed, but rarely get any water at all.

Figure 314 is a sectional elevation of one stall of a urinal for which several authorities have a preference. Several of this

FIGURE 314.

kind will be found in railway stations, both in town and country, and to have to use them is simply abominable. The slate apron B is generally covered on both sides with urine as well as the sides of the stalls, the back only being washed by the sparge pipe C. Bad as this evil is, in some cases it is made worse by having the slate apron piece fixed so high that people of short stature eject on the outside, at D. The first great evil in all the conveniences described is allowing the urine to spread over a too large surface by having more exposed than is really necessary. For instance, if the divisions were cut off at about the height of a person's knees, little or no matter would be splashed on them, and if there were no aprons there would be about 14 *square feet less of surface in each stall* to get fouled, thus leaving only about 6 feet at the back to be kept clean.

In some public places has been introduced a new kind of urinal. Figure 315 is a sectional elevation of one stall. This is

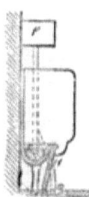

FIGURE 315.

a great improvement on the old-fashioned open stalls, but it is far from being a thorough success. One great advantage is, the urine passes directly into a trough, E, of water, and at short intervals of time the flushing-tank, F, automatically discharges itself into one end of the trough with sufficient force to displace the contents and leave a body of clean water behind. A branch water pipe is so arranged as to wash out the foot-channel, G, at the same time. Those the writer has seen are very much furred with urinary deposit, in spite of the violent scour of clean water sent through it. The trough is fouled down the front side, at H. The writer has

URINALS. 249

watched visitors, and noticed that some of them stand too far away, as if afraid of soiling their dress, and boys can scarcely reach high enough. In addition, the front edge of the trough appears to be too wide, and, being rounded, any drippings on it sometimes runs outward.

Figure 316 is a plan of a public place. Although the smells are offensive, they are not nearly so bad as some of the others described. In this case the slate divisions are placed at an acute angle to each other, and a large cast-iron bowl with three lips is placed in the centre. The

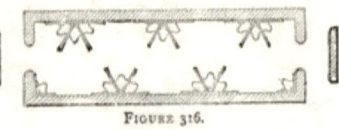

FIGURE 316.

stalls are cut off at the level of the top edge of the bowl, and, as they do not project far, users are obliged to stand so close to the basin that the lip catches most of the drippings. A constant supply of water is laid on to the bowl, so that it is always full and the contents in continual motion.

Figure 317 is a section through the basin showing the overflow pipe, which can be lifted out when necessary to empty the contents for cleaning. The bowls are generally made of cast-iron, but a few have lately been fixed made of vitrified stoneware.

Figure 318 is a sketch plan of another arrangement of the same kind of urinal. The frame-

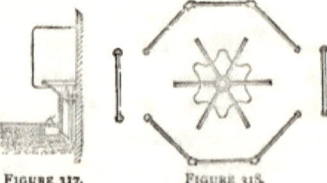

FIGURE 317.   FIGURE 318.

work of the enclosure is of iron, with slate panels. A skeleton or lattice-iron roof is fixed over all, and the bottom of the centre post forms the waste or overflow pipe. On the top end of the post is a gas-lamp, with red-coloured glass, on which the word "Gentlemen" is embossed. On the floor, at K, in Figure 317, an open grating is let in to catch drippings. This is the most objectionable part of the whole arrangement, as the space beneath is generally charged with offensive smelling matter, such as urine, street-driftings, cigar-ends, and a host of other things.* This

---

* Since writing the above, several improvements have been made in this arrangement.—J. W C.

grating has an advantage: by catching any wet that may fall, the user's boots do not become so saturated as to be offensive and objectionable to others besides himself, and neither does he leave wet foot-marks behind him. It would be an advantage if a provision was made so that a small stream of water would be distributed under the grating, so that the space beneath would be washed clean.

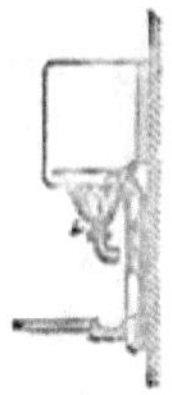

Figure 319 is a sketch section, showing a trough-urinal fixed in a London street, and which appears to act very well and give off very little smell. The trough is porcelain-enamelled iron. A constant supply of water is laid on to it and passes through perforations in the bottom, as shown at L, and the overflow is through perforations near the top edge, at M. As urine is considerably heavier than water it sinks to the bottom, hence the above system of flushing has reason on its side.

FIGURE 319.

When plain stall-urinals are ordered to be fixed, there are a few improvements that can be made on the way in which they are usually fitted up. For instance, when flushed with a sparge pipe, a great deal of water splashes outward and on to visitors' boots, &c. If the sparge pipe is perforated, as shown in fragmental section, Figure 320, the water-jets will sometimes rebound from the back slab, as shown by the arrow, to A, and more especially when the face of the back slab is rough. If the pipe is perforated, as shown by dotted lines, at B, the splashing is not quite so bad. The jets of water being thrown upward, in falling meet and break the force of the water issuing out of the holes in the pipe. If the perforations are made in this way an evenly-distributed sheet of water runs down the back slab, but in the ordinary way of arranging the holes the water will run down in streams. A ledge is generally improperly left, as at C, from which the water will sometimes rebound, as shown by the arrows. Sparge pipes should be made of copper. When iron is used the metal rusts, some of the holes get filled up, and others grow larger and very much out of shape, so that the water is not evenly distributed. All sparge pipes should have a small hole in the bottom near one end, so that when out of use the

FIGURE 320.

whole of the water may drain out. In winter this water will get frozen, and perhaps burst the pipe. When an apparatus is fixed so that the flushing is intermittent, it generally happens that when the force is spent the last portion of the water will run out of the holes and to the under side of the pipe, as shown at D, Figure 320, and fall on to the floor, or into the channel, and splash very much. In some cases a metal sheathing has been fixed over the pipe, as shown at E, Figure 321, to prevent this. A great deal of the splashing of water may be avoided if the back slab is fixed with a slope, as shown at F, Figure 321.

In some of the principal railway stations and a few public places in London white earthenware basins have been fixed, as shown at Figure 322. This is a step in the right direction, as urine is not splashed over such a large surface as in the other kinds described. Most of these basins have flushing-rims, so that the incoming water washes over the whole of the inner surface.

FIGURE 321.   FIGURE 322.

Figures 323, 324, and 325 are sectional plans of three of the commonest kinds of basins.

Figures 324 and 325 are considered the best shape for catching drippings. The waste-holes in Figure 325 are the best arranged, as they are partly in the back as well as the bottom. Sometimes a cigar-end or piece of cigarette is thrown into the basin, which, on unfolding, will cover the bottom waste-holes and prevent the water from running away. It is very rarely that the back waste-holes get stopped in this way.

All the basins described have straight-down waste pipes. Some others are also made with back outlet waste-arms to connect to the waste pipe behind the back marble or slate slab, instead of as Figure 322, where the waste pipes are shown on the front of the slab, and discharging into an open channel. Urinal-basins, when connected to the waste pipe instead of into the open channel, should be

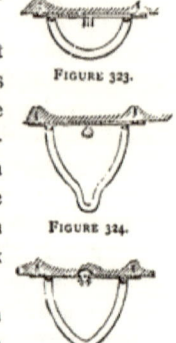

FIGURE 323.

FIGURE 324.

FIGURE 325.

trapped. This is generally done by means of a lead trap soldered in the waste pipe—in some instances one trap to the range, and others a trap to each basin, this latter being the right thing to do.

Figure 326 is a vertical section of a basin and trap in one piece of earthenware, so constructed that water is retained in the

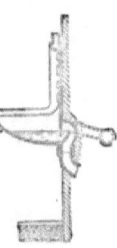
FIGURE 326.

bowl. The lip, G, hangs so that whatever runs down will drop on to the floor instead of running under the basin and down the back slab. This evil is generally unnoticed, although a frequent source of smells near these places. There are other basins, with earthenware traps to them, not constructed to retain any body of water. Some sanitary plumbers will fix a range of ordinary urinal basins, and a valve and overflow pipe, so that water is retained at the same level in all the basins.

Figure 327 shows how this is done, the basins and slabs being omitted for clearness. H H are as fixed for back outlet, and I I

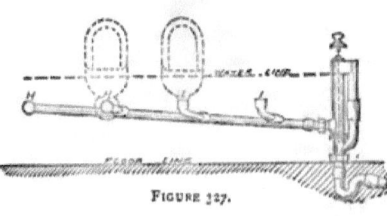
FIGURE 327.

are for straight-outlet basins. When fixed in this way a constant dribble of water should be running, so as to keep the contents of the bowls, &c., in motion, and also to prevent an accumulation of urine in the basins. It is a further advantage to fix an automatic flushing-tank, so as to periodically send a good scouring flush through the pipes, &c. If this is done, precautions should be taken to have the overflow large enough to prevent the basin overflowing. A moveable panel should be fixed in front of the discharging apparatus, so that it can be readily got at for cleaning should it become furred and choked up. The basins can easily be removed should the branches or the horizontal waste pipe require cleaning.

When several basins are flushed with one apparatus, great care is required in so arranging the service pipes that each basin will get its fair quantity of water. Figure 328 shows a good way of doing this K is the flushing-tank fed by a small tap, L.

URINALS. 253

A dribble or continuous flush can be arranged by fixing the pipe and stop-cock, M.

In some cases, where the supply of water is limited, what is commonly called a "treadle-action" apparatus is fixed to each basin. There are several kinds of these treadle actions. They mostly consist of a small hinged platform, either above or level with the flooring. Some are inlaid with fancy tiles, and others have iron gratings. The treadles are so arranged that a person stepping on them causes a valve fixed beneath to open, or act upon a lever connected by means of wires to a valve in a specially-arranged cistern. In some cases the valves are arranged so as not to open until the person steps away. These valves are specially made to run one gallon of water and then close themselves, but do not allow any water to pass during the time the platform is depressed.

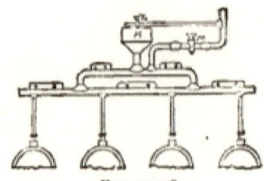

FIGURE 328.

These treadle-action apparatus are a common cause of complaint by reason of the smells that escape from them. They are mostly found to contain a quantity of stale urine, and when the floor is swept, or washed, other matters get in to add to the evil. When fixed in a wooden floor, or on brickwork, or stone, into which the offensive matter soaks, no amount of cleaning will get rid of the smells. One very good description of treadle apparatus in the market has an enamelled-iron tray in which the valve is fixed, and a small perforated pipe is so arranged that jets of water escape at each opening of the valve and thoroughly wash out this tray at the same time as the basin. But even this is a nuisance if improperly fixed.

Figure 329 represents in section one the writer had to alter some time ago. The cause arose from the blunder the plumber made when fixing it of branching the waste pipe from the tray into the trap in such way that water, &c., laid in it, and sometimes rushed up into the tray in such a volume that it did not get thoroughly washed out again.

The reader is again referred to Figure 322. The stalls or divisions should be clear of the floor

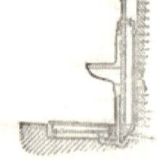

FIGURE 329.

so that it can be swept clean, and have no corners for dirt to accumulate in. The stalls are generally about 2 feet apart. This gives a fair amount of room, and at the same time prevents users from standing on one side of the basin so as to use it improperly. The divisions generally project about 2 feet from the back. This is a great mistake, and it causes users to stand too far from the basin, and this occurs more particularly in railway stations, where travellers perhaps have rugs and travelling-wraps on their arms. The stalls should only project about 12 inches clear of the back, so that visitors can stand close to the receivers.

The iron gratings placed over a hollow space in the floor should in all cases be removed, as they only become receptacles for filth. If these places (urinals) are properly fitted up, very little will fall on the floor, and hence it is unnecessary to provide for something that ought not to happen, and which eventually becomes a source of evil. Veined marble is a good material to use for urinal-stalls, and costs very little more than enamelled slate. Light coloured enamelled slate looks very nice when quite new, but it is objectionable for reasons that have already been given in another place. In some cases the slate has been enamelled black, in others to imitate granite and marbles; but there are so many mischievous people, amateur artists, about who, when they cannot

FIGURE 330.

use a pencil, do not hesitate to take the point of a pocket-knife to gain their object. The front lip of the basin should not be more than 2 feet above the floor, and should be 1 inch lower rather than higher. In one public institution in London the urinals are arranged as shown by Figure 330, which is a plan.

FIGURE 331.

No basins are used; a constant stream of water is kept running down the Λ-shaped backs, and also down the sunken part in the floor. This floor-flushing is shown by Figure 331, which is section across Figure 330, at A B. N is a perforated pipe hidden beneath the foot-stone. These places are in the charge of attendants, whose duty it is to keep them clean, but in spite of all the care taken there is always an unpleasant odour near them.

CHAPTER XXXI.

URINALS—*continued*.

In some hotels and clubs urinals are fitted up similar to Figure 322, but, instead of the iron gratings on the floor, perforations are made in the slate foot-stone, and an earthenware receiver, with flushing-rim and water-supply attached, is placed to catch drippings.

Figure 332 is a plan of one basin showing this. This drip-pan has a waste pipe connected, and is flushed at the same time as the urinal-

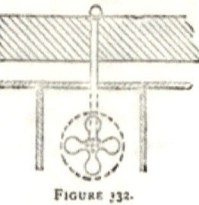

FIGURE 332.

basin. One large London club has, by the advice of their sanitary engineer, removed the fixed urinals of this kind and substituted small portable utensils. A slop-sink, as shown at Figure 230, is fixed for emptying the bowls into, and a hot-water tap fixed over the sink for rinsing purposes. An attendant is told off for the special duty of keeping these places clean. The man's wages is a bar to a general adoption of this system, but where this is not objected to the cause of smells is removed. In another London club a range of six urinals having been fitted up, the plan of each stall being as Figure 332, was a cause of complaints by

FIGURE 333.

reason of the smells being driven out at the bottom end of the waste pipe each time the automatic flushing-cistern was discharged.

Figure 333 is a sectional elevation. The stench was so bad

that the grating, at A, had to be removed, a solid cover bedde over the gulley-trap, and a vent pipe fixed to a considerable height to carry away the unpleasant odours. The new vent pipe is shown by dotted lines.

Figure 334 is a sketch, drawn from memory, of a urinal that was shown at the International Health Exhibition, held in London,

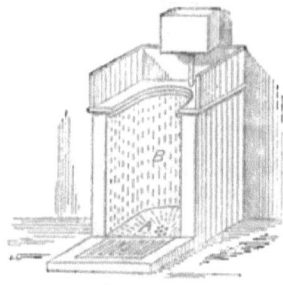

Figure 334.

in 1884. A was a white porcelain bowl, rounded to fit the back of the stall, and the front part extended beneath the iron grid on which the users would stand. The back, B, was made of one piece of thick glass, bent as shown in the sketch, the whole being enclosed with white veined marble. The flushing was done by means of an automatic flushing-tank arranged to empty itself at regular intervals of time through a sparge pipe bent to fit the back. The grid, C, was easily removable for cleaning the bottom basin. Although this urinal was shown by itself, there is no reason why it should not be fitted up in a range.

There are sanitary engineers who prefer to have a narrow step fixed to urinals, for the reason that users must then stand close and so avoid spreading urine over a larger surface than necessary. An opposite argument may be used, that people in a

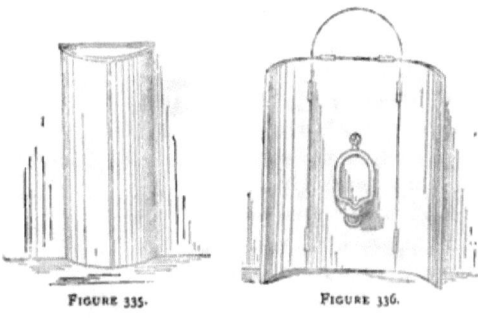

Figure 335.   Figure 336.

hurry would stumble at the step, especially when fixed in a dark situation.

One large firm of sanitary engineers in London fit up a urinal with a mahogany enclosure, so arranged that upon lifting up the top of the enclosure the sides open at the same time by means of the necessary brass couplings, and by an ingenious piece of mechanism a valve is opened so that water streams over the whole of the inner surface of the basin during the time the place is being used. On closing the lid the sides shut up and the water is turned off.

Figure 335 shows the fitting closed up, and Figure 336 when it is open. It is spoken of as being suitable for offices and billiard-rooms.

There are two or three makers of sanitary fittings who fit up a wash-hand basin and urinal in the same enclosure. Figure 337 is an illustration of one. The urinal-basin is fitted to the door, on opening which water begins to flow. The waste pipe is connected to that of the wash-hand basin by means of a hinged and telescopic joint.

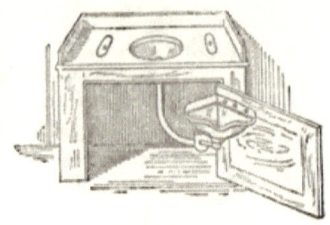

FIGURE 337.

There are also one or two folding-urinals in the market. These are hinged on the back edge on to a cast-iron frame fixed over a recess made in the wall. The basin is pulled down for use, after which it is closed by lifting up the front edge, when the contents are tilted into a kind of hopper and so run away down a waste pipe.

Figure 338 is a sketch showing one that is spoken of as being compact, and suitable for offices, ships, and other places where space is limited. A round basin and slab, in one piece, fitted up in a similar way as Figure 338, makes a compact wash-hand basin. In this case the rod which connects the basin to the key of the supply-cock should be omitted, and bibb-faucets, having jointed nozzles for pushing back out of the way, used so that hot and cold water can be turned on at pleasure. A great many of the fittings that have been described are very ingenious

FIGURE 338.

and compact, but they all have the disadvantage that unless they are well looked after and get the necessary attention they soon become offensive-smelling. For this reason urinals should always be situated in a well-lighted place, and, if possible, away from the dwelling. When, of necessity, they must be fixed inside the house, the room in which they are situated should be thoroughly well ventilated, and an attendant instructed to thoroughly cleanse the basin, and as much of the waste pipe, &c., as can be got at as often as possible—even once a day would not be too often.

In some public museums and exhibitions urinettes are fitted up in ladies' cloak-rooms. Figure 339 is a sketch of one. These are mostly arranged with a valve beneath, so that when the seat

FIGURE 339.

is depressed a stream of water flushes the basin. There is no doubt that all urinals, whether stalls or basins, should have a constant stream of water running over the parts exposed to the action of urine, but in some places the supply of water is limited, either by scarcity or by the rules of the Water Companies, to a discharge through a water waste-preventing valve or cistern of about one-half to one gallon to each basin; but it frequently happens that even this limited quantity of water is not used—people come and go and never think of flushing the place after them. To ensure a periodical flushing of the places under discussion automatic flushing-cisterns are in great favour, as when once started they require no further attention beyond making good any of the working parts that may wear out. Most of the Water Companies in London permit the use of these cisterns, but some of their inspectors put seals on

FIGURE 340.

the regulating-cocks after testing that a not too extravagant quantity of water is used.

Figure 340 is a sketch of a very simple automatic flushing-cistern holding about two gallons. If required to empty itself about every eight or ten minutes a small tap can be fixed and regulated to fill the cistern in that time; but if it is intended to discharge about every ten or twenty minutes, it is sometimes necessary to fix a reversible ball-valve. The action is as follows: The bottom parts of the pipe-coil inside the cistern retain a small quantity of water, the upper parts being charged with air. This makes the coiled pipe what is commonly called air-bound. The water has to rise in the cistern so as to cover the coil a few inches so that the weight of the water is sufficient to overcome the resistance offered by the pent-up air inside the bent pipe. A small supply-cock, set so that the water dribbles very slowly into the cistern, will not fill it quick enough, so that the water will dribble away down the pipe as fast as it comes in. By adding a reversible ball-valve and regulating it so that when the cistern is partly full the floating ball will open the valve and let the water run in at full bore, it will head up so quickly that any small quantity dribbling away has no effect on the ultimate results.

Figure 341 is a sketch of one which acts precisely as Figure 340, but a great deal depends upon the depth of the bag part, A.

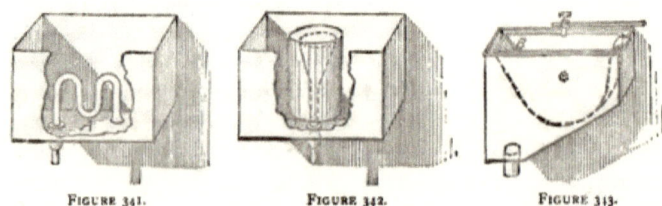

FIGURE 341.   FIGURE 342.   FIGURE 343.

Figure 342 is a patent automatic cistern used for the same purpose. Another kind is shown by sketch, Figure 343. In this case the inner chamber is hung on bearings fixed slightly out of the centre, but so shaped that when quite full of water the part B becomes the heavier, so that it falls down and allows the contents to escape into the outer chamber and down the pipes. When empty, the other end being the heavier, it falls back to its original position. Indiarubber buffers have to be fixed for the tumbling chamber to knock against, otherwise it is very noisy in its action.

Figure 344 is another description of automatic flushing-tank. In this case a syphon has to be fixed as shown at C, so as to retain the air in the syphon pipe, D, so that the water will head up to the dotted line, when it will have sufficient weight to drive out the air and start the syphon. These cisterns are also made to a large size and used for flushing drains, &c.

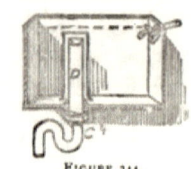

FIGURE 344.

CHAPTER XXXII.

SOIL PIPES.

AT a very old historical building in London a retiring-place was found to be made at the top of a two-storied building, and a shaft built in the walls for the purpose of conveying excreta to an opening leading into the River Thames. Figure 345 is a

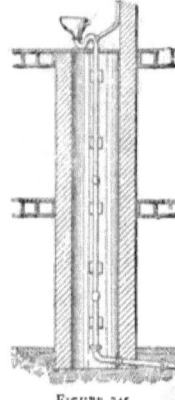

FIGURE 345.

sketch section showing the shaft. A plain seat was originally fixed over the top of the shaft, until the stench that escaped became unbearable, when a water-closet basin and trap were introduced. As this did not improve matters very much, it was finally decided to fix a lead soil pipe, as shown, the men and necessary materials being lowered from the top. A drain was also fixed from the bottom of the soil pipe to an adjoining sewer. There is no doubt the shaft remains an evil, as the walls were partly covered with excreta, and it is more than probable that the smells from this can pass through into the adjoining rooms.

Some people are under the impression that anything in the shape of a tube will do for conveying soil from a water-closet to a drain, and all sorts of schemes are practised with that object, but with varying degrees of success. In some modern cases common drain pipes have been fixed as soil-conduits. In other cases drain pipes have been used and fastened to the walls of houses with pieces of hoop-iron. In another case common drain pipes were built in a party-wall between two houses to act as a ventilator to the house drains. In this case a week was spent in fruitless search for the cause of smells in the drawing-room, when it was decided to take down the wooden skirtings. In doing this the wooden plugs, driven into the joints of the brickwork for fixing the skirtings to, came out, and a loose brick was also found. On removing this brick the source of evil was discovered.

Figure 346 is an illustration showing this. Drain pipes, no matter how well the joints may be made in the first instance, are not to be trusted to either for the conveyance of soil or for vent pipes to drains.

FIGURE 346

Zinc is not a proper material for soil pipes. In some suburban residences, built by jerry builders, zinc pipe has been used, and, in less than twelve months, holes have been eaten through by the gases emanating from sewage. In some cases D-traps made of zinc have been discovered beneath water-closets; one exhibited at the Parkes Museum of Hygiene, in London, is literally all in pieces. In spite of the knowledge that zinc is not a good material to use, there are a great many cases where it is being put in by scamping builders.

A common way of fixing the soil pipe at small residences is shown at Figure 347, which is a section of a back part of a house. Ordinary iron rain-water pipe is fixed up to A, when a short piece of lead pipe is inserted, and a lead branch pipe carried through the wall to the trap of the water-closet. Above this the vertical pipe is continued to eaves-gutter to receive rain-water from the roof. It is very rarely that the joints of the pipes are made air-tight, and, even if they were, smells escaping from the top of the vent and

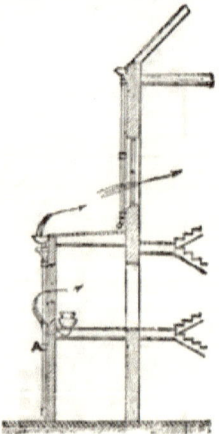

FIGURE 347.

rain-water pipe can pass into any open window, as denoted by the arrows. The joints of the pipes, when made at all, are made of red-lead cement. If in the sun, these joints soon become defective. The expansion and contraction of the metal pulls the cement out of the socket, as shown in section, Figure 348. In some cases when the cement has been finished flush with the top edge of the socket, two or three days' sunshine will cause it to stand up from $\frac{1}{4}$ to $\frac{3}{4}$-inch, as shown at A, leaving a crack through which any smells can escape. The sun has no effect on iron pipes when fixed inside the house, but it is almost impossible to make sound joints to them. Out of some hundreds of tests made, the writer has never yet found an iron pipe of this description to stand either the peppermint or smoke tests, defective joints being the rule and not the exception.

FIGURE 348.

At a first-class (?) house in the West-end of London the soil pipe was of the same description as mentioned above, and also acted as a drain ventilation pipe, the top end being connected to the gutter of the roof to receive the rain-water. It was complained that none of the back windows could be opened on account of the smells that escaped from the pipes.

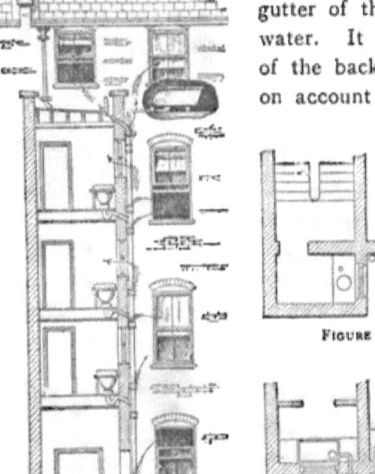

FIGURE 349. 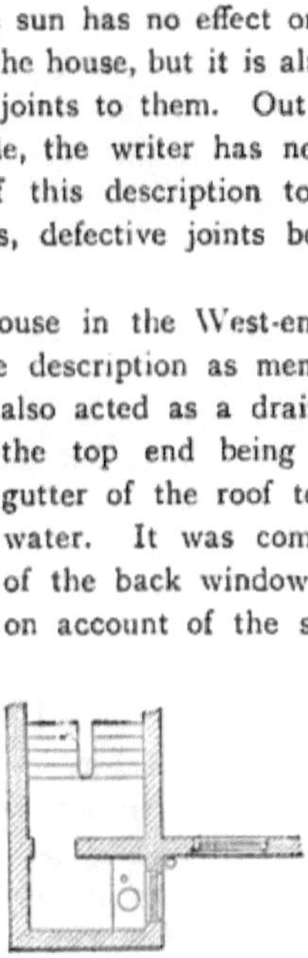   FIGURE 350. 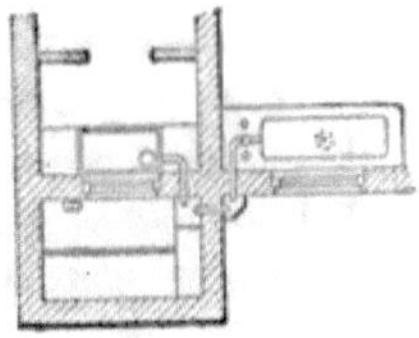   FIGURE 351.

Figure 349 is part of the back elevation of the house, with the projection built for the water-closet shown in section. Figure 350 is a plan of one floor, and Figure 351 of the roof over projection, bathroom, and housemaid's closet. The sketches speak for themselves, the arrows denoting the evils complained of. The waste pipe from the bath and the sink, which also received

SOIL PIPES.

chamber slops, discharged on to the roof, which was offensive from the splashings of the slops. In addition to the defects pointed out, there were no traps beneath the sink or bath, so that the smells from the drains *were laid on* to the house by means of the waste pipes.

The writer could give numerous examples of this class of work, and cases where purchasers have tried to save a few pounds by not employing a sanitary expert to advise them, before purchasing a house, as to its sanitary arrangements, preferring rather to run the risk of being duped by unscrupulous builders.

Another example is given at Figure 352 of one house in a street, all the others being arranged in the same careful (?) manner. In this case the soil pipe was of lead up to the point B, and the joints were properly soldered, but, above that, iron pipe was fixed to take the rain-water from the roof. The joints were not made air-tight, so that smells could escape. The head on the top end was dangerously near a window,

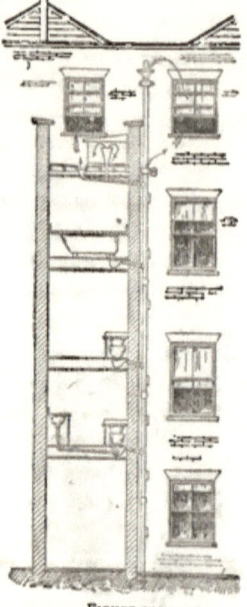

FIGURE 352.

but the greatest evils were the branch pipe to carry off rain-water from the lower roof, which was only 2½ feet from a bedroom window, and the overflow from the cistern, which supplied drinking-water for the household, connected to the branch rain-water pipe. The arrows show the defective arrangements, which, perhaps, were the cause of the illness of the inmates, and which led to the examination and discovery of the evils.

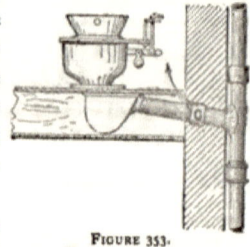

FIGURE 353.

Figure 353 is a fragmentary section showing a defect that recently came under the writer's notice. This was in the house of a medical officer of health. The stench in the water-closet was so great that it was thought advisable to take away the old pan water-closet and D-trap, and fix a better kind of trap and an apparatus of a more sanitary description. The floor of the water-closet was taken up for access to change the trap, when it was discovered that a hole was eaten through the lead by sewage gases, so that it was necessary to change the branch soil pipe. On cutting away the brick wall for that purpose a slip-joint was found, as shown at C. The discovery of this defect may almost be termed accidental, as it was hidden in the brickwork and could not be seen until the wall was cut away.

This is an example of how some builders plumb their houses: They will let the plumbing, piece-work, to some journeyman plumber, as unprincipled as themselves, at a price that would scarcely pay for good materials. The plumber, to make the work profitable, will make several T-pieces, and solder short pieces of soil pipe on to D-traps, which are generally made of five-pound lead, at his home. The T-pieces are sent on to the job and fixed, perhaps, by the bricklayer. A few days afterward the plumber will bring the traps, socket them in the branches of the T's, as shown at Figure 353, and, after the carpenter has laid the floor, will fix the water-closet apparatus and pipes to flush them, and the job is completed. The writer has known dealers of materials to go round and buy up old pan water-closet apparatuses, do them up, and sell them cheap to the above class of builders.

It is to be hoped that the long-talked-of registration of plumbers in England will soon be a fact, and, in addition, that properly-qualified inspectors of plumbers' work will be appointed, and laws passed to make it a criminal offence to do any plumbers' work in such a way as to be injurious to the health of the poor victims of people who have to live in the above kind of houses.*

---

* Since writing the above, some hundreds of plumbers have been registered by the Plumbers' Company, and better things are now to be hoped for.

CHAPTER XXXIII.

SOIL PIPES—*continued.*

ENOUGH has been written to show the danger of fixing light iron soil pipes, with the rain-water leaders connected to them, on the outside of the house, but there remains to be told the evils of *fixing the same class of work inside the house.* Figure 354 is a very common example. One case was tested with a smoke-rocket, when smoke escaped from each joint of the pipes, and also through the wall into an adjoining bedroom. This will be more clearly understood by referring to Figure 355, this being a plan of one water-closet, the arrow showing the way the smoke passed through the wall, which, at this point, was only about 4½ inches thick. The mortar was very poor, and as the chase for the pipes had been cut after the walls were built, the bricks were loose. The plaster on the walls was very much cracked, thus leaving little or nothing to act as a barrier against smells passing through. The soil pipe was continued to the small flat roof to receive the rain-water from that and also the upper roof. *The overflow pipe from the cistern was fixed so as to discharge into the top end of the soil pipe.*

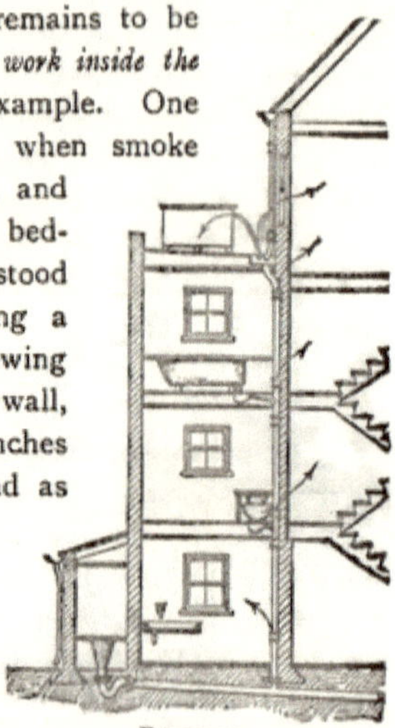

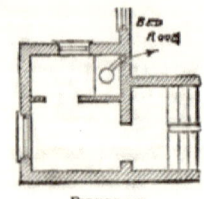

FIGURE 354.

FIGURE 355.

Another class of house is shown in section at Figure 356. One water-closet was on the second floor, and the soil pipe from it was fixed inside the house in an angle of the drawing-room on the first floor and the dining-room on the ground floor. The pipes were of light iron and the joints defective. Each time the water-closet was used the noise of the water was distinctly heard in the above rooms, so that it was deemed advisable to remove the whole affair to a less objectionable position, and the soil pipe fixed outside an external wall.

In another case the water-closet was fixed over a drawing-room with a decorated ceiling. The space round the trap was packed with sawdust to deaden the sound of rushing water. The housemaids had been in the habit of bringing all bedroom slops to this water-closet, and, although the floor beneath the apparatus was partly covered with lead, water splashed over, and the ceiling beneath was disfigured with stains. In spite of all protests as to the unsuitability of the position for a water-closet the owner insisted on retaining it in its present place. This is in a house the rental of which is about £150 a year. There is not the least doubt the plaster of the ceiling below will become so loose that, should a pipe leak or water by any means get splashed over the water-closet safe, a large portion of the ceiling will fall down and perhaps do several pounds' worth of damage to furniture or whatever may be near, and then there will be another growl at the plumber who fixed the water-closet there.

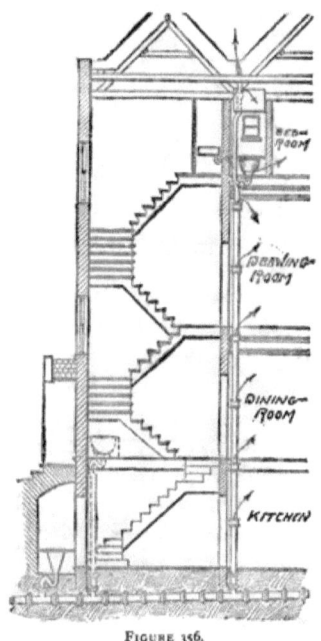

FIGURE 356.

There are several houses near Hyde Park, in London, that were built some twenty-five years ago, that are very badly designed with regard to the sanitary arrangements. Figure 357 is a section across the two upper floors and roof of one house the writer had to make several alterations to some years ago. The soil pipe was continued to the small flat roof next the party-wall of the next house. Beneath the floor of attic an open trough, made by lining between the floor-joists with lead, was made to carry away the water from the front gutter into the soil pipe. In addition to the smells escaping out of the end of the trough and entering the attic window, the floor-boards over the trough were so shrunken as to allow any smells to freely escape into the bedroom in which maidservants slept.

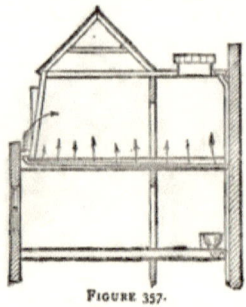

FIGURE 357.

The arrows denote this escape. The way this was improved was rather interesting. Space was made by the side of the soil pipe to fix a new lead pipe to receive the rain-water from the small flat roof. The lead trough beneath the flooring was taken out, and, as the space was only 2½ inches deep, a 6-inch lead pipe was flattened, as shown in section, Figure 358. 10-foot lengths of pipe were used, which necessitated three joints being made when the pipes were in their position. This was got over by cutting open the pipes about 18 inches each side of the joint, which was then soldered on the inside of

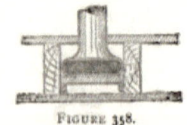

FIGURE 358.

the bottom and sides. After doing this the pipe was closed and the other part of the joint wiped on the outside. A seam was then wiped over the slits, paper being pasted over the soldered joint at the sides of the seam to prevent that being melted. One end of the 6-inch pipe was bent upward, as shown by Figure 359, and soldered into the bottom of the gutter, and the other end soldered into the vertical rain-water pipe fixed from the upper roof as predescribed.

FIGURE 359.

Innumerable cases could be given of the evil of attaching rain-water leaders to soil pipes. Figure 210 is a plan of a house, one of the soil pipes in which was used as a rain-water leader. The top end was finished as shown in section, Figure 360. A is a skylight over a staircase, and which also gave light to several bedrooms. The sides of the light were louvred for ventilation, but, unfortunately, the skylight was in a valley between two high roofs, and it did not matter which way the wind

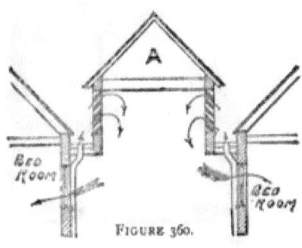

FIGURE 360.

blew there was always a draught into the house, carrying with it any smells that escaped from the combined rain-water and soil pipe. In this case a separate soil pipe was fixed and continued to a good height above the roof as ventilation to drains and soil pipe.

At a country house taken by a lady and family for the summer months, the servants sat by their bedroom window the whole of the first and only night they were there, and could not stay in bed because of the abominable stench that appeared to be immediately beneath. On an examination being made it was discovered that the rain-water leader was connected to a drain leading into a cesspool, no trap of any kind being fixed to prevent smells escaping. Figure 361 illustrates the evil. Smells escaping through the pipe passed under the

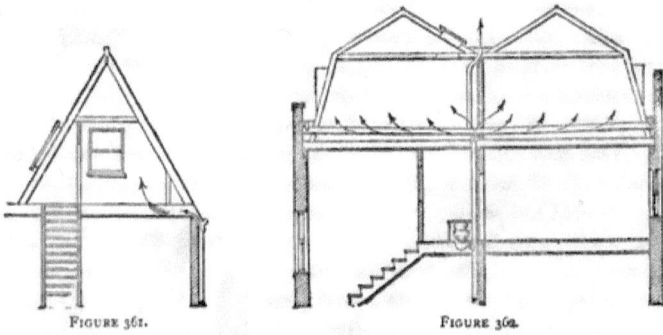

FIGURE 361. FIGURE 362.

eaves of the slates, between the floor-joists and through the joints of the boards beneath the bedstead.

One more illustration is given of defects of construction often met with in a certain class of London houses. In this case a servant-maid had fever, which led to an examination for the cause being instituted. Figure 362 is a fragmental section of the third and attic floors of the house. The soil pipe, which was of lead, with good soldered joints, was connected at the top with a leaden gutter as shown. The attic rooms were used for sleeping in, and the above gutter was continued through each attic to receive the water from the back and front roofs. *The only protection to prevent smells passing into these bedrooms was a board loosely laid over the trough gutter.* In addition to this abominable state of things, appearances lead one to think at times this board had been removed so that bedroom slops could be emptied down, and thus add to the already insanitary state of things.

I should like to diverge from the subject of soil pipes for a few minutes, to show that it is not always the journeyman plumber who is to blame for bad work. No matter how skilful he may be, he cannot make a good job with bad or improper materials, and, when working to some one else's dictation, it is not fair the plumber should be charged with the dictator's mistakes. Further, the public—that is, those interested in houses as owners or tenants—are very much to blame for a great deal of bad work. As an illustration: In a great number of cases where the writer has been sent for to make an examination of a house, he has been met by the person interested, and it is really ludicrous the pains that are often taken to impress upon him (the writer) that there never has been any illness in the house, and there is nothing the matter. In these cases one is almost tempted to ask the person: "Why send for a sanitary man, then?" Novices at making examinations of houses would be influenced by the above class of people, and not thoroughly test the drains, &c., while old hands at it would simply smile *and make a thorough examination.*

A case occurred where the writer tested the drains of a house with smoke and found them very defective, the smoke escaping between the joints of the stone paving in passages and floor-boards of rooms. In spite of this proof the owner was not satisfied until he had called in another adviser, who made a report similar to the first one's. The services of both advisers were dispensed with, and a jack-of-all-trades employed to stop up the cracks in the flooring with mortar and putty.

Very often after an examination of a house has been made

and the report sent into the owner, the sanitary man has to submit to all sorts of interrogations, such as "Is it as bad as you say? What will it cost to put right? Are you sure it is defective? How long will it take to do? I don't think it is so bad as you make out! Can it be done without the family leaving the house? You must have made a mistake, as no one has been ill! Could not the holes be stopped with putty?" After about an hour, and sometimes two, of this misery, when one wishes he had never been called in, he is dismissed with the remark, "I will think about it and let you know." This promise is very often not kept, but some talented expert, who can do wonders with paint and putty, is called in, with the result that very often the doctor's bill is considerably more than would have paid for that which would have prevented the illness of the family.

But to return to our original subject. In several cases of testing, smoke has been found to escape beneath the seat of an upstairs water-closet, and this sometimes when trying to find cause for smells in some other position. In taking down the wood enclosure of the water-closet it is often found that a small trap is placed beneath the lead safe to take away any water that overflowed the basin, the outgo of the trap being branched into that from the water-closet trap. The water has become syphoned out of the small trap, as the soil pipes were not ventilated, and the small pipe, called the "weeping pipe," which is arranged so as to recharge this trap with water at each usage of the water-closet, has become choked, so that no water can pass through to recharge the trap. Figure 363 is a plan of the two traps and soil pipe, and Figure 364 is an elevation on A B showing the apparatus fixed and the weeping pipe, at C. In these cases the water-closets are generally flushed by

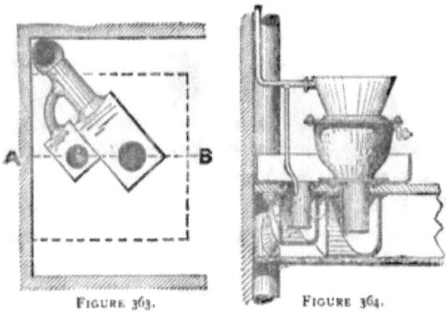

FIGURE 363.   FIGURE 364.

SOIL PIPES.

means of valves and service-boxes fixed in the cistern over. In other cases, where a valve and regulator flushing apparatus has been used, the weeping pipe has been branched into the pipe between the valve and arm of the basin.

Twenty to twenty-six years ago the writer assisted to fix large numbers of water-closet traps in the manner above described. At that time it was considered to be first-class work, and was done in all high-class houses. One case comes back to memory where the traps for a range of eight water-closets were arranged, as shown on plan, Figure 365. So much value was not attached

FIGURE 365.

to ventilation pipes at that time as now, ¾-inch and 1-inch pipes being considered quite large enough for the purpose of preventing syphonage of traps. In some cases the D-traps were made very large—that is, 10 inches or 11 inches deep instead of 9 inches which was the usual size, and 7 inches wide, the dip pipe being kept a little distance from the heel, as shown in section, Figure 366. By doing this, space is made for a larger quantity of air to enter, with less displacement of water

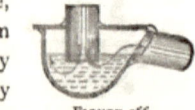

FIGURE 366.

than is the case with a small size D-trap, in which the heel and sides are close to the dip pipe. But traps of that size have now fallen into disuse, being often described as small cesspools. Large soil pipes were also in much favour, as it was more difficult to fill them with water so as to start a syphonic action on the water in the traps. These precautions were taken in the case of the work shown in Figure 365, which, no doubt, was a thorough success when judged with the experiences of that age, but it was also found necessary to fix ¾-inch flushing pipes to the water-closets, a larger size with a good head of water often upsetting all calculations on the point of trap-syphonage. In cases similar to Figure 365, which have been renewed within this last ten years, it has often been found that the safe-traps have been quite empty, partly by syphonage, and the rest by evaporation; the weeping

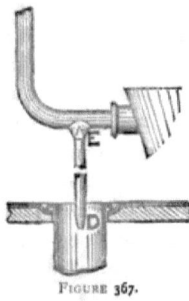

FIGURE 367.

pipes in these cases having become useless by being choked at D, Figure 367. This is not so likely to occur if the bottom end is left open and the other end tapered, as shown by dotted lines at E. Any dirt lodging on the top end of the pipe would be removed by the scour of the water as it passes toward the basin, whereas in the first instance anything getting into the end D would become further jammed by the water-pressure above it. In the case of stoppage in the soil pipe, the safe-trap, when arranged as shown at Figure 363, becomes perfectly useless. The outgo being branched into the soil pipe, any stoppage in the soil pipe affects this trap as much as that under the water-closet.

In some cases the waste pipe from the safe has been branched into the cheek of the water-closet D-trap, as shown at Figure 368.

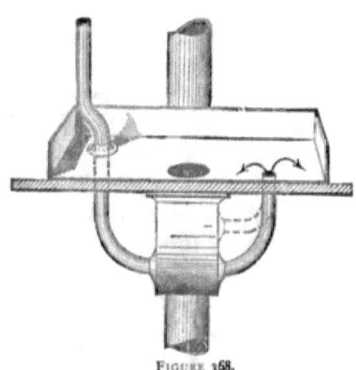

FIGURE 368.

So long as the water is in the trap to the necessary height no smells can escape from the soil pipe. But the waste pipe is a serious evil in another way. Little puffs of air are driven out, as shown by the arrow, each time the water-closet is used, and sewage-matter is also driven out and lays in the bottom of the safe to an extent often injurious to health. There are several cases on record where this safe waste pipe has been branched into the trap above the water-line, as shown by dotted lines, and others where it has been branched into the soil pipe, and no provision whatever made for keeping smells from escaping. On the left hand of Figure 368 is shown a fragment of the cistern-waste branched into the water-closet D-trap. Several years ago this was considered the right thing to do, and the writer has done them that way. When newly fixed, as shown, it is highly dangerous to pull out the cistern stand, or cleaning-out, pipe as the water will rush down the waste pipe through the trap, and play up as a fountain through the safe-waste pipe. After being fixed for a few years the ends of the pipes in the traps become furred up so that no water whatever will pass, thus rendering these pipes perfectly useless for their purpose. Most sanitarians carry the waste pipe from the safe out of doors into the open air, with a hinged flap on the end to prevent any inward draught and to keep out birds, &c. Where this cannot be done, the lead safe should be continued in front of the water-closet enclosure, so that if a leakage occurs it can be seen.

CHAPTER XXXIV.

SOIL PIPES—*continued.*

PLUMBERS' work done about twenty years ago, and which recently has had to be taken out, has very often been found to be of good materials and workmanship, but badly arranged. In first-class work it was usual to branch waste pipes from all kinds of fittings into the water-closet trap. Figure 369 is a sketch of a trap with four ends of waste pipes attached that was taken out a short time ago. A was the waste pipe from the lead safe on the floor under the water-closet, B was that from the cistern fixed over the water-closet, C was the waste pipe from a wash-hand basin, and D from a small sink fixed on the floor above.

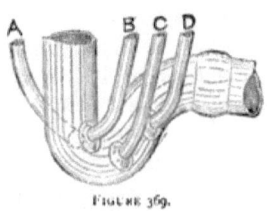

FIGURE 369.

Figure 370 is a sketch of some old work recently taken out of a building in Lincoln's Inn Fields. It transpired that this job

FIGURE 370.

never had worked properly, although a plumber, about two years ago, took out an old D-trap from under the water-closet and fixed one of a better description. In the figure, E is a sink in the scullery; L, a bath; F, a small sink on floor of staircase landing; G, waste pipe from water-closet safe; I, waste pipe from cistern over water-closet; H, the water-closet trap; K, the joint of trap to branch soil pipe, soldered on the top side, and red-lead cemented on the under side. The whole of this work was fixed in the caretaker's apartments on the top floor. Every time the water-closet was used puffs of foul air were sent out of the various waste pipes, and when the scullery sink was used, dirty water would flow up into the bath and the sink on the floor. The caretaker, to prevent this, corked up the bath and sink waste pipes, the corks being removed when those fittings were used. After doing this he was continually sending for the plumber to unstop the waste pipe from the sink E. Very little comment is necessary; the veriest tyro will see at once the stupid arrangements. It may be added, however, that when the waste pipes from L and F were corked up, the horizontal waste pipe would become air-bound by reason of its being trapped and retaining water in the bagged parts. Before the branch pipes were corked the pent-up air could escape, and thus allow the waste water to escape past. Several instances of this kind of botch-work could be given, but the writer refrains, thinking that perhaps his readers would be under the impression that their credulity was being drawn upon.

To return to soil pipes. Figure 371 is a plan of horizontal soil pipes for water-closets which had traps attached to the apparatus. The writer was working with several other plumbers, about eighteen years ago, at a large public hospital where several ranges of water-closets had the soil pipes fixed as above plan.

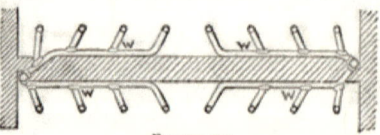

FIGURE 371.

The vertical soil pipes were of lead 6 inches in diameter, and continued to roof as ventilators. The horizontal pipes were 5 inches, and the branches were 4½ inches in diameter. The floors were fire-proof, and, so that they should not be impaired or weakened by cutting any part away, it was decided to fix the branches as shown, in preference to having one horizontal pipe to each range of water-closets, for the reason that little or no fall could be given to a long length of pipe unless a step up was made to the water-closets. Steps should be avoided as much as possible to all water-closets, more especially in a hospital for sick people.

Figure 372 is a plan of several soil pipes that were fixed in the same hospital where it was required to have two water-closets

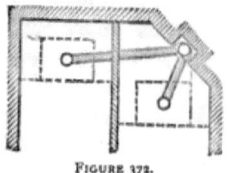

FIGURE 372.

side by side. In this case there were four floors, the same arrangements being carried out on each. The seats were arranged as shown by dotted lines. That on the right-hand side had to be kept forward, otherwise the flap would not remain open because of the splayed angle of wall.

Figure 373 is an elevation; Figure 374, a plan; and Figure 375, a section on A B of a range of water-closets fixed several years

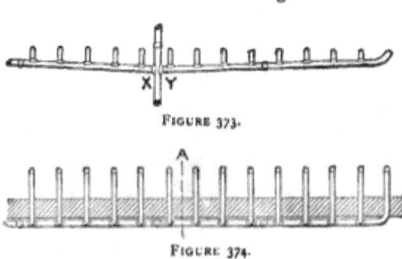

FIGURE 373.

FIGURE 374.

ago at a London railway-station for the use of the passengers. The water-closets were on a level with the platform, which was about 40 feet above the level of the street. The reader is referred to the branch-joints, which

are all at right angles to the main soil pipes. The branches at X, Y, Figure 373, are badly arranged, as what came down one

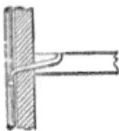

FIGURE 375.

horizontal pipe would doubtless rush up the one opposite, and perhaps lay there until a discharge from one of the water-closets on that branch would again wash it up into the first one. Similar work was being done at another station by different men, when the above evil was foreseen and precautions taken to prevent what has been described from occurring.

Figure 376 is a plan, and Figure 377 an elevation, showing how,

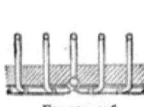

FIGURE 376.

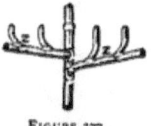

FIGURE 377.

in this other case, the branch-joints were made good to the vertical pipe. It will be seen that it was impossible for matter to rush down one pipe and up

the one opposite. In addition to the branch pipes being jointed to the vertical pipe at different levels, they were arranged so that they entered at the front side, or nearly so. At Z, Z, Figure 377,

SOIL PIPES. 277

is shown how the branches were made good to the horizontal soil pipe, the bottom ends being bent so as to direct the current in the proper direction. There is no doubt that those shown in Figure 373 would cause the stream of water and fæcal matter to be directed on the bottom of the horizontal pipe in such a way that part would be driven up the pipe, where it would lay until a discharge from a fitting higher up would send it down again. The illustrations, Figures 371, 372, 373, 374, 375, 376, and 377, are all of soil pipes prepared to receive water-closets that were made of one piece of earthenware—that is, the trap and basin were combined

Referring again to Figure 371, it will be noticed that the short pipes branched into the horizontal soil pipe have the connections made in such a way as to reduce to a minimum the liability of the discharges from the water-closets running back up the main soil pipe. The branch-joints, when made in this way, require to be carefully fitted, or little spurs of solder will be found inside when finished. They also require about half as much more solder, and a thin cloth to be used, so that as much solder can be wiped out of the throat of the acute angles, W, W, as possible.

Some little time ago the writer was acting as foreman on a large job, and instructed the plumbers to make the branch-joints of a soil pipe for a range of water-closets in the same manner as shown at Figure 371. Unfortunately, he fell sick, and was away for some few days, and when he came back, found part of the work done and fixed, the joints being made as shown at Figure 378. It was very difficult to make the men understand that the object sought was entirely lost by bending the end of the branch

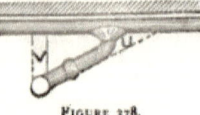

FIGURE 378.

soil pipe in such a way that it entered the horizontal pipe at right angles, and that the work would have answered just as well, and the labour to the bend saved, if the pipe had been branched, as shown by dotted lines at V. It is scarcely necessary to add that the joints were intended to be made as shown by dotted lines at U.

A case of faulty construction occurred where complaints were made of an abominable stench issuing from a water-closet near a bedroom. An examination was made for the cause, when it was found that the waste pipe from the slop-sink—shown in sectional elevation at A, Figure 379—was connected to the side of the water-closet trap, on the floor below, in such a way that whatever was thrown into the sink rushed down the waste pipe, through the water-closet trap, up the safe waste pipe, and lay in the lead safe at B.

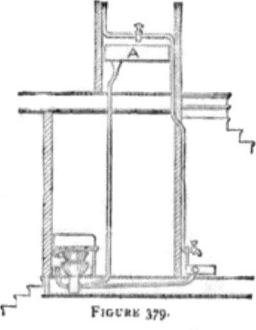

FIGURE 379.

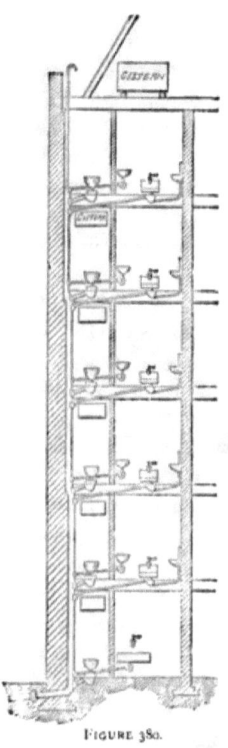

FIGURE 380.

Another sink was fixed beneath the stairs, as shown at C, and sometimes when the water-closet was used liquid matter would wash back up the waste pipe, and could be seen to knock up against the grating which is soldered over the end of the waste pipe in the sink.

Figure 380 is a sectional elevation of a wing of a large building in London which is occupied as offices. The vertical stack of soil pipe is fixed inside the building and is 4½-inch diameter lead pipe. The water-closets are of the valve description, and fixed over lead D-traps. In the lobby outside each water-closet is a wash-hand basin, a urinal-basin, and a lead sink on the floor for drawing water into pails for cleansing purposes. A 1¼-inch lead ventilating pipe is carried from the top end of the soil pipe to the roof. It transpired that for years there had always been complaints of smells escaping from somewhere, but hitherto no one had been able to discover where. Sometimes the smells were found on one floor, and at other times on another, and sometimes in a water-closet, and sometimes in the adjoining lobbies, or in the offices behind. Chemical and smoke tests failed to prove any defects in the soil or waste pipes, or that the seals of any of the traps were broken. It was found impossible to break the seals of the traps by syphonage, for the reason that the branch waste and soil pipes had not sufficient fall, and in spite of violent tests the D-traps always retained sufficient water to seal the ends of the dip pipes, but in some cases only to the extent of $\frac{1}{8}$ to $\frac{1}{4}$ of an inch. The examiner having called in two other persons to assist him, it was found that when two water-closets on the upper floors were used at the same instant of time, the air in the lower portion of the soil pipe would become sufficiently compressed as to burst through the traps fixed on the floor below. After the air had escaped, the water would fall back into the traps, thus leaving them sealed again. It was also found that another reason the water was not syphoned out of the traps sufficiently to break the seals was, that so many traps were connected to the same soil pipe, that each would allow a small quantity of air to pass, which, in the aggregate, was sufficient, when added to the air entering through the $1\frac{1}{2}$-inch vent pipe, to fill the soil pipe, and thus prevent the vacuum being sufficient to start a syphonic action in the other traps. Air currents up the pipe-casings would sometimes carry a smell from one floor to another at a higher level. This is often very troublesome when making examinations, and the engineer is sometimes misled by this means as to the source of smells. In the case under consideration the walls were found defective, so that smells could sometimes pass through to the annoyance of the people in the offices. It is proposed to ventilate all the traps and enlarge the vent pipe at the top of the soil pipes, also to take precautions to prevent smells from passing from one place to another.

At a large building near the Bank of England, a difficult case came under the writer's notice. Several experts had been called in at various times to discover the cause of an abominable smell that was intermittent. The fact of the smells not being continuous added to the difficulty, as no vapour test would betray the cause in the same manner as if the smells were constant. Each sanitary man had taken away the water-closets and fixed others that he had a preference for. The one who preceded the writer had the syphon-traps taken out and D-traps placed beneath each water-closet, but did not succeed in his object.

Figure 381 is a sketch diagram, showing the water-closets and soil pipes which were continued to the roof full size, and the first floor was the place where the smells were complained of. After applying vapour tests and finding nothing defective in the materials or appliances, the water-closets were taken up, and also the flooring, so as to be able to make a closer examination of the traps and pipes. Nothing being discovered to account for the smells, men were sent to the upper floors. The handles of the closets, A and B, being simultaneously pulled up, and the writer stooping over the trap of the water-closet, C, to watch the result, was anything but agreeably surprised to have the contents of the trap blow up into his face. This solved the mystery at once. On discharging the contents of these two water-closets, air was driven downward in each soil pipe. The two columns of air meeting at the first-floor level burst through the traps of the water-closets at that point. Several other fittings, such as urinals and wash-hand basins, were attached to the same soil pipes, but they are omitted for the sake of clearness. The remedy applied was to take two ventilation pipes from the traps of the first-floor water-closets and continue them to the roof. On using any one water-closet on the upper floors no evil resulted, as the air driven down the soil pipe by the falling water, &c., could freely escape up the other one. The above evils have frequently been found in smaller houses. Figure 382 is a sketch of the back of a very common description of house. This kind of house has been referred to in an earlier chapter, where was pointed out the evils of connecting the rain-water leader to the soil pipe. In great numbers of cases a trap is fixed at the foot of the soil pipe to prevent

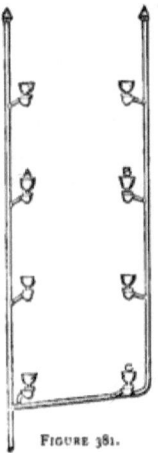

FIGURE 381.

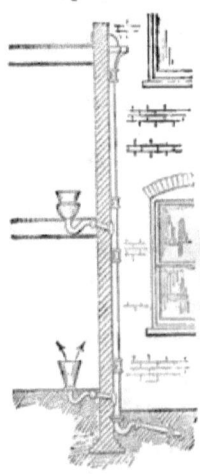

FIGURE 382.

any bad air from the drains passing out at the top of the soil pipe, &c., and into any open window. This trap seals the bottom of the pipe, with the result that the air cannot freely escape when driven downward by discharges from the water-closets. On using the top water-closet, the air in the soil pipe is driven downward, and will frequently burst through the trap of the lower water-closet. Innumerable cases could be given of defective arrangements of soil pipes. Those that have been illustrated were simply those out of several that came first to the writer's mind, and which were given as typical examples to show that even with good materials and skilled labour, the whole affair may be rendered a failure for want of technical knowledge as to what will be the results when completed.

The water-carriage system of conveying sewage matter from the dwelling is one that requires a great deal of thought and study. The most elaborate water-closet apparatus is perfectly useless without the necessary water to cleanse it, and float the matter deposited in it away to a suitable place. The best kind of traps are of no value if they have no water in them. Drains or soil pipes may be *made* of the very best materials and yet be sources of serious evils unless the joints are both air and water-tight when in or near the dwelling. All the above evils may be guarded against, and, at the same time, an error of judgment in the arrangement or setting out of the work may be committed with dire results. In olden times the plumber was simply a manipulator of lead, but now he is called upon and expected to be a highly-trained scientist. There is not the least doubt he will rise to the occasion, and that in the future he will not commit the same mistakes as were made by his predecessors in the craft.

CHAPTER XXXV.

SOIL PIPES AND TRAPS.

The commonest kind of trap used a few years ago, and which has been illustrated, written, and spoken about in such a way one would think that the last had gone into the melting-pot never to return, except in another form, is still very frequently used by a certain section of people. The D-trap, Figure 383, is referred

FIGURE 383.

to, and it is to be hoped that our lately-organized Plumbers' Registering Committee will keep their books clear of the names of those who advocate its use in preference to all others. Some fourteen or sixteen years ago a plumber, who had advanced ideas, saw some of the evils attached to the D, and designed a trap as illustrated in Figure 384. This was a great improvement, as the scour through this trap kept it much

FIGURE 384.

cleaner than the old-fashioned one could be, and, in addition, if the gases emanating from sewage corroded the trap it could be discovered by water leaking out of the holes; whereas, in the D-trap, the dip pipe, being exposed on both sides to the action of the gases, was soon eaten through, and thus rendered the trap useless for keeping smells from escaping. Being out of sight, these holes can very rarely be discovered until the smells become so bad as to lead to the water-closet being taken up for the trap to be examined.

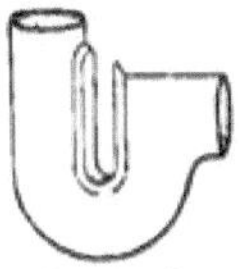

Figure 385 is an illustration of a good old trap. One is shown at the Parkes Museum of Hygiene. The writer has seen a few taken out of a certain locality in London, and as they are not generally met with, it leads to the presumption that they are all the work of one man.

FIGURE 385.

From appearances one is led to the conclusion that the body of the trap was bossed up and then the throat-piece wiped in afterward.

A friend of the writer's, and a man of great experience, used to make all traps for fixing beneath water-closets as Figure 386. The traps were made in two halves out of sheet-lead, and I have assisted to make them, on iron moulds, and then soldered together afterward by wiping a seam on each side. The inlet and outlet ends were 4 inches, but the body of the trap was made 5 inches in diameter. They were made in this way with the object of reducing the syphonic action that takes place more or less with all unventilated traps when fixed under water-closets, and with a soil pipe attached to the outgo. Another reason was to break the impetus of the water discharged from a water-closet. The impetus given to the water when falling into the trap often being sufficient to carry it right through an ordinary round pipe trap, scarcely any water remaining behind to form the proper seal. In the north of England great numbers of traps are made as shown at Figure 386, but of equal diameter throughout, the seams at the sides being made with fine solder and copper-bit. This seam is not nearly so strong as the wiped one described in Figure 386.

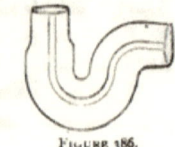

FIGURE 386.

Some years ago commenced a new era in trap-making, when they began to be cast instead of being made by hand. The old patterns, with very little variation, of the D and round pipe P traps were made, and then later on a patent was taken out for casting traps on the principle of that shown at Figure 384. Figure 387 is an illustration of this trap, which is now being superseded by that shown at Figure 388.

In Figure 387 the body of the trap is much larger than the inlet, but in Figure 388 the sizes are in reversed order, the waist of the trap being much smaller than the inlet. The inlet is

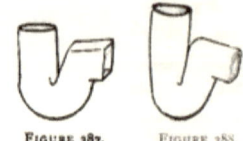

FIGURE 387. FIGURE 388.

4 inches, and the body of the trap 3 inches in diameter. Those the writer has fixed have been found to be quite free from fur after two years' usage, and he has seen one, fixed nearly seven years

ago, which had pieces of glass fitted to the sides. Although the water-closet over this trap is much used, the accumulation of fur is so very small that a person standing on one side of the trap can see the light of a candle held on the other side. The patentee claims that this trap is proof against momentum, or the water rushing through the trap, as described when alluding to Figure 386, and certainly the trap fulfils the claim.

In addition to cast-lead traps, a great many are being used that are made in the same manner as drawn-lead pipe—that is, the lead is forced through dies which gives it a round section, but by manipulating the press, the pipe, as it issues from the dies, is made to curve round in the desired shape to form traps to use in various positions. To show how much under control is the press for making traps, one kind is shown at Figure 389, which is of one piece, no solder or other means of joining lead together being used. Figure 390 is a trap differently shaped, but made by the same means as the other one. Figure 390 is made especially to the order of a master plumber, who claims that it is proof against momentum.

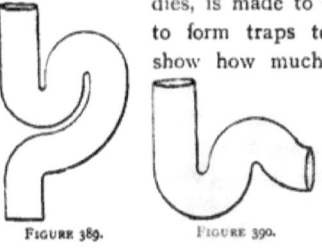

FIGURE 389.   FIGURE 390.

A great many other examples of traps could be given, but it is quite unnecessary, as they are mostly made on the lines of those that have been described in this and earlier chapters. It may be added, however, that all traps for fixing under water-closets must have a clear waterway through them, and that mechanical traps such as those with hinged flaps or valves, or balls, either floating or heavy, so arranged as to form an effectual seal, cannot very well be used, as paper and other matters that pass through a water-closet would cling round the working parts, and so render them useless. In some cases certain matters would accumulate round the valves, &c., and in time form a complete stoppage.

The subject of traps cannot be dismissed without a few remarks on trap fixing. It is admitted by all advanced sanitary engineers that traps are necessary under water-closets and other fittings that require waste pipes. There are some few people who think otherwise, but they are mostly amateurs, and those of limited experience. If we are to accept the dictum that traps are necessary, it follows that they should be fixed in such a way that they are not rendered useless by the way of fixing them. One of the first things to guard against is fixing the traps in such a way that the seal is broken. The writer has several times found traps fixed as shown in sketch section, Figure 391. D-traps, and also the P-traps, have been found to have the seal broken by improper fixing. In some instances this appears to have arisen from branching the trap outgo too low down in the vertical soil pipe when the trap has been pulled up after fixing, so as to be in its proper position for the water-closet. It may also have occurred in another way. If the piece of pipe from the trap to the vertical soil pipe has been cut too short, the P-trap has been strained open and so distorted in shape, in the endeavour to make it longer, that the water-seal is broken, as shown by Figure 391. In other cases the vertical soil pipe has been insufficiently supported, with the result that it has broken away from its fixings, and slid downward, causing the branch pipe to drag down the outgo-end of the trap, thus breaking the water-seal. The straining and bending of traps in their position after fixing is a natural sequence with those men who work by rule of thumb. I will illustrate this: When a plumber is going to fix a trap and soil pipe, he will often stand a length of soil pipe in the position intended for fixing it. He will then place the trap in its position, and take a short piece of soil pipe and cut and rasp the ends to fit between the trap and vertical pipe. The next operation is to remove the whole of the work to the bench in the workshop, when the hole is cut in the vertical pipe, and the necessary soiling and shaving done to the parts intended for soldering. The work is again placed in its position and the parts fitted together. After doing this the joints are "tacked"—that is, some hot solder is splashed on the joints and left until it has set. The work is now again moved to the bench

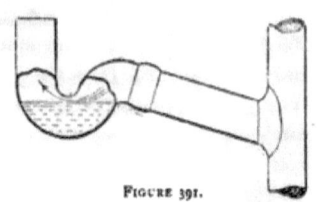

FIGURE 391.

and the joints made. During the operation of making the joints it frequently happens that so soon as the tacking solder is melted, the work will shift and so get out of its proper position. On again moving the piece of work into its position, it will be found not to fit; but lead being a very soft metal, it is easy to "spring" it—that is, bend and strain it—until it suits the intended position.

This rule-of-thumb business should be strongly condemned. In the first place, a great deal of time is wasted when moving the work up and down between the shop and the intended position. The shop is generally in the basement, and the work may perhaps be going to be fixed on the third, fourth, fifth, or sixth story of the building, and thus a great deal of the plumber's time is misspent. In spite of all care taken when fitting the work, it frequently happens that it does not "fit" when completed. The proper way of setting out this kind of work is as follows:— Let Figure 392 represent a plan of the opening for a single water-closet. Now water-closets, such as the valve and others,

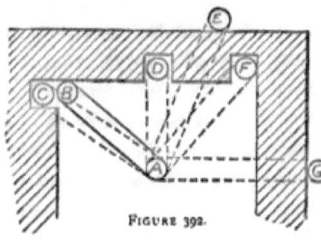

Figure 392.

which require a trap beneath them, and which have the outlet in the centre of the apparatus, require the trap to be placed central—that is, equidistant from the side walls. The distance from the back wall is governed by the depth of the water-closet seat. Some architects will have the seats 1 foot 9 inches from back to front, and others specify them to be 2 feet 6 inches, so as to leave more room for the user's dress, and also to leave room for hanging the perforated part of the seat as well as the flap. To take a medium, and the most commonly used, width of seat—say 2 feet—the trap should be fixed with the centre of the inlet-end 1 foot 3 inches from the back wall. To mark out the work, first of all set out full size on the bench or floor the side and back walls, as shown at Figure 392. Find the position of the trap A, and with a pair of compasses describe a 4-inch circle—that being the usual size of the inlet of traps. If the soil pipe is 4 inches, and going to be fixed in the

SOIL PIPES AND TRAPS. 287

angle as shown at B, describe a similar circle at that point. If the soil pipe is to be 3½ inches in diameter, or any other size, describe a circle equal to the end section of the pipe. If the soil pipe is to be fixed in any of the positions shown by dotted lines at C, D, E, F, G, or any other position, the circle should be made in that position. The next thing is to take a piece of strong lath—say 1 inch thick by 2 inches wide—and cut it to the exact length between the two circles. All the lines can now be rubbed out as being of no further use. The next operation is to set out a pair of parallel lines, as shown at H, J, Figure 393, the distance apart being equal to the diameter of the vertical soil pipe. Now take a long rod and place the bottom end on the drain-socket if it is for the first length, or on the top end of the last length of

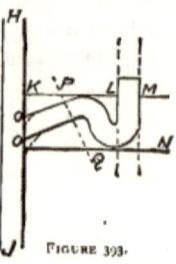

FIGURE 393.

soil pipe that was fixed, if for an upper floor, and mark on the rod the top of the floor-joist where the trap is going to be fixed. When a rule is used for taking a dimension, a mistake may crop in, but by using a long rod the liability to error is minimized. Now lay the length of soil pipe between the lines H, J, and measure from the bottom end with the rod, allowing half-an-inch for a joint at the bottom end, and transfer the mark of the top of the floor-joist from the rod to the pipe, and also on the setting-out marks, as shown at K, Figure 393. With the small rod that was cut to an exact length, set out the distance K to L. 4 inches away make another mark at M. Between these marks lay the trap so that the crown of the outgo is below the line of the floor-joist. Scribe or mark on the bench the shape of the trap, and draw parallel lines from the trap outgo to the lines representing the vertical soil pipe, giving these lines a declination from the trap, but not allowing them to come below the line N, which represents the bottom edge of the floor-joist. The reason for this being that the pipe would look unsightly if seen below the ceiling. When small soil pipes are used it is necessary to bend the end of the branch soil pipe as shown by dotted lines, for reasons given in an earlier chapter.

Marks should be made on the long length of pipe at O, O, and then the hole for the branch pipe opened between those marks. The branch pipe can be cut the exact length, and with the end at the proper angle for connecting to the other pipe, and if properly done will not require any rasping or fitting in any way beyond what is necessary to keep the solder from running through when the joint is being made. . The other end of the branch pipe should be cut half-an-inch longer than the mark P, Q, to allow for the outgo of the trap to enter a short distance. As a rule, the joint on the outgo of a P-trap is straight. After soiling and preparing the ends, this joint should be made. The branch joint should then be prepared and placed in position for making, as shown at Figure 394.

Wood blocks should be fixed inside at R, as described in an earlier chapter, to support the weight of the trap and branch pipe.

The piece of lath that was cut to the exact distance between trap and vertical pipe should be placed as shown at T, Figure 394, to insure that they are the proper distance apart, and should be left there until the branch-joint is made, to prevent the trap falling forward, and also help to support the weight of it. The sides of the pipe laying on the bench should be scotched to prevent its rolling, and a piece of 1-inch pipe bent to a U-shape and laid across the trap, with the ends resting on the bench, will be found to be all that is necessary for fixing the work until the joint is made; or two clout-nails can be driven into the edges of the bench, and a stout piece of string passed round the trap, the ends being fastened to the nails for the same purpose.

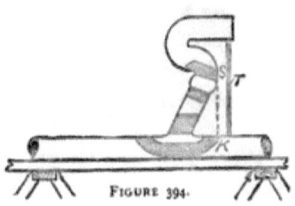

FIGURE 394.

Before making the branch-joint, a set square should be placed on the bench-mark K, Figure 394, and should touch the crown of the trap at S. If a D-trap is being fixed, a straight-edge laid on the flat top of the trap should touch the bench-mark, K—that is, if the trap is properly fixed. The above way of setting out traps and soil pipes is not new, it being taught to the writer when he was a lad. If taken as a problem in geometrical projection, it would be set out as Figure 395; the dotted lines being the plan, and the firm lines the elevation.

To show the value to plumbers of a knowledge of drawings, the writer a few years ago had a set of drawings of a house being built at Shanghai, in China, given to him for his guidance, as to the positions of the cisterns, water-closets, sinks, &c. The whole of the work was set out full size on the workshop floor, in London, and made and put together ready for placing in position. Brass unions were soldered on to the pipes where, of necessity, they had to be made in sections for convenience of removal or stowing away in packing-cases.

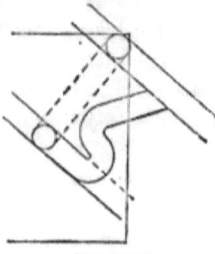

FIGURE 395.

Each part of the work was labelled as to its position, and the unions were all numbered for the guidance of the workmen who fixed them. The whole of the work was found to fit its intended position, and the only hitch was a delay in replacing some marble slabs for wash-hand basins and urinal backs and stalls that got broken in transit from England to China.

Figure 392 represents a plain setting-out for a water-closet trap and soil pipe, but sometimes difficulties present themselves. Figure 396 is an example. In this case the water-closet was fixed on the staircase landing, and so as not to spoil the appearance of the elevation of the house, the window was continued to the top of the stone landing on which the water-closet was fixed. In addition to the bend shown on the plan, Figure 396, another bend was required, as shown by dotted lines, Figure 397, so as to avoid cutting away the skew-back of the brick arch outside and the bearing of the wooden lintel inside.

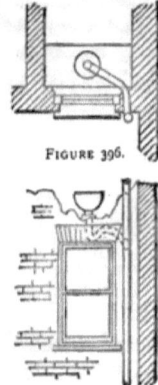

FIGURE 396.

FIGURE 397.

There are architects who bestow some thought on the arrangements of water-closets and soil pipes, but they are the exception rather than the rule. The above fittings are generally the last consideration when designing a house, and any small

U

spare space that may be left after forming the rooms is generally appropriated for a water-closet or sink, or something of the kind. Very often no spare space is left, and the result is the water-closets are introduced into all sorts of unsuitable positions, thus presenting all kinds of difficulties for the plumber to overcome. In one case the writer had to make three bends in a short length of branch soil pipe to avoid going through a chimney-flue. In another case five bends had to be made to avoid cutting away the main supports of a floor in the centre of a building. A case occurred where 3 inches were cut out of the top edge of a 9-inch rolled-iron joist, and another where 4½-inch holes had to be cut through the web of a riveted plate-girder. It is unnecessary to describe these details any further, but close this chapter by drawing the reader's attention to Figure 398, which is a fragmental plan showing how the water-closets are arranged in a large hotel in London. A is a shaft in which are fixed all the soil, gas, waste, main, and down service pipes. Also the hot-water circulation pipes; these last being a valuable auxiliary for creating an up-current of air. The water-closet chambers are ventilated into the shaft, and may be said to be fairly free from local smells which are generally found in these places.

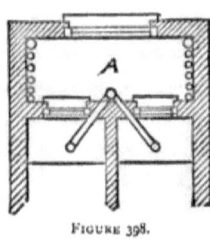

FIGURE 398.

Another advantage of the above-mentioned shaft is, the pipes are all easy of access for repairs or any other purpose, such as removing stoppages, &c.

CHAPTER XXXVI.

WATER-CLOSETS.

It would be a waste of time to give a history of water-closets, and no real advantage would be gained by filling pages with descriptions of those kinds which are now obsolete and, it may be said, forgotten. The writer had thought of describing all those in use at the present day, but had to give up the intention for two reasons: 1st—that, instead of a short chapter or two exhausting the subject, it would fill several volumes of books, and take more time than he has at his disposal; 2ndly—there are so many that would have to be named and condemned in the same sentence as being far from good or suitable for their intended purpose, that is from a hygienic point of view.

One reason for their unsuitability probably arises from the fact that there are very few water-closet apparatus that have been designed or invented by plumbers, or those answerable for the healthy condition of our houses. A great many have been invented by others than plumbers, such as—watchmakers, locksmiths, carpenters, doctors, truss-makers, and such-like people. It is only just to some of these people to say that in some cases a really good fitting has been designed. Take for example the valve water-closet, which was invented by Bramah, and bears his name. At the present time it is made by all the most important manufacturers of closet fittings we have, and is still looked upon, by advanced sanitarians, as the best closet in the market. Several manufacturers have made improvements in the details, but none have invented anything to supersede it. Bramah was a cabinet-maker, although in some of his specifications, for patents, he describes himself as an engineer.

Figure 399 is a sketch of Bramah's valve water-closet, here shown to illustrate and compare with Figure 400, which is a modern valve water-closet with several patented improvements.

Figure 399.

In Figure 399 the handle is placed on the left of the sitter. The basin is very large, the water-flush is at the back and distributed round the basin by means of a metal fan or spreader. If we may judge from the drawing, the size of the flushing pipe was much too small to thoroughly flush the basin. This evil would be aggravated if the head of water was low. In several old Bramah closets taken up by the writer, the water has only dribbled down the back side of the basin, the front part getting none at all. These

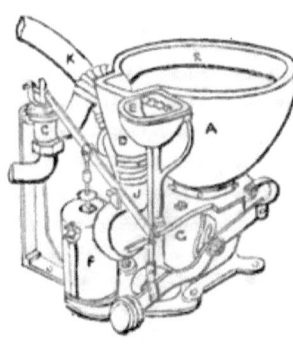

Figure 400.

closets are invariably found to be flushed by means of cranks and wires attached to a valve fixed over a service-box in a cistern. In some cases the cisterns are some height above, or some distance away from, the water-closet. When this is the case, the handle has sometimes to be held up for several seconds before the water comes into the basin. The common result is, the handle is dropped before the water comes in and the paper in the basin gets jammed in the discharging-valve at the bottom, thus keeping it open and allowing the water, that should be retained in the basin, to dribble away down the soil pipe.

The overflow pipe from the basin is very small and connected with the valve-box under the basin. This overflow is trapped, but no provision is made for keeping the trap charged with water, in which case it becomes useless. It is just possible that when the handle of the apparatus is lifted, that a small quantity of water rebounds off the bottom valve into the pipe which connects the trap to the valve-box, but this is objectionable, as fæcal matter may be carried into the trap in the same manner, where it would lay and give off unpleasant odours into the basin. The framework to which the apparatus is attached is objectionable. Being made of wood, it at times gets saturated with splashings and then gives off offensive smells.

The closet shown at Figure 400 is without all the above objectionable points. The handle is on the opposite side, and instead of the raised handle, has a sunk dish, E, made of porcelain. Or the closet can be had with a knob-pull, so arranged that the flap of the wooden enclosure can be closed before the handle is raised, and thus deaden the sound made by the inrush of water of which some people complain. The basin, A, is much smaller, has less surface to be kept clean, and has a hollow rim, R, from which the incoming water streams down over the inner surface of the basin on the front and sides equally with the back, thus exercising a cleansing force over the whole of the inner surface of the basin. The flushing-valve, C, is attached to, and forms a part of, the apparatus. By this arrangement water enters the basin immediately the handle is raised, this being a very great advantage. The valve slowly closes by means of the regulator, F. The size of the valve should be according to the head of the water-cistern, varying from 1 inch with a head of 20 to 25 feet, and $1\frac{1}{2}$ inch with a head of 3 to 8 feet. A closet, to be properly flushed, should have at least from two to three gallons of water enter the basin in four or five seconds, and it is impossible to get this through $\frac{3}{4}$-inch supply-valves which is the size made by most makers. In Figure 400 the overflow, D, is large enough to take away the water as fast as it comes into the basin. Should the supply-valve get jammed by any means, so that it does not close when the handle is dropped, the overflow will take away the surplus water. On referring to the drawing it will be noticed that the overflow arm, D, is open at the top so that a small mop could be pushed down to clean the inside. Water-closet overflow pipes generally smell offensive by reason of their dirty inside. The hollow rim of the basin is continued round the overflow arm, so that it and the trap beneath gets thoroughly flushed at the same time as the basin. The trap, J, to the overflow-arm of the basin is connected with a pipe, K, one end of which is continued out into the open air, and the other end connected with the valve-box, G, under the basin. The valve-box is porcelain-enamelled inside so that no corrosion can take place, and it is also made as small as convenient so that no large amount of bad air can be retained in it. A vent-arm and pipe, K, are fixed so that when the contents of the basin fall down into the trap they displace the air out of the valve-box into the open air through the vent pipe. Without this vent the pent-up air would escape upwards through the valve-opening of the basin. In some cases this air is forced through the overflow-trap. This is a frequent cause of complaint, which, no doubt, led to the adoption of the vent-arm and pipe. Another use for this pipe is, that with a poor supply of water fæcal matter will sometimes lay in the trap below and give off certain gases which could pass through the pipe into the open air, instead of into the house. The framework of the above water-closet is cast-iron.

There are great numbers of makers of valve water-closets who each have their own ideas as to what constitutes a good apparatus; but there are also several makers, who, in their endeavour to undersell other people, turn out poor, flimsy things, and which become a permanent tax, when fixed, for repairs on those who buy them.

Some sanitary engineers have valve water-closets made to their own especial design. Two or three have them made with a small weeping pipe from the supply pipe to the arm of the basin to the trap fixed beneath the overflow-arm so as to keep it charged with water.

Other engineers will not have any trap to the overflow from the basin, but retain the overflow-arm. In this case, if the basin fills too full the overflow discharges into the safe under the apparatus. In a great many cases the lead safes are found to have large pools of dirty water laying in them, and this occurs especially when fixed on the bedroom floors where slops are thrown into the water-closet. In a recent case, when searching for a cause for smells to which a doctor attributed a case of sore throat, the safe was found to be full to overflowing of slops, and which was giving off an abominable stink. The waste pipe from the safe was choked up with pieces of paper.

The writer once had to see to a valve water-closet that was much used by children. The overflow from the basin had originally been made to discharge into the safe, and complaints were made that the place always smelt offensive. In addition there was frequently a quantity of water dripping from the safe waste pipe and down the walls of the house. To prevent this dripping of water being such a nuisance, a pipe had been fixed from the end of the safe-waste and continued to the level of the ground outside. This new pipe had become so foul inside that the air which passed through it escaped beneath the water-closet enclosure and added to the cause for complaint. In another case the overflow pipe from the water-closet basin had been continued outside and connected with a rain-water pipe, but this had to be altered because of the offensive smells, and also because of the unpleasant draught of cold air that passed through.

One or two engineers have valve water-closets without any overflow-arm to the basins. When fixed in this way the plumber is being continually sent for to regulate the supply of water, as the basins fill too full or not full enough. In one case the maid-servants were taught by the engineer how to adjust the valve-regulators when it was found necessary. No doubt these maids would make valuable "mates" should they ever become wedded to plumbers. After several years' experience, and with nearly all kinds of valve water-closets, the writer prefers those fitted and fixed that are similar to that shown at Figure 400.

When inspecting and reporting on the sanitary condition of houses, the writer very rarely condemns the closets if they are of the Bramah-valve description. Even the very worst valve water-closets, if well flushed and fixed over a good description of trap, are much better than a great many other kinds that have been introduced these last few years.

The traps beneath valve water-closets should always be fixed beneath the floor. Sometimes **valve closets** have been made with the trap attached to them, but they are not nearly so good as the others for the reason that when the handle is dropped the discharging valve will pick up any paper, &c., that may be laying in the trap. Figure 401 is a sketch of an apparatus with the trap

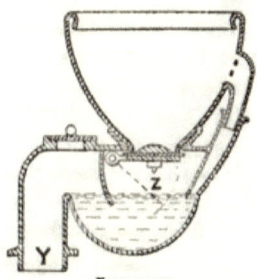

FIGURE 401.

attached, on referring to which it will be noticed that the basin-valve, Z, when opened dips into the water in the trap. Neither can the joint, Y, be made secure or trustworthy. With a lead trap fixed beneath the floor a soldered joint can be made with the lead soil pipe. At an examination of a large building, recently made, the valve water-closets were found to be fixed without any traps at all, the engineer, doubtless, thinking that the basin-valves would keep back any smells. But on testing the soil pipes with smoke it freely escaped at the junction of several of the apparatus with the soil pipes. There is not the least doubt that the joint, Y, Figure 401, is liable to the same defects. At the examination above referred to, one of the valves was found so defective as to allow all the water to leak out of the basin, and smoke freely escaped through and into the house. Traps may be considered objectionable, and plumbers may try all they can to do without them, but they are the only trustworthy barrier for keeping smells from escaping from soil or waste pipes. But the traps must be good ones, and properly fixed and ventilated to prevent the water-seal being broken, as described in an earlier chapter.

In some cases the writer has found that on filling the soil pipe with smoke—the pipe being left open at the top as a ventilator—that on pulling up the handle of a trapless valve closet, smoke has freely escaped through the discharging-valve, even in defiance of the downward rush of water. But with a ventilated water-trap beneath the apparatus this cannot occur. In several cases where trapless valve closets have been fixed, they have been afterwards taken up and traps fixed beneath them.

CHAPTER XXXVII.

WATER-CLOSETS—*continued*.

SOME years ago an eminent master plumber designed and patented a plunger closet, as shown by Figure 402.

The basin and trap were made in one piece of white earthenware, and water was retained in the basin by means of a hollow plunger, fitted in a side chamber over the outlet where it is connected with the trap. If the basin filled too full the water overflowed down the hollow plunger. Closets of more recent make have a solid plunger and a separate overflow-arm. Since the patent ran out several other firms make slightly-varied copies of this closet. Although several people prefer this kind of closet, the writer still thinks that

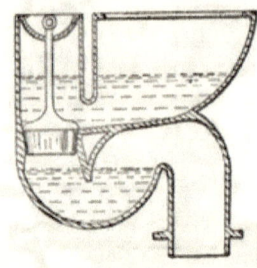

FIGURE 402.

they are not nearly so good as the valve closet shown at Figure 400 for the reason that those he has seen were always very dirty and offensive-smelling in the chamber in which the plunger and overflow are situated. Neither is the joint of the outgo of the trap to be trusted, for reasons given further back when speaking of trapless valve closets.

Those closets that retain a body of water in the basin are considered the best because the fæcal matters are at once immersed and thus prevent any bad odours escaping from them. In addition, immediately the handle of the apparatus is raised, the contents of the basin are floated away by the accompanying body of water. Neither does the basin get stained in the same way as some of the other descriptions of water-closets which do not retain a body of water in the basin.

Some people believe that the "pan closet," Figure 403, is a very good one because it retains water at the bottom of the basin.

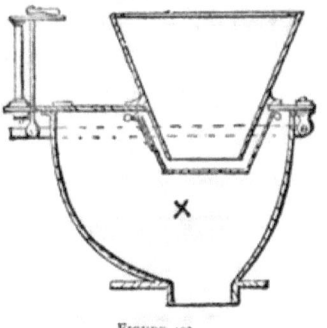

FIGURE 403.

But it is now abandoned by all good plumbers as being, compared with other kinds, the most offensive in its action, and which no amount of ventilating or flushing will prevent giving off foul vapours each time the handle is raised. These vapours emanate from the inside of the large chamber, X—the whole of the inner surface of which sooner or later gets coated with filth.

It being generally acknowledged that water-closets should retain water at the bottom of the basin, various designs have been invented to gain that object and yet do without discharging-valves for emptying the contents of the basins. With valve closets the contents fall or gravitate from the basins, but those referred to, and shown at Figure 404,

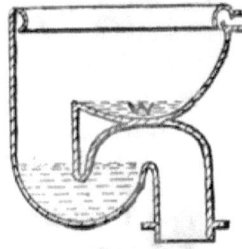

FIGURE 404.

require some power to dislodge the contents and carry them away. This power is supplied from the flushing-water, which has to come into the basin with such force as not only to cleanse the basin but dislodge the contents collected in the hollow at W. This kind of water-closet is made, almost without exception, by every English maker of water-closets, and at the present time it is the most popular closet of the day, being sold and fixed by thousands. This is rather strange, as amongst the hundreds that the writer has seen in use, he has seen scarcely any that were clean in their action, or could be kept clean without the daily use of a brush. The water in the hollow is not deep enough so that fæces can be immersed, and the stains left after being used are in some cases as bad as if the basin had been quite dry before using.

Another-shaped basin, shown at Figure 405, is much cleaner than that shown at Figure 404. With this basin the deposit drops into water which is about 4½ inches deep, and there is considerably less amount of basin-surface to be kept clean. These basins have flushing-rims, and, in addition, a jet is placed near the rim so that the incoming water will play on the floating matter in the basin and drive it through the trap and down to the drain.

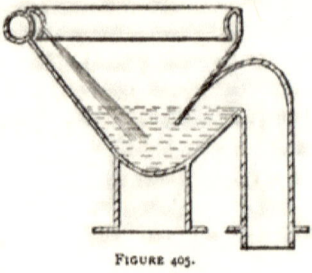

FIGURE 405.

Figure 406 is a basin and trap made in one piece of earthenware, and in its action is similar to Figure 405, excepting that it has no water-jet to assist the flushing-rim in washing away the contents of the basin. These closets are very much liked, and great numbers are found to be fixed. In the writer's opinion it is not so good as Figure 405, in that it has considerably less surface of water at V, and the larger the water-surface the lesser liability there is for the basin sides to get fouled.

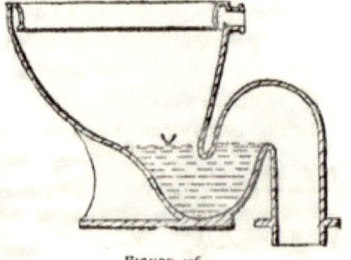

FIGURE 406.

The last three kinds of water-closets that have been described are sometimes made to an ornamental pattern or otherwise decorated. They are then fixed without any enclosure, but with a rim-seat, as shown at Figure 407. A great many sanitary engineers have a dislike for urinals, as it is so very difficult to keep them clean, and fix the water-closets, as shown, so that they may be used for either purpose. If there is no enclosure to water-closets, the whole

FIGURE 407.

of the space around them is lighted and ventilated, and there is less liability for any slops or filth to accumulate in the manner so often found with enclosed water-closets. One eminent firm makes valve water-closets to fix without any enclosure.

With water-closets fixed as above described, it is important that the floors beneath them should be made of impervious material. In some cases fancy tiles have been laid on a concrete bed. Although they look clean and smart they are not so good as a close-grained slab of stone. In a large public institution in London several tiled urinal and water-closet floors were taken up, and, as the tiles smelt so strongly of urine, it was decided to relay the floors, where necessary, with slate slabs, in as large pieces as possible, so as to avoid having more joints than could be avoided. At a large theatre in London the floors were relaid with black asphalte for the same reason.

There are several makers of closets of the kind shown by Figure 408, and for a good, plain, and cheap basin, they are very serviceable when well flushed, and have flushing-rims. Although they are all good it is advisable to choose those with the outlet of the basin well back, as shown by the sketch. When the outlet is in the centre, fæces will fall on the back side of the basin, but with those suggested as the best it falls directly into the water in the trap.

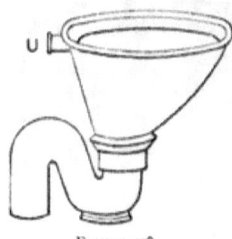

FIGURE 408.

The plumber should always be careful how he attaches the flushing pipe to the arm, U, so that the incoming water will pass in equal quantities each way round the rim, and then what escapes near the front part of the basin will meet in such a way as to turn over with a kind of cascade action, and fall into the trap, driving the contents before it. When the connection is improperly made more water will pass round one side of the basin than the other, and thus form a vortex or whirling motion which fails to clear the contents out of the trap. In some cases a small defect in the construction of the basin will cause the water to whirl round, when it does not exercise such a cleansing force as when it streams downwards from the hollow

WATER-CLOSETS.   301

rim of the basin, converging at the outlet. A good way to test how a water-closet basin will flush is to crumple up a piece of soft water-closet paper between the palms of the hands and throw it into the trap and then flush the basin in the usual way. This is a very severe, though simple, test, as the paper is very buoyant and difficult to send through the trap. At public exhibitions of sanitary goods, the attendants, when showing off water-closet basins, will place pieces of paper all round the inside of them. On flushing the basins these pieces of paper are carried down by the water, and, getting over the outlet, are forced through the trap by sheer weight of water. In use a basin is very rarely filled with paper in the manner described, so to get to know how one is cleared by flushing, the applied test should be as nearly as possible what would occur in ordinary usage. It appears strange at first sight, but it is more difficult to get rid of one piece of crumpled-up paper than six or eight pieces opened out flat in either wash-out or wash-down water-closet basins.

Figure 409 is the common long hopper with the side-inlet flushing-arm. As the incoming water whirls round the basins they are never so clean as those with a flushing-rim. Three or four flushes of water of about two gallons each will very rarely drive one piece of crumpled-up paper through the trap, and in use these closets are invariably found to have fæces and paper laying in the trap.

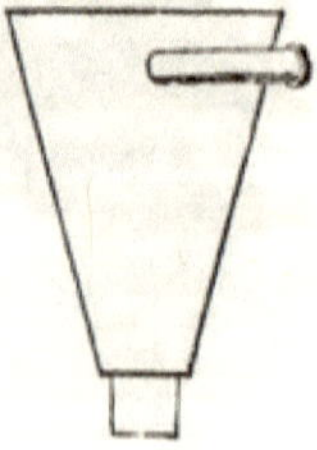

In public schools and institutions, where there is a large number of inmates, it is difficult to provide water-closets suitable for their purpose, and, at the same time, to be free from offensive smells. If basins are used they are constantly being broken, or the traps choked up by boys' caps, jackets, and other articles of dress. Or books and pieces of stick are poked down. In these cases trough water-closets have to be fixed as shown at Figure 410.

There are several objections to these water-closets, one of the

FIGURE 409.

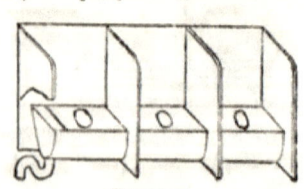

FIGURE 410.

principal being that the sides get splashed at the back with fæces and at the front with urine, and no provision is made for flushing the sides. It is difficult to scrub them with a broom, as the seats have to be fastened down to prevent children opening them and falling in. One seat-holder can splash his neighbour, and a mischievous boy has been seen to splash another boy by means of a stick. Further, a pyramid of matter will accumulate under each seat and remain there until an attendant raises the discharging-plug and empties the contents of the trough down the drains. It is also necessary to scrub the seats with water and broom two or three times a day as they get fouled by people standing on them. Figure 411 is a section across a trough water-closet, showing

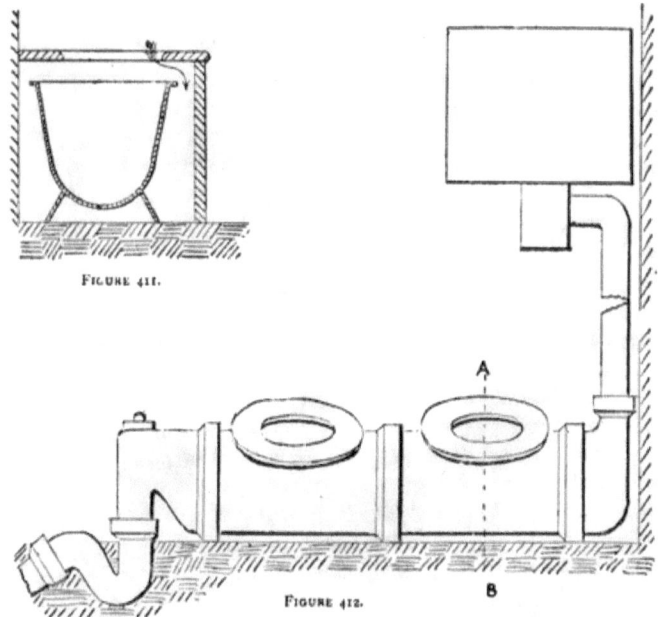

FIGURE 411.

FIGURE 412.

what has been found in several cases. Nearly all the urine is discharged over the top edge of the trough, as shown by the arrow, and falls on the floor.

WATER-CLOSETS.

Ranges of closets of a better description have been designed and fixed these last few years. Instead of U-shaped slate or iron troughs those of a cylindrical shape are much used. They are also made of vitrified stoneware, and are so constructed as to retain a body of water in them. The flushing is arranged to take place at regular intervals of time by means of an automatic discharging cistern which is a great improvement on the older system of having it done by an attendant. Although there are several patented kinds in use, an illustration of one only, which the writer considers the best, is shown at Figure 412. Some the writer has recently fixed, for the use of the outdoor poor at some Union offices, had no enclosures of any kind, and the seats were simply rings of wood bolted on to flanges made on the fitting.

They were fixed sloping towards the front so that they could not be stood upon, and the space from the crown of the cylindrical part to the back wall was filled up with portland cement concrete, the surface being worked up to a smooth face so as not to harbour vermin, &c. Figure 413 is a section on A B, showing what is meant.

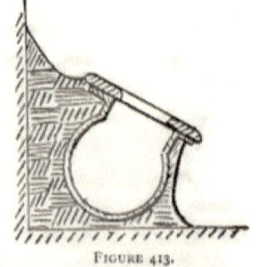

FIGURE 413.

The writer was shown, at a large public institution near London, some water-closets invented by a civil engineer several years ago, and it seems rather strange that they are not more generally used than they are. Figure 414 is a sketch drawn from memory. Only two basins are shown, but each range had about eight or ten. Each basin dips about $\frac{1}{2}$-inch into the water in the trough, and is flushed by means of a valve

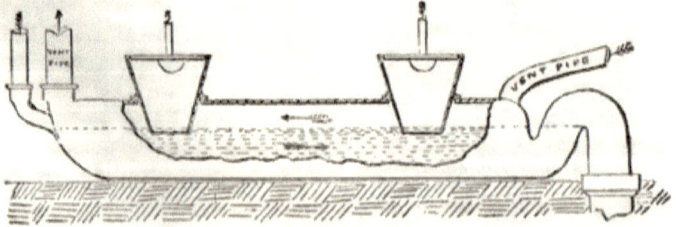

FIGURE 414.

worked by the opening of the door to each stall. At stated intervals of time a valve is opened so as to send a stream of water through the trough and thoroughly flush it out. Ventilating pipes are attached as shown, so that a current of air passes through the trough.

All the water-closets that have been described may be accepted as typical of those in general use, the principles being the same but varied in detail. For this reason it is not necessary to pursue the subject of water-closets any further, excepting to say that the best kinds of apparatus are useless unless a good water-supply is attached so as to float the matter away. Human ingenuity has not yet invented an apparatus that will eject what is deposited in it into the public sewer or wherever the matter has to go.

All water-closet enclosures, when used, should be made so as to be readily taken down or opened. The writer had to take out thirty-two (32) screws one day when making an examination before he could take down a water-closet seat to see what was beneath it.

As there is always a liability of the woodwork round a water-closet being splashed with slops, &c., it is a good plan to paint, varnish, or otherwise render the woodwork impervious. This applies more especially to the perforated part over the basin, the under side of which is invariably found to be saturated with urine and to smell very offensive.

As stated in an earlier chapter, water-closets should never be fixed in or near sleeping apartments, and the room in which a water-closet is situated should always have a provision made for constantly changing the air. Two pipes, one to let fresh air in and the other for the vitiated air to escape, are necessary. A back spring should always be fixed to the doorway, so as to keep the door closed and allow for ventilation to take place independently of the rest of the house.

This door-spring is of great importance, for if several air pipes are fixed they will generally be found to act as air-inlets so long as the door is open, and thus blow any smells in the water-closet into the house.

CHAPTER XXXVIII

HOT-WATER BOILERS OR WATER-BACKS.

It is only in large hotels and public institutions in England that a special provision is made for heating the water which is so necessary for the various purposes required, such as for supplying baths, lavatories, sinks, and for other domestic purposes, and in some instances for radiators and coils for heating the various corridors or rooms.

In ordinary mansions and private dwelling-houses it is usual for the boiler or water-heater to be attached to, and form an integral part of, the kitchen range or cooking-stove.

Figure 415 is a sketch representing a front view of an open range, such as is generally used in the kind of houses above referred to, and which has a large fire surface in front, before which joints of meat, game, &c., can be hung for roasting. In the sketch, A is a side oven sometimes placed as shown, and used for baking pastry, &c., in. Where no provision is made for supplying hot water to other places in the

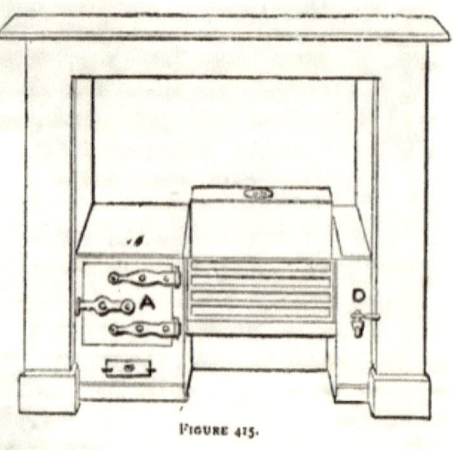

FIGURE 415.

house it is usual to have a large boiler, sometimes made of cast and sometimes of wrought iron, placed at the back of the fire and extending the whole length of it. Where this, which is commonly called a low-pressure boiler, is used, a draw-off tap is usually fixed as shown at D. A small-sized feed-cistern with a ball-valve and the necessary cold-water supply pipe is fixed in a convenient position, and a pipe connection made between the feed-cistern and the boiler. Figure 416 is a sketch showing how this is done. At E is shown a bag in the connecting pipe, made to

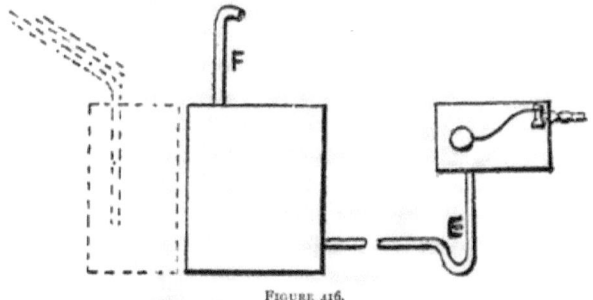

FIGURE 416.

prevent the water circulating back into the feed-cistern, making the contents of that hot. But for this bag, or trap as it is sometimes called, the water in the feed-cistern would get very hot and give off steam to a very inconvenient extent. This bag, under certain conditions, must be made rather deep. When a large fire is kept up and not much water drawn from the boiler the water will expand in bulk, with the result that although a circulation may not be set up between the boiler and feed-cistern, yet the water will ebb and flow, so to speak, between them, and so raise the temperature of that in the feed-cistern. In some cases the boilers have tops bolted down to them and a hole cast in the top, over which is placed a loose cover. When the water in the boiler is so hot as to give off steam it will escape through the loose cover, the latter making an unpleasant clattering noise the whole of the time the steam is escaping. To prevent this noise it is usual to fix a steam or vent pipe from the boiler, as shown at F, Figure 416. The top end of this pipe is generally carried up the chimney-flue a few feet, some of the escaping steam passing upwards, but a great deal condenses and falls down on to the fire, sometimes bringing soot with it. This is very objectionable, especially where cooking has to be done over the fire. Cases have occurred where the steam pipe, when fixed as described, has had to be altered and continued to a place out-of-doors, or turned into a condensing-chamber.

Where the boiler under consideration has been of wrought iron, with a manhole bolted or screwed on tight, the steam or vent pipe is indispensable. Water would not run into the boiler if it was air-bound, and when steam was generated it would force the water back into and overflow the feed-cistern. Figure 417 is a sketch showing an arrangement sometimes met with, but not by any means to be recommended. In the sketch, G is a cold-water cistern supplying the boiler, H, and also the cold-water draw-off tap, J. In a case that the writer had to alter, the work was fixed as shown. From appearances it was seen (and was confirmed by report) that originally only one pipe was fixed to supply the boiler and also the cold-water tap over the sink. Whenever the kitchen fire was alight a considerable quantity of warm water had to be run to waste before cold water could be drawn at the tap, J. As it was found that hot water passed from the boiler to the cold-water cistern, which was about 40 feet (forty feet) away, the writer introduced an ordinary feed-cistern at L, so as to break the continuity of the feed pipe to the boiler. At the same time the expansion pipe, K, had the top end lowered so that if the water in the boiler got very hot and expanded it would overflow and run to waste, instead of forcing its way back into the feed-cistern. The tap, I, was fixed over the scullery-sink, and was the only means of drawing hot water from the boiler. In spite of the alterations made the job was a poor one, and the system is not to be recommended for anyone to adopt.

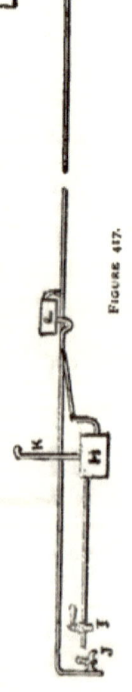

FIGURE 417.

When repairing or altering work fixed some years ago, one often finds a low-pressure boiler and feed-cistern as shown at Figure 416, and a coil of lead, sometimes iron, pipe inside the boiler, one end of the coil being connected with a pipe leading from a cold-water cistern fixed on an upper floor, and the other end with a pipe leading up to a bath or sink. By this means cold water is simply passed through a reservoir of heated water, and in its passage absorbs and takes away a certain amount of heat. This system is not a good one and is now very rarely practised. One of the greatest objections is the coil has to be removed each time the boiler is cleaned out. Because of this the boiler does not get cleaned so often as it should be, and consequently soon wears out. When the coils are made of iron, oxidation takes place inside and outside at the same time, and they soon corrode.

A great many kitchen ranges of modern construction have two boilers at the back of the open fire, the large one, as shown by firm lines, Figure 416 and a smaller one, as shown by dotted lines. The former is called the low-pressure boiler, the draw-off cock being generally as shown at D, Figure 415. The other one is called the high-pressure boiler, and is connected by means of circulation pipes with a hot-water cistern or cylinder fixed near or at some higher place in the building so that hot water may be drawn at the sinks, baths, &c., on the various floors.

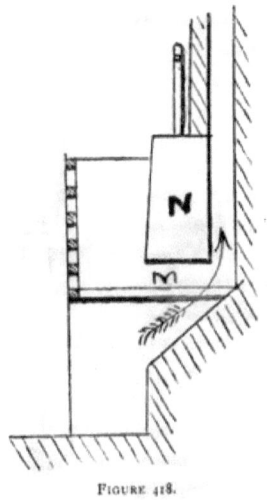

FIGURE 418.

A great many hot-water engineers now omit fixing the two boilers in smaller houses, and arrange for the high-pressure boiler to provide all the hot water that is required. In clubs, hotels, and some of the largest mansions, the low-pressure boiler is fixed and fitted up slightly different, and is used to generate steam to be used for cooking purposes.

In some cases where very large fires are used, say 6 or 7 feet long, so that several joints of meat can be roasted at the same time, the fire lays against the iron boiler, and that is found sufficient to heat the water, but

HOT-WATER BOILERS OR WATER-BACKS.

in smaller ranges a flue is constructed to pass beneath and up the back side of the boiler so as to get a larger heating surface. It is always necessary, when flues are fixed, to have dampers so as to regulate the draught.

In some cases the fire-bars are continued beneath the boiler, as shown in section, Figure 418, at M but the writer has often found it necessary to put a dead-plate over that part immediately under the boiler to prevent cold air instead of heat passing up the flue behind. This cold-air draught is denoted by an arrow.

The shape of the boiler for open ranges is generally as shown at N, Figure 418, that being an end view, but for a close range, or what is commonly called a "kitchener," a boiler, technically named a "boot-boiler," is used. This is shown at Figure 419. A manhole for cleaning out and other purposes is generally placed at P, but the best makers have another manhole at Q. The reader will, no doubt, notice that from the manhole, P, access cannot very well be had to clean the toe of the boiler where most of the fur accumulates, and which it is necessary to remove at intervals of time. The boiler, Figure 419, has fire-bricks placed as shown at r, r, leaving an open space

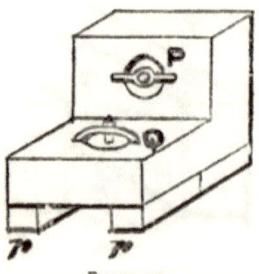

FIGURE 419.

for the heat from the fire to pass up the back flue. A great many boilers are now made and used as shown at Figure 420, a flue being made in the boiler itself. These, sometimes called saddle-boilers, are bedded on a solid base of fire-bricks. A few boilers have a tube through them as shown at Figure 421, the fire passing

FIGURE 420.

FIGURE 421.

through the tube and then up an ordinary flue behind the boiler. Where the kitchen fire has been expected to provide or heat sufficient water for a very large house, a double boot-boiler has sometimes been fixed. This is shown at Figure 422. The fire lays between the two feet of the boiler, and an under and back draught is provided for, as shown.

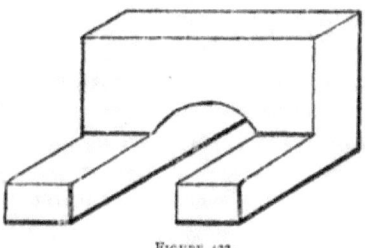

FIGURE 422.

In some cases water is circulated from a boiler fixed in a "hot plate." This hot plate sometimes has ovens also, and may be described as a horizontal kitchener. The shape of the boiler is usually as shown at Figure 423, the fire being in the centre at X.

There are a great many other-shaped boilers used, but the few illustrated are sufficient for our present purpose. A great many of them are rivetted together, but this last few years the welded boilers have come more into general use.

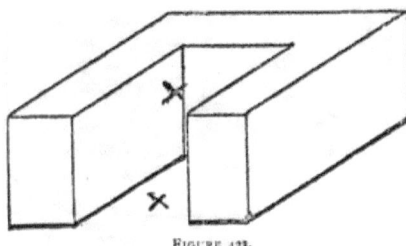

FIGURE 423.

CHAPTER XXXIX.

CIRCULATION BOILERS.

A GREAT many blunders are committed by people who fix hot-water boilers. There are so many intricacies that no one, unless he has had a very varied experience, can always be certain that his job is going to be a success.

The mistake is frequently made of people attending public exhibitions and such like places, where they see something they take a fancy for and straightway purchase it, or a facsimile of it, have it taken home, and then send for someone to fix it. After it is fixed, grumbling commences, even when the fixing has been properly done. In the case of ranges, one person will grumble that it burns too much coal, no doubt thinking that all cooking, &c., can be done by the range without any coals. Another person will grumble that the ovens will not bake, or the water in the boiler will not get hot. Perhaps the real fault may have been that the range, although a good one, was not suited for the particular class of work required to be done. It is to be recommended that, before buying, an expert should always be consulted as to the suitability of any particular fitting for its purpose. The above evils are aggravated if the range is not properly fixed. To confine our attention to the boiler—it is necessary the flue should not be too small, or it will choke up with soot, and if too large, a great deal of heat will pass away in a useless manner up the chimney. The division between the flues and those from the ovens should be built of fire-bricks, or made of other material that will not crumble away by the action of the fire, or otherwise be broken. If this happens, there can be no control over the heat from the fire, as it will pass away round the ovens when it is most wanted for the boiler, and *vice versa*. The flues should also be continued to a fair height inside the chimney, and more especially when the building is low, so that the chimney is short, and perhaps so situated that the draught is sluggish, and there is a certain amount of difficulty in getting the smoke to go upwards. On the other hand, in some cases the flues should not be too long, or the fire will burn coals to waste, and more heat be generated than is required.

Where the draught is too strong, and the fire is raised to almost a white heat, the water in the boiler will not get hot so readily as with a fire just above a red degree of heat. The reason for this, no doubt, is that an excessive heat acting on the iron repels the water on the inside of the boiler, leaving a space, and, as the water is heated by conduction, it necessarily follows that it must be in actual contact with the iron, or whatever the boiler is constructed of, to absorb the heat conducted by the iron. Experiments prove that iron heated to redness will repel a globule of water when dropped on it, but a drop of water placed on a piece of iron moderately heated is almost immediately evaporated, as in this case the water is in actual contact with the iron, but in the other case the water is repelled and a space left between, the water dancing about for some considerable time before evaporation is complete. If a piece of red-hot iron is plunged into a vessel of cold clear water, the red colour of the heated iron can be seen for some considerable time, until the temperature of the surface of the iron is so reduced as to allow the water to come into actual contact with it, when it soon cools.

Before the boiler is fixed in its position the plumber should always drill the necessary holes. Some men will chip the holes with a hammer and hand chisel, or use a diamond-pointed chisel, but it is better to drill them, as the holes are made truer, and there is not such a large burr left on the inside edge of them, which requires to be afterwards filed away.

The position of the holes should always be considered. No holes should be made near the bottom of the boiler, as any connection made at that point would, in some cases, become choked with sediment. Another reason is, that if a draw-off pipe is taken from the bottom, the boiler could be drained empty. When empty the fire would act on the iron, oxidizing it to a serious extent, and should water enter the empty boiler when very hot, the sudden conversion of water into steam would cause an explosion, and the destruction of the boiler. Neither is it a good plan to connect the pipes to the boiler on that part exposed to the action of the fire. In some cases the pipes have been fixed through the flues, but this has generally led to their early destruction by external corrosion, besides being an obstruction when the flues require to be cleaned.

One of the most important points to be considered is the way in which circulation pipes are connected with the boiler, and also that the boilers themselves are fixed so as to allow all air to escape when water is entering.

Figures 424 and 425 show boilers fixed unlevel, so that air is pent up at the highest points. Steam will generate in these air-spaces, and eventually escape up the flow pipe with such violence as to alarm anyone near and lead them to think the boiler was going to burst. These noises in the boiler and pipes are worse

FIGURE 424.

FIGURE 425.

still when the circulation pipes are too small in the bore, or have any contractions so as to impede the free circulation of the water, so as to convey the heated water away from the boiler as freely as possible.

Figures 426 and 427 illustrate bad or improper connections made between the circulation pipes and the boiler. In each case an air space is left between the water and the top of the boiler.

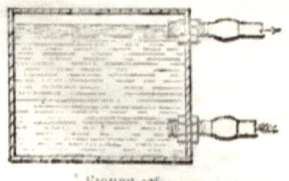

FIGURE 426.

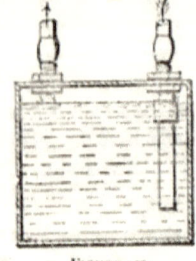

FIGURE 427.

The return pipes in Figures 426 and 427 are denoted by the arrows pointing to the boiler, and may be fixed as shown, but the flow pipes should be connected so that they do not project inside the boiler. Figure 428 is a fragmental section showing how this is usually done by first-class workmen. The hole is tapped with a

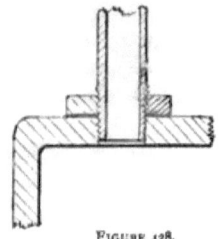
FIGURE 428.

thread to suit the pipe, and a back-nut screwed on outside, as shown, so as to prevent the pipe being screwed in too far. This connection is a good one when the substance of the iron will allow for the threads being tapped, but when the boilers are made of iron of a light substance, a boss should be rivetted on for screwing the pipe to, as shown in section, Figure 429.

Different kinds of waters have different effects on boilers. Hard waters will deposit lime to a very serious extent. The writer has

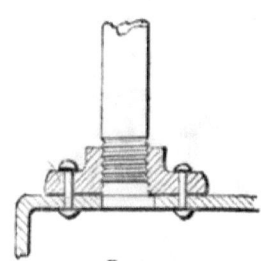
FIGURE 429.

found them with a lining exceeding an inch in thickness, and which could only be removed by chipping with a hammer and chisel Other waters leave a deposit inside the boiler in flakes of a reddish brown colour. This has been found nearly 2 inches thick, built up of thin flakes, and had accumulated in less than two years.

Some little time ago the writer cleaned out a kitchen boiler that had no fur whatever on the inside, but he took out nearly a pailful of what can only be compared to gravel, the sizes of the granulated matter varying from a grain of rice to a large-sized horse-bean.

Other boilers have been found with a kind of slimy coating inside, and others have rusted so that flakes of iron oxide came off the sides.

It is a very common complaint, when iron boilers and pipes are used, that the hot water, when drawn, has a very objectionable dirty red colour.

Sometimes, when galvanized-iron or copper pipes have been used, the water has still been very much discoloured. This being traced to the boiler, a coating of lime-white or a thin wash of portland cement being applied to the inside has prevented any further oxidation of the iron boiler and discolouration of the water.

When cleaning out boilers, it is generally difficult to get the furred matter out. In the first place the fur is very hard, and adheres very closely to the boiler. In the next place the manhole is generally too far away from where the fur most accumulates; it is also so very small that a man can only get one hand in at a time, and he can very rarely see what he is doing. In some cases special-made chisels, bent for the purpose, have to be used. Most of the boot boilers have only one manhole, and that is situated, as illustrated, at P, Figure 419.

These manholes should always be made water-tight. The importance of this was recently brought to the writer's notice. A house was changing owners, and the whole of the plumbing-work was being overhauled, when it was found that the kitchen boiler, which was working under a head of about 35 feet water pressure, was corroded all round the manhole, and so thin that an *accident* (?) would have occurred at an early date. Figure 430 is a drawing of a section of the manhole and part of the boiler, showing the position of the corrosion. It ap-

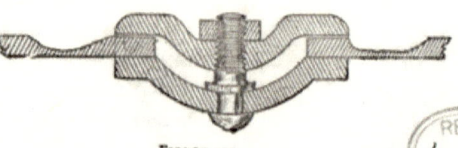

FIGURE 430.

peared that the manhole had leaked, but as the corrosion did not extend more than 1½ inch at the widest part, and only for about ¾ inch round the manhole, it is highly probable that a galvanic action had set up between the iron and the red lead cement which was used for bedding on the cover. The boiler was of wrought iron, and the manhole plates of cast iron. The wrought iron only was affected.

Figure 430 represents a manhole as usually fixed to boilers of kitcheners, but for an open-range boiler it is generally made as shown at Figure 431. The cover being

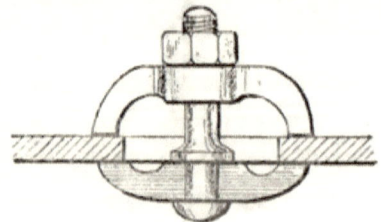

FIGURE 431.

fixed inside the boiler, and a bolt passed through an iron bridge-piece, fixed outside, and screwed up tight with a nut. The hole in the boiler and the plate are made elliptical. If made round it would be impossible to get the plate through the hole.

CHAPTER XL.

HOT-WATER FITTINGS.

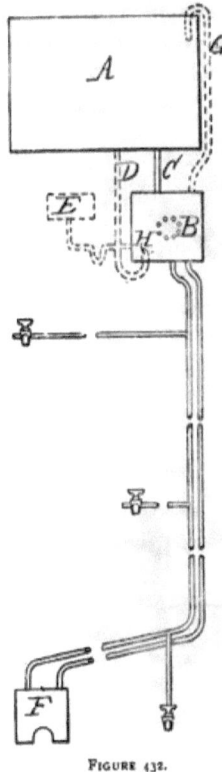

FIGURE 432.

AFTER the boiler is properly fixed the next thing is to consider the circulation pipes. Figure 432 is a sketch diagram showing how they are fixed in some of the London jerry-built houses. The pipes are usually of ¾-inch bore, and of the gas pipe description which is very light in substance. The bends are often carelessly made and buckled in their throats, so as to offer serious impediments to the free flow of water through them. The pipes are rarely coated inside, or galvanized, to prevent oxidation, with the result that frequently all the hot water has to be run to waste, as it is so highly coloured with iron rust. In some slightly-better houses 1-inch bore pipes are used, and in first-class work it is usual to fix 1¼-inch galvanized-iron or tinned-copper pipes. The iron is what is called the lap-welded steam pipe, and is very strong. The copper pipes have a lapped and brazed seam, but the last few years a great deal of patent seamless drawn copper tube has been used. Copper pipe should always be tinned when quite new and bright. When kept in stock for some time before being tinned, it is often found that the insides, which cannot be seen, do not get coated over the entire surface, patches of copper being left untinned.

Referring again to Figure 432, A is the cold-water and B the hot-water cistern. C is a pipe connecting the two. The writer has frequently met with this stupid arrangement. In some other cases the feed pipe has been connected as shown by dotted lines at D. Both of these arrangements are bad, as the hot water will sometimes work back into the cold-water cistern, rendering the contents into a heated state. This evil is more serious when the cistern, A, has to supply cold water to the sinks, &c. This can be prevented by fixing a light flap or valve on the end of the connecting pipe, as shown at H, Figure 432, but, as this valve is liable to set fast, it is not a good plan.

In some cases a small feed-cistern is fixed at the side of the hot-water cistern, as shown at E, Figure 432. This is a better arrangement, as it breaks the direct communication between the cold and hot-water cisterns, but the blunder is frequently made in fixing a $\frac{1}{2}$-inch—in some cases a $\frac{3}{4}$-inch—ball-valve in the feed-cistern. There is some little risk of damage to the boiler attending this arrangement. The low head of pressure on and the smallness of the ball-valve allows the water to dribble in very slowly, so that if a tap is left running, say into a bath, and another tap is opened at a level with the boiler, it is possible to empty the cistern, pipes, and boiler; and, it need scarcely be added, this is a very dangerous thing to do when the boiler fire is alight.

In some cases the draw-off taps have not been connected with the circulation pipes, but to a separate pipe fixed from the hot-water cistern. This is not to be recommended, as in the case of a high house a long length of pipe would have to be emptied of the stagnant cold water contained in it before hot water could be drawn. And again, the pipe would have to be connected to the upper part of the hot-water cistern, the lower strata of water being often quite cold. This would limit the available supply of hot water.

The writer has seen a few cases where the feed pipe was fixed direct from the cold-water cistern to the boiler. In one case the pipe was only $\frac{1}{2}$-inch in bore, which was much too small. The feed pipe should always be so large that the cold water will run in as fast as the hot water can be drawn off.

Hot-water cisterns generally have steam-tight covers, or manholes, and have a pipe fixed from the top to the outer air above the level of the cold-water cistern. In some cases this, which is commonly called expansion pipe, is turned over the top edge of the cold-water cistern, as shown at G, Figure 432. In one case the writer saw several years ago, when the cold-water cistern was full the water covered the end of the expansion pipe, and the connection thus made created a circulation between the water in the two cisterns.

It is very important that all pipes connected to a boiler should be kept free from any stoppage, or a possibility of the water being frozen. There is no more frequent cause of boilers bursting than pipes being frozen. In some cases a patent plug, with a thin disc of copper in the centre, regulated to break when an extra pressure has been brought to bear, has been attached to the boiler. In other cases a safety-valve has been attached. Some of the safety-valves are similar to those usually attached to steam boilers. Another kind much used is called a "dead-weight safety-valve," and is shown in section at Figure 433. The top of the inner tube is conical and ground into the outer cap. Weights, A, A, A, are added in proportion to the pressure the valve has to resist, care being taken not to add more than is really necessary. This valve should always be connected with the boiler by means of a separate pipe. If connected with the circulation pipes, as is sometimes done, and common keyed bibb-faucets are used for drawing off hot water, water-hammer in the pipes, caused by sudden closing of a cock, will sometimes cause the safety-valve to jump, so to speak, and allow a small quantity of water to leak out. This kind of valve is a very good one, and is preferable to some that have the weights piled on the top, making them top-heavy and more likely to leak if they should be knocked against or jarred.

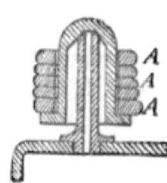
FIGURE 433.

It is an open question if safety-valves are to be trusted in cold countries, as it is possible the valve itself may become fixed if the water inside gets frozen.

Fusible plugs are sometimes fixed to boilers, so that if by any means—such as shortness of water or insufficient supply—they

HOT-WATER FITTINGS.

become empty when the fire is burning, the plugs will melt and thus relieve the pressure brought to bear by the sudden expansion of water into steam should it come in when the boiler is hot.

In first-class houses of modern construction the old way of fixing the hot-water cisterns at the top of the house, or on an upper floor, is not so generally practised, but what is known as the cylinder system has been adopted.

Figure 434 is a sketch diagram showing one of the simplest arrangements as fixed in some of the middle-class houses. J is the boiler; K, the hot-water cylinder; L, the cold-water cistern; and M, M, M, the draw-off cocks. In this system the water circulates between the boiler and cylinder, and a pipe is fixed from the top of the cylinder and continued above the cold-water cistern. From this pipe branches and draw-offs are fixed in the various positions as necessary. The only drawback to this system is, that the water in the rising pipe being stagnant, and at times nearly cold, it has to be drawn off before hot water can be had at the cocks. In some cases a return pipe has been fixed as shown by dotted lines, and as it was only necessary for the water to circulate slowly, this return pipe has been fixed of a smaller size than the flow pipe. For best work the return should always be the same size as the flow pipe, or the circulation of the hot water is very much impeded.

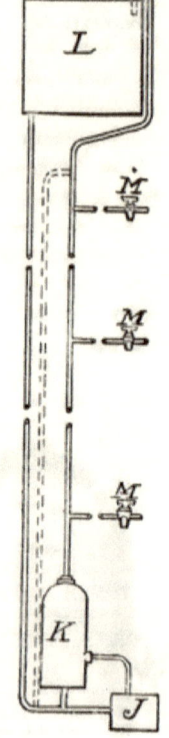

FIGURE 434.

In some cases the hot-water chamber has been made of iron plates rivetted to angle-irons, and made as shown in sketch, Figure 435. But even when made of ¼-inch iron plates it has been found necessary to have several stay-rods inside to prevent the

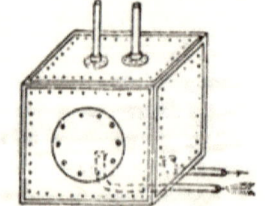

FIGURE 435.

sides and top bulging out by the internal pressure. This bulging sometimes causes the joints of the pipe connections to leak, and it is nearly always the case that the manhole leaks. This manhole must, of necessity, be large enough for access for cleaning out the chamber, which is rendered more difficult by reason of the complication of stay-rods inside.

The manhole plate is of wrought iron, and bolted over the manhole with a series of bolts tapped and screwed to an inside strengthening-plate. Should the manhole-plate require to be removed for cleaning-out purposes, the bolts are generally difficult to get out by reason of their having rusted in. Sometimes the bolts break off and have to be drilled out, necessitating new bolts and retapping the holes for them. It is very rarely done, but it is better to have gun-metal or copper bolts instead of the iron ones.

Hot-water cylinders are made of galvanized-iron plate, varying from about ·094 to ·25 of an inch, or of sheet-copper of about ·035 to ·120 of an inch in thickness. In some cases the copper cylinders are tinned inside.

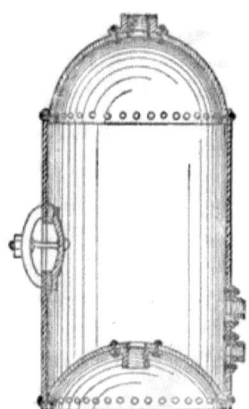

FIGURE 436.

The shape of the iron cylinders is generally as shown in sectional elevation, Figure 436. The sketch shows the position of the manholes and connecting bosses, and speaks for itself. The bottom is domed upwards, as shown, as being more convenient for rivetting, and also because if a flat bottom is put on the pressure of water is found to make the bottom bulge downwards, so that the cylinder will not stand firm on a flat shelf or platform. Copper cylinders are sometimes made with the bottom domed downwards, and, instead of being rivetted up, the seams are brazed together. When made this way, and the cylinder is to stand upright, a hollowed seating or bracket must be made for the bottom to rest in.

Some copper cylinders have the straight seams dovetailed and brazed, and the domical ends soft-soldered on, but as after a time the soldering comes away from the copper it is necessary to solder, or sweat as it is sometimes called, an outer ring of copper over the soldered seam, as an extra precaution against a leakage or breaking asunder of the domes and the body of the cylinder.

Figure 437 is an elevation showing the copper cylinder as described, and with a piece broken out so as to show the section of the joints referred to.

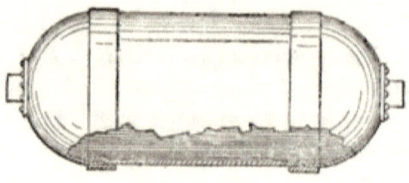

FIGURE 437.

The round cylinders are generally made with bridged manholes, as described for the boilers, for access to the inside for cleansing, &c.

Cylinders, as a rule, should stand above the level of the boiler, and should not be fixed on wooden shelves or brackets, but should have iron stands or brackets. Brick bases are sometimes built to support the cylinders, but the brackets are best, as more room is left for access for making the connections or giving any attention to them that may be necessary.

CHAPTER XLI.

CYLINDERS AND HOT-WATER CIRCULATION.

HOT-WATER cylinders should always be enclosed so as to avoid unnecessary waste of heat by radiation. They may be fixed in an enclosed recess, or covered with a non-conducting material such as "fossil meal," and other similar preparations.

A wooden casing may be fitted round them, with a space left between to be afterwards filled with a suitable material.

Dry hair-felt has been much used for this purpose, but the writer has recently had to take away all the felt that was placed round some hot-water pipes, as moths were found to propagate in that material. In the case referred to the moths propagated to an extent that would appear almost incredible, and then spread into the woollen furniture of a house and had played sad havoc, some being almost entirely ruined and spoilt. In another case fleas were found to an unpleasant extent, and the hair-felt had to be removed because it was found to harbour them.

For ordinary use "slag-wool" has been found very efficient to prevent radiation of heat from cylinders, and is free from the objections referred to above, but it is necessary to have an outside covering. As the cylinder is usually fixed in the kitchen, anything unsightly should be avoided. For first-class work nothing looks smarter than a polished mahogany lagging, with bright polished brass bands or hoops to hold it together, and if the hoops are tightened up with screws the whole enclosure can easily be taken to pieces should it be necessary to do anything to the cylinder or the connections.

I think I have read somewhere that in New York it is the practice to supply the hot-water system direct from the public water-main. This must be a very dangerous practice. Although the writer has no experience of this kind, still he cannot help thinking that there is danger of the boiler bursting, besides the annoyance of being without hot water should the mains break down or be emptied for repairs to them or any branch connections. Or the pressure from the mains would be seriously reduced should all the water be required to supply the engines when a fire broke out in the neighbourhood. In the writer's practice he always advises that the water be laid on from the main to a cistern or cisterns which will hold at least enough for twelve hours' consumption.

We will now return to the question of hot-water circulation.

The pipes between the boiler and cylinder should never be less than 1¼ inch in diameter, but 1½ inch is much better. Or, if smaller pipes are used, the connections to the boiler should be of the sizes stated, as it is not at all uncommon to find the ends in the boiler partly filled with fur, and especially where very hard waters are used. The cylinder may be fixed in any convenient position, but it is usual and best to fix it as near the boiler as possible.

Figure 438 represents a very good way for connecting the pipes, and a great many men fix them so.

Figure 439 is another way much practised. In this figure the cold-water service pipe is connected with the return pipe to the boiler, for the reason that it has been thought not advisable for cold water to enter the boiler to replace the hot water that has passed away into the system of pipes. By connecting it with the return pipe a certain quantity of warm water from the cylinder mixes with the cold, thus not submitting the iron boiler to sudden expansions and contractions. In Figure 438 the cold water is shown as passing directly into the boiler. In Figure 438 the manhole is shown as usually fixed, and in Figure 439 it is shown near the bottom of the

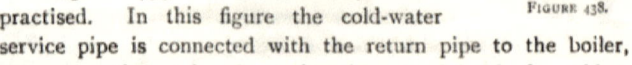

FIGURE 438.

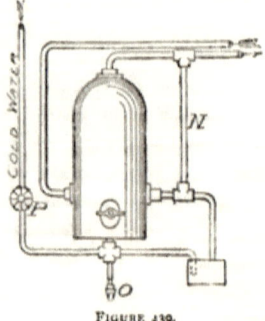

FIGURE 439.

cylinder, so that any mud or sediment can be more easily removed. In Figure 439 the pipes from the cylinder are shown as running horizontally. It is found in practice that the water will not circulate so freely in horizontal as in vertical pipes. In cases similar to the one quoted it has been found necessary to fix the pipe, N, from the boiler flow to the cylinder flow pipe, so as to accelerate the circulation in the cylinder pipes. It may here be mentioned that no draw-off cocks should be fixed or connected to the boiler or cylinder, but should be branched into the flow pipe from the cylinder. The reason for this is, that should the cold-water supply be exhausted or cut off for any purpose, it would not be possible to empty the boiler, &c., for reasons it is not necessary to repeat. A special cock should be fixed for the plumber to empty the cylinder when necessary for cleaning out or making any repairs. This cock should have a square head, or else be locked up so that servants could not tamper with it or make use of it. This cock is shown at O, Figure 439. In some instances a pipe is connected with this cock so as to discharge the contents of the cylinder into a drain or waste pipe. This is not a good plan, as the cock may be left open so that the water can run to waste and not be noticed. It is much better to fix a bibb-cock as shown, so that the water can be drawn into pails. This gives a little more trouble to the plumber, but as it has to be done only once or twice in a year, the trouble is not worth considering. A stop-cock should always be fixed in the cold-supply pipe as shown at P, Figure 439, so that in case of accident the water can be shut off at once instead of the person in charge having to run to the top of the house, or wherever the cold tank is fixed, to shut off the supply. No expansion pipe is shown in the figures, as it is usual to fix that from the highest point of the circulating pipes.

The circulating pipes from the cylinder should never be less than 1 inch diameter; $1\frac{1}{4}$ inch is better, and for large mansions the writer has used $1\frac{1}{2}$-inch pipes. In addition to iron, galvanized iron, and copper pipes—as described in an earlier paper—lead and lead-encased tin pipes have been used for hot-water pipes. Lead pipes have now fallen into disuse; indeed, the writer has not fixed any lead circulation pipes for this last fourteen years. Hot and cold water passing alternately through lead pipes causes them to expand and contract to such an extent as to cause the pipes to break, and this occurs more when the pipes are too firmly fixed. Branch pipes frequently break near the joint to the main pipes. On looking closely at the ends of the fractured pipes the lead looks as if its particles were disintegrated, and there is no doubt that it is the want of tenacity in the metal that will not bear a tensile strain when shrinking that is the cause of the pipe being pulled asunder. The writer has used lead-encased tin pipe for hot-water work, and will not readily forget his first job. The great difficulty was to make the joints. A plumber's wiped joint was found to melt the inside lining of tin, leaving a thin shell of lead, the tin melting and running into a mass in the bottom of the inside of the joint. Blow pipe and copper-bit joints were tried and found to be failures. Next was tried a lining of pasted brown paper inside the joints and then an ordinary wiped joint made. This answered fairly well, but the paper was found to wash off, and several pieces would congregate in some bend or branch-joint and cause a stoppage. The ends of the pipe were packed with whiting, but that was not a success. Tinned sheet-iron nipples inside the pipe ends before wiping the joint were fairly successful, but it was found that gun-metal coupling-unions were the best means for connecting the pipes.

Figure 440 is a section of a joint made as described. The pipe ends were passed through gun-metal linings and tafted or flanged over. These linings had hexagonal ends at R, R, and spanners were used for screwing them into the socket, S. T is an asbestos or indiarubber washer, but if the flanged ends of the pipe are properly done and left with a smooth

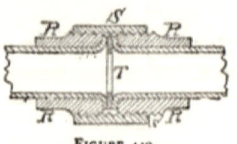

FIGURE 440.

face, they can be butted together and screwed up so as to be water-tight without any washers. These joints require periodical examination and further screwing up, especially when the pipes are in long straight lengths. Horizontal pipes should always lie on wooden fillets fixed to the wall. If fixed on hooks the pipe will bag down between the fixings. The wooden fillet should be rather wide. Although it does not add to the appearance of the work, it is best not to have the pipes too straight so that they will have a direct strain on the screwed coupling connections, but any bends in horizontal pipes must lay flat on the fillet, or air will accumulate in any high parts. When pipes are quite straight the joints leak by the alternate pull and thrust motion which is continually taking place at each change of temperature of the water inside them. It is often lost sight of, but a fact nevertheless, that hot-water pipes when in use *are always in motion*, and cannot be said to be still for more than a few seconds at a time.

Great care should be taken when making branch connections for draw-off pipes. Figure 441 is a section showing the T-union

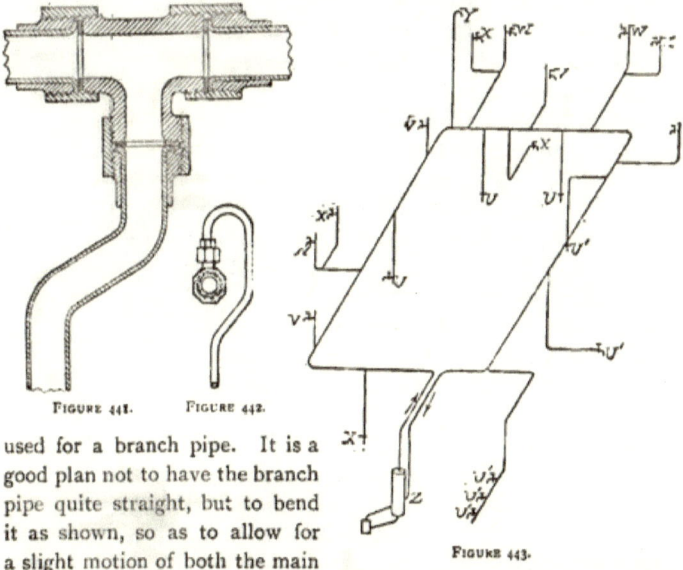

FIGURE 441.   FIGURE 442.

used for a branch pipe. It is a good plan not to have the branch pipe quite straight, but to bend it as shown, so as to allow for a slight motion of both the main

FIGURE 443.

and branch pipes, and thus render the liability of a strain on the joint to a minimum. In some cases the writer has fixed the branch joint upwards as shown at Figure 442, so as to allow for the motion predescribed.

Figure 443 is a bird's-eye view of a cylinder and the circulating pipes fixed to a nobleman's mansion in the country (England). The whole of the pipes were lead-encased tin. U, U, were draw-offs to sinks on the ground floor. V, V, were similar draw-offs on the first or chamber floor. W, W, were baths. X, X, were wash-hand basins, and Y was the expansion pipe. This was placed at the highest point of the circulating pipe. Doorways and similar obstacles compelled this to be fixed where shown. The return pipe was taken back and branched into the cylinder, as shown at Z. This, no doubt, caused a slight check on the circulation of the water, but it was deemed advisable to make the connection as shown. It was thought that by reason of the long length of pipe and the consequent friction of the water passing through that any water drawn at the cocks, U', U', would most likely be cold instead of hot if it had been connected in the usual way.

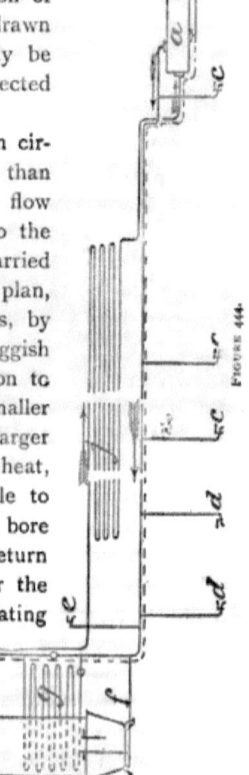

FIGURE 444.

Sometimes, as an economy, the return circulation pipe has been of a smaller size than the flow pipe. For instance, a 1¼-inch flow pipe has been fixed from the cylinder to the furthest point away, and an inch pipe carried back to the cylinder. This is not a good plan, as reducing the size of the pipe causes, by the extra friction of the smaller pipe, a sluggish circulation of the hot water. In addition to the extra friction of the water in the smaller pipe, there is also, proportionately, a larger radiating surface and consequent loss of heat, so that for these reasons it is advisable to have the circulating pipes of the same bore throughout. Another view of a small return pipe may be mentioned—viz., the smaller the pipe the larger, proportionately, is the radiating surface, so that the temperature of the return-water is much more reduced, and, theoretically, this ought to cause the water to circulate more freely.

In some cases the hot-water cylinders are fixed horizontal instead of vertical.

Figure 444 is a sketch diagram of a horizontal cylinder and the circulation pipes, as recently fixed under the supervision of the writer. In the sketch, A is the cylinder fixed on brackets over a doorway; B is the boiler; C, C, C, are taps over the scullery, servants' hall, and butler's pantry sinks; d, d, are taps over lavatories; e is fixed to supply hot water to the housemaid's sink on chamber floor, and f to the spray and plunge bath; g is a coil for heating the bathroom, and is arranged so that towels, &c., can be dried or warmed; h is a coil on upper floor to warm a water-closet room and adjacent lobby. The waste heat is made use of by carrying the return pipe through the housekeeper's room and the linen-room, the pipe being bent to form a radiator, as shown at j. It will be noticed that all the draw-off taps are branched into the flow pipe, which is carried as direct as possible, and the coils or radiators are connected beyond the furthest tap so that the water may be drawn as hot as possible at the taps. The coils are so attached to the circulating pipes that by closing three or four stop-cocks the water will flow round without passing through the coils, heat not being wanted in the rooms in the summer time. K is the cold-water supply-tank, and the service to the boiler is shown by dotted lines. The whole arrangement works very well. In this, and similar cases, it was found that after drawing sufficient hot water for a bath the heat from the coils fell considerably, and some little time had to elapse before they got hot again. For this reason it is an open question if the pipes, &c., for supplying hot water to sinks, &c., and also for heating corridors or rooms by radiators is a good one, as it is almost impossible to keep one constant heat.

In some instances a hot closet has been fixed in a kitchen, for warming plates and dishes, and connected with the boiler circulation pipes. For want of the proper knowledge some of these have been found to be failures. The writer had to inspect one that had been fixed but a few months, but no matter how hot the water was in the boiler the closet was always quite cold.

Figure 445 shows the arrangement. The man who fixed it said he would give it up and own he was beaten, but when asked if water or air was in the coil, Z, he at once saw where he had made a mistake. He then made the necessary alterations so as to allow the pent-up air to escape from the highest end of the horizontal coil. Acting on advice, he also connected both ends of the horizontal coil to the flow pipe from the boiler, and put a small perforated disc of copper at Y so that the water did not circulate past and miss the T-junction. This air-binding of pipes is of frequent occurrence and in a great variety of ways. In a row of six mansions a stupid blunder was committed in each house. The storage capacity for cold water was too little, with the result that the cisterns would frequently get emptied. Every time the cisterns were emptied and filled again it was found impossible to draw water at the hot-water taps.

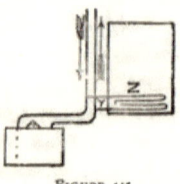

FIGURE 445.

The reader is referred to Figure 446, which shows what was found. The cold-water supply to the boiler was connected to the bottom of the cistern, and, about 6 feet away, the pipe was lifted about 2 inches, so as to fix it with a pipe-hook to the wooden bearer on which the cistern was placed. On taking out the hook in each house, and lowering the pipe, air came bubbling back into the cistern, after which the water ran freely into the boiler and could be drawn at the various taps. It may be further explained that the bottom end of the supply pipe was sealed by the water in the boiler so that the air could not escape or be driven out at the bottom, although the cold-water cisterns were about 60 feet above the boilers. Great numbers of similar blunders could be given, but the writer does not consider it necessary to weary his readers by repetition of cases.

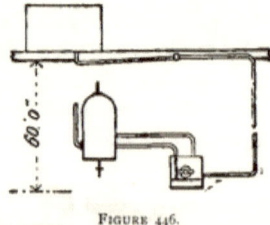

FIGURE 446.

A common cause of complaint is that such a quantity of cold water has to be run to waste before hot water can be had at the taps. This occurs chiefly with long branch draw-off pipes to taps some distance away from the circulation pipes. In a great many cases it is possible to fix return pipes to these branches so that the water will circulate through them and thus avoid the evil referred to. Where it is impossible to circulate through the branches, the size of the pipes should be reduced as much as possible, so that less water will have to be run to waste. This applies more particularly to wash-hand-basins, where only a small quantity of hot water is required. For baths it is not so important to reduce the size of the pipes, as a few pints, or even a gallon, of cold water having to be drawn does not much matter, but if the pipe is reduced, the bather's patience would perhaps be exhausted by waiting for the bath to fill from a small pipe. In some cases it has been found to be an advantage to have two, or even three, sets of circulating pipes carried from the cylinder to the various parts of a house, especially when the shape of the house is large and straggling, and the draw-offs are wide apart. No matter how many sets of circulating pipes are fixed, it is always necessary to have a vent pipe from the highest end of each to allow the air to escape. In some cases a draw-off pipe has been fixed near the highest end, on opening which any air could escape, but this is not by any means a good system to adopt. When air pipes are fixed, it is scarcely necessary to add they must be carried to a level above the cold-water supply-cistern.

A reason was given in an earlier chapter for connecting the return pipe to the cylinder instead of branching it into the boiler or the cold-water service pipe to the boiler. To prevent water from being drawn from the return or cold, instead of the flow or hot-water pipe, some makers will fix a small flap-valve inside the circulating pipes, that opens in the direction of the flow of water when circulating, but closes should the circulation by any means become reversed, or to prevent back-water being drawn when a tap is opened for drawing hot water.

A great many hot-water men fix stop-cocks in the circulation pipes, but this practice cannot be too strongly condemned. A few years ago a serious accident (?) occurred through this. The stop-cocks were closed while a new draw-off cock was being fixed, and it was forgotten to reopen them, with the result that when the fire was lighted the water in the boiler expanded sufficiently to burst it.

Elbows in circulation pipes should always be avoided as much as possible, it being much preferable to have bends of an easy sweep or radius. Some first-class men will always insist on the pipes being heated and bent to suit the various positions, instead of using those bought at the manufacturers, the reason being that every bend, especially when made to a small radius, causes a retardation of the free flow of water, and, as the force in the form of heat which puts the water in motion is very small, all obstructions and unnecessary friction should be avoided as much as possible. When galvanized-iron pipes are heated and bent the zinc coating is melted off. For common work no notice is taken of this, but for best work the pipes are bent first and galvanized afterwards. When done this way the bends should be carefully examined, as they have sometimes been found to be partly or entirely choked with zinc. Some men, who take a pride in their work, will make very nice bends, and when two or three are fixed together will make them match, so to speak, as shown at Figure 447. Figure 448 is an elbow drawn for comparison with the bends. Mention has been made of the expansion and contraction of hot-water pipes. If no allowance is made for this motion of the pipes the result will be that some of the joints will leak, and this occurs more especially with long straight lengths. As

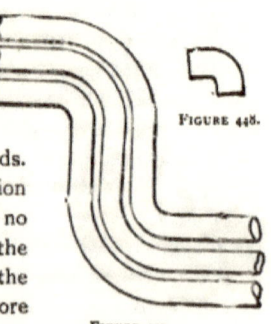

FIGURE 448.

FIGURE 447.

expansion and contraction are, to a certain extent, irresistible forces, it follows that provision should be made to allow for them. The introduction of bends, as shown at Figure 447, will allow for a great deal of motion of the pipes, and is a very simple and easy way of overcoming the difficulty. Figure 449 is a gland-joint that can be used for the same purpose, the lining sliding in and out of the gland-box.

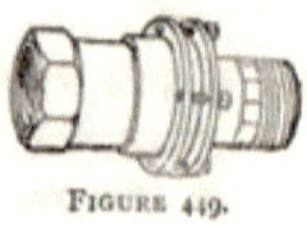

Figure 450 is, I think, an American invention, and appears to be a simple and easily-applied expansion-socket, but not

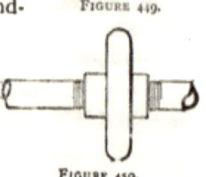

FIGURE 449.

FIGURE 450.

having used them the writer does not know how long they will last without breaking. But perhaps they are made of very tough but flexible iron, and are as strong as the other parts of the fittings. In addition to expansion-joints it is important that the circulation pipes should not be fixed by any branches, but this was referred to when writing on lead-encased tin pipes.

When hot-water pipes are fixed round and inside a room, they should not be too tightly fixed at the return-ends, but should be kept at least one inch clear of the walls. Figure 451 explains

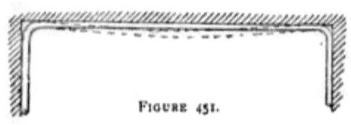

FIGURE 451.

what is meant, the dotted lines showing how the pipes will bulge out in the centre portion when expanded by heat. If the pipes are fixed clear of walls, as suggested, this would not occur, as the expansion would be expended at the ends. Neither should the pipes be too rigidly fixed by means of wall-hooks. Horizontal pipes should lay on brackets, which, although supporting, will not prevent them freely moving backwards and forwards. Lead pipes should lay on wood fillets, as stated in an earlier chapter. Figure 452 is a neat brass

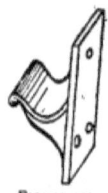

FIGURE 452.

or gun-metal bracket which can be screwed to a wooden block or lining-board fixed to the walls. Long vertical pipes should be supported in the centre of the length, so that the expansion may be equally distributed at the ends.

To carry out hot-water work in a thoroughly efficient manner, a man requires to know all the principles which govern his work. If he does not know these principles he is always liable to blunder in the arrangement of the fittings and pipes. A man may be a good tradesman, and fix new work that will answer first-class, and yet very often be at a loss to know how to alter other work that has been improperly done by someone else.

Figure 453 is an illustration of a job near Regent's Park, in London. A is a ½-inch expansion pipe; B, a ¾-inch flow pipe; C, a ½-inch return pipe; D, a 30-gallon cylinder; E, a ¾-inch draw-off pipe; F, a ½-inch cold-water feed pipe; and G, a small feed-cistern with a ⅜-inch ball-valve. The top of the feed-cistern was fixed nearly level with the top of cylinder. The talented expert who arranged this work was ignominiously dismissed, and another man was called in who spent several days in altering the work, but still could not see how it was that no water could be drawn at the cock, H. Eventually he fixed another expansion or vent pipe, as shown by dotted lines at I, when he succeeded in getting

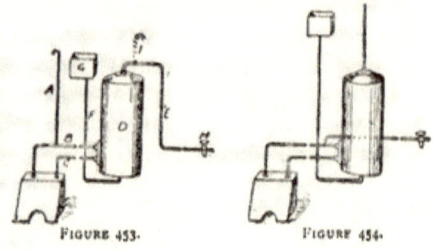

Figure 453.   Figure 454.

a small dribble of water at the bibb-cock. A large fire had been kept up, with the result that sometimes the water was forced out of the feed-cistern and steam drawn at the bibb-cock when opened. This was before the second expansion pipe was fixed.

The second man having been sent away, the owner thought perhaps it would be an advantage to go to a respectable firm of engineers, instead of employing cheap men who knew nothing about the principles which govern the work. He was very much surprised when he was told that the whole of the pipes would have to be changed to a larger size and differently arranged.

Figure 454 shows the alterations, which may be enumerated as follows: The pipe leading to the draw-off cock was branched into the flow pipe from the boiler to the cylinder. The pipes between boiler and cylinder were made 1 inch in diameter. The expansion or vent pipe was fixed from the top of the cylinder, and the feed-cistern was fixed at a higher level. A larger service pipe and ball-valve were fixed, and also a larger pipe from the feed-cistern to the bottom of the cylinder. A damper was also fixed in the boiler-flue so as to regulate the draught. On being told that there was plenty of hot water and enough to supply a bath, the gentleman was so pleased that he had one fixed in an adjoining room.

Figure 455 is an illustration of another blunder which was, if possible, worse than the last one described. This was

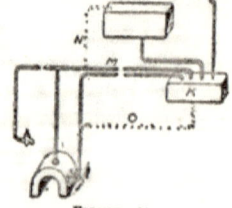

Figure 455.

fitted up in a nobleman's mansion in London by a village blacksmith, in whom his lordship had so much confidence that he paid railway and other expenses sooner than trust to anyone in London. The strange part about this arrangement is, that the water in the cistern, K, did sometimes get hot, but was never to be depended upon for heating. At times it was found necessary to open the cock, L, as air would accumulate in the horizontal pipe, M. The only alteration made in this case was to fix a vent pipe as shown by dotted lines at N, and connect the return pipe from the hot-water cistern to the boiler as shown by dotted lines at O.

Figure 456 was a rather strange experience of erratic circulation. A boiler was fixed for heating water to supply sinks in which bottles were washed at a wholesale wine merchant's. Two radiators were fixed, as shown in the diagram, for heating the offices. The coil, P, got very hot, but there was no heat in the coil, Q, so long as there was a fire under the boiler. But, on allowing the fire to go out, the water would at once begin to circulate backwards, and the coil, Q, got moderately warm, but on again lighting the boiler fire this coil would gradually get cool again. As the whole of the work appeared to be done in good and workmanlike manner, some little thought was necessary before it could be understood why one coil got hot and the other did not when the boiler fire was alight, and yet on putting out the fire the circulation was reversed. The pet-cocks on the coil-heads were opened, but as water came in each case there was proof that they were not air-bound.

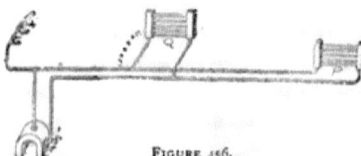

FIGURE 456.

Two mistakes suggested themselves. One was the pipe from the boiler was too small, being only 1¼-inch, and the branch to the coil, Q, was at right angles, *but horizontal*. The return pipe was branched into the main return pipe as shown in the sketch. This branch being upright accounts for the back action in the circulation. The alterations made were as follows: 2-inch pipes were fixed from the boiler as far as the branches to the coil, Q, and the flow pipe branch was turned upwards as shown by dotted lines. The alterations, when made, proved to be successful, as both the coils heated equally.

In another case a long branch pipe was fixed to supply a pantry-sink. This pipe was returned so that water would circulate through it, and thus avoid having to empty the pipe of cold water before hot could be drawn. It was found that no circulation took place. The way it was branched is shown at Figure 457, S being the main flow and T the main return pipe, and U the way the draw-off pipe was branched so that any pent-up air could escape when the bibb-cock was opened. Figure 458 shows the alteration made,

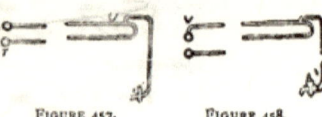

FIGURE 457.   FIGURE 458.

the branch flow being taken off the top of the main flow as shown at V. The issue was successful and hot water freely circulated through the branch pipe, which was about 40 feet in length.

The last two problems in hot-water circulation are typical of a great many mistakes made by men who may be compared to pieces of machinery, who can execute work in a fairly good way, but who lack the necessary experience or knowledge to properly plan or arrange the scheme before commencing to carry it into execution. Other examples could be given, but those cited are sufficient to set the reader thinking, and, if he is a practical man, may perhaps prevent him making similar mistakes.

In some instances fire insurance societies insist that no hot-water pipes shall be in actual contact with, or within several inches of, any woodwork. Where, of necessity the pipes must be fixed close, the woodwork has to be covered with sheet-metal or be otherwise protected.

When pipes are carried beneath wood flooring, and rooms, &c., are beneath, it is always advisable to construct a proper channel for the pipes to lay in, and line the channel with sheet-lead. This metal lining not only keeps the pipes away from the woodwork, but should the pipes leak the water is caught, and, running to the lowest end, the bottom of the channel being laid to the same declination as the pipes, can be conveyed away by means of a waste pipe fixed from the channel to a suitable position for discharging. It is also a good plan to lay the hot-water pipes on small rollers, which can be made of remnants of iron pipe, so as to keep the circulation pipes further away from the woodwork, and also allow them to move freely when expanded or contracted by differences in temperature. These rollers should not be too far apart or the pipes will sag down between them, thus allowing air or vapour to lodge in the high parts and cause a partial obstruction.

It is important that the cold-water supply-cistern should be fixed a fair height above the highest draw-off cock, and, as stated in another chapter, that the cold-water supply-pipe should be a good size—that is, larger than any of the branch or draw-off pipes. This has again been brought to the writer's notice at a house where complaints were made that very often no hot water could be drawn on the upper or chamber floor for a minute or two at a time, when it would begin to run at the tap and perhaps immediately afterwards stop again. On seeking for the cause of this it was found that sometimes one or two taps were opened on the ground floor for drawing hot water at the scullery or pantry sinks, during which time none could be drawn on the floor above.

Figure 459 is a sketch diagram showing the position of the cold-water cistern, cylinder, boiler, tinned-copper circulation and draw-off pipes, &c. To remedy the evil, all the cocks on the ground floor were pea'd—that is, a copper disc, with a small perforation, was inserted in the bosses,

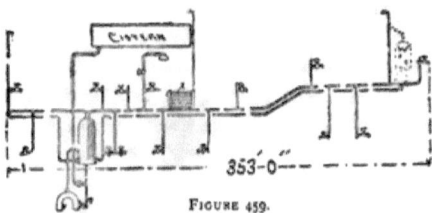

FIGURE 459.

so that the water did not run quite so fast when the cocks were opened. A 1¼-inch pipe was fixed from the cold-water cistern to the boiler. The cold-water cistern was only 4 feet above the level of the highest draw-off cock, and this cock was one of great importance, as it supplied the principal bath-room, which was used by a nobleman. The above alterations were found to improve matters very much, but at times no hot water could be had at the above-named cock, so it was deemed advisable to fix a 20-gallon tinned-copper cylinder, as shown by dotted lines at W, the circulation pipes being so connected that the cylinder could only be emptied by opening the cock attached to it. This arrangement was to avoid the contents of the cylinder running back into the circulation pipes when they were being drained by cocks at a lower level, and was found to be fairly successful, although some time had to elapse before the lower strata of cold water was heated to the same extent as the upper portion of the contents of the cylinder.

In the diagram, X was a coil made of 1¼-inch copper pipes, nickel-plated, and was used for warming the bath-room, and also the bath-towels.

FARQUHARSON ROBERTS & PHILLIPS, LIMITED, Printers, 7, Upper Thames St., London, E.C.

ORIGINAL MAKERS.   ESTABLISHED 1844.
THOMAS GLOVER & CO.

THE SIX MEDALS AWARDED TO
THOMAS GLOVER'S PATENT DRY GAS METERS.
*The latter being the Highest Award for
Dry Gas Meters at the Paris Exhibition,* 1867.
Since then we have not Exhibited FOR PRIZES.

PATENT
DRY GAS METERS

Are a Remedy for all the Defects of Wet Meters.
Are suitable for all Climates, whether Hot or Cold.
No Loss of Gas by Evaporation.
Cannot become fixed by Frost, however severe.
Are the most accurate and unvarying Measurers of Gas.
Prevent Jumping and unexpected extinction of the Lights.
Cannot be tampered with without visibly Damaging the Outer Case.
Are upheld for Five Years without Charge.

THOMAS GLOVER & CO.,
214 to 222, St. John Street, Clerkenwell Green, London, E.C.
TELEGRAPHIC ADDRESS—"GOTHIC, LONDON."

INDEX.

A

ACCESS to openings in drains, 154.
Access to pipes for repairs, 101.
Acids act on seams of soldered pipe, 38.
Action of oxygen on lead, 20.
Action of sewage gases on solder, 81.
Action of sun on cast and sheet-lead, 24.
Action of waters on boilers, 314.
Additional hot-water cylinder fixed on upper floor, 337.
Aero-hydrogen blow-pipe, 38.
Air-binding of drains, 127.
Air-binding of hot-water pipes, 329.
Air-binding of pipes, 95.
Air blown through water-closet traps, 279, 280.
Air compressed in sewers, 123.
Air-current with water-current, 166.
Air-current through vent pipe evaporates water out of trap, 191.
Air-extractor cowls, 169.
Air-fans driven by water, 170.
Air-fans fixed in drain vents, 170.
Air flushing for drains, 166.
Air in boilers, 313.
Air in drain warmer than outside, 167.
Air-inlets to drains under sidewalks, 177.
Air-inlets at each end of a drain, 183.
Air-inlets to drains, 171, 173, 174, 175, 176, 177.
Air-inlet pipe continued to roof, 166.
Air-inlet valves, 176.
Air-testing of iron drains, 152.
Air-tight covers to manholes, 175.
Air-pump to test pipes, 153.
Allow for expansion of lead in sinks, 118.
Alloys of lead, 21.
Alterations to improperly arranged hot-water work, 333, 334.
Amateur artists and urinals, 254.
Amateur objects to complication of traps, 188.
Angles of cisterns shaved too wide, 108.
Angle-shaving filled up with solder, 109.
Appliances for casting sheet-lead, 22.
Arches through walls for iron drains to pass through, 150.
Architects and pipe chases, 100.

Area traps, 193.
Arrangement of scullery-sinks, 192.
Asbestos joint, 164.
Asbestos washers, 325.
"Asphyxiator" as drain tester, 127.
"Asphyxiator" as vermin exterminator, 127.
Astragals, casting, 105.
Astragals on rain-water pipes, 104.
Attendants at exhibitions of water-closets, 301.
Autogenous soldering, 38.
Automatic flushing apparatus for urinals, 258, 259.
Automatic flushing-tank used, 173.
Avoid use of tools when possible, 109.

B.

BACK spring to closet doors, 304.
Bad arrangement of cold and hot water cisterns, 317.
Bad arrangement of scullery-sink, 185.
Badly-arranged manhole, 136.
Badly-shaped joint, 73.
Bad position of water-closet and soil pipe, 266, 267, 268.
Bad work and the public, 269, 270.
Baggy parts in lead pipes, 95.
Ball-traps, 240.
Band on joint, 92.
Banker's house, defective drains at, 130.
Base to trap, 140.
Basins for urinals, 251.
Baths, 208.
Bath cocks and valves, 209, 210.
Bath connected with hot-water cistern, 217.
Bath cradled, 208.
Bath enclosures, 219.
Bath, needle, spray, &c., 221.
Bath overflow pipes, 211, 216.
Bath-room decorations, 218.
Bath-room furniture, 219.
Bath-room heated with coil, 337.
Bath safes, 215.
Baths, shower, 220.
Bath utensils, smell offensive, 217.

INDEX.

Bath-valves and water companies, 215.
Bath waste connected to slop-sink-waste, 216.
Bath-waste-pipe, position of, 210.
Bath-waste-valves, 212, 213.
Bedrooms and water-closets, 304.
Bench-made bends, 40.
Bends v. elbows, 57.
Bending dresser, 51.
Bending pipes, a modern invention, 57.
Bends heated with gas, 46.
Bends in heavy lead, 28.
Bends in 1-inch pipe, 52.
Bends in two halves, 55.
Bends larger than the pipe, 51.
Bends loaded with sand, 53.
Bends made by cutting wedged-shaped pieces out of throat, 56.
Bends made by slitting throat, 56.
Bends made on bench, 40.
Bends made quickly, 49.
Bends made to a large radius, 53.
Bends made with a water core, 54.
Bends made with dummy, 47.
Bends on pipes of square section, 67.
Bends out of thick lead, 56.
Bends should be heated, 40.
Bends should be drawn on bench, 56.
Bell-traps as air-inlets to drains, 194.
Bell-trap in dairy floor, 172.
Bell-traps in floors and sinks, 185, 186.
Bell-traps in a wine cellar, 197.
Bell wires lay side of lead pipes, 100.
" Bent-hook " for cisterns, 115.
Best materials for drains, 146.
Best tradesmen make sound work, 146.
Bevels set for bends and elbows, 57.
Bidet-pan, 223.
Bird's-eye view of hot-water pipes in a mansion, 326.
Block-joint, 78.
Blow-down cowls, 169.
Blow-pipe joints, 91.
" Blow " through solder, 115.
Blunders in fixing hot-water work, 332.
Board to catch solder, 72, 86.
Bobbins and ball, 44.
Bobbins and followers, 44.
Bobbins through pipe with several bends, 46.
Boiler air-bound, 307.
Boiler, cold-water supply to, 323.
Boiler flues, 309, 311.
Boiler manholes, 315.
Boilers, various kinds. 309, 310.
Boiler top corroded outside, 315.
Boilers and fusible plugs, 318.
Boilers and kitchen range, 303.
Boilers and pipe connections, 313.
Boilers and safety valves, 318.
Boiler difficult to clean out, 314.

Boilers being emptied when fire is burning, 312.
Boilers fixed unlevel, 313.
Boilers lime-whited inside, 314.
Boilers with furred coating inside, 314.
Bolts for manhole-plate to hot-water cistern, 320.
Bolt for opening pipe-ends, 69.
Boot boilers, 309.
Bottles in drains, 154.
Bottom length of soil pipe should be left out, 88.
Bottom lengths of pipe get battered, 103
Bowl for washing crockeryware, 198.
Bower-Barff'd pipes, 145.
Boxes of charcoal, 123.
Brackets for hot-water pipes, 332.
Brackets for wash-hand basins, 241, 242.
Brad-awl to close joints, 83.
Bramah's patent for pipes, 30.
Bramah, the inventor of valve water-closets, 291.
Branch connections for hot-water pipes, 326.
Branch drains, 135.
Branches for iron drains, 156.
Branch hot-water pipes, 330.
Branch joint, how to fix, 88.
Branch joints welted, 85.
Branch soil pipes, 277.
Branch soil pipes bent near joint, 84.
Branch-wiped joints, 81.
Brass face-plate for cocks, 207.
Brass gratings in sinks, 120.
Brazed cylinders, 320.
Breakage of lead pipes by hot water, 202.
Brick drains, 128.
Brown pasted paper, 98.
Builder threatened with an action at law, 165.
Bungling way of reducing size of iron drain, 158.
Bungling way of cutting iron pipes, 147.
Burning, effect on eyes, 38.
Burning lead seam on pipes, 27.
Burnt seams, 38.
Burnt seams cleaned off and burnished, 56.
Bursting of drain pipe sockets, 133.
Bursting pressure for a 2-in. pipe, 202.
Butler's pantry sink, 195.
By-pass in vent pipe, 170.

C

CABINET wash-hand basins, 225.
Cap and screw on branch-joint, 84.
Cap over waste of wash-hand basin, 235.
Cappings on sinks, 119.
Carbonate of lead, 20.
Carbonic acid in drains, 168.

INDEX.   341

Cardboard models of elbows, 64.
Card-wire, use of, 70.
Card-wiring bends, 47.
Careful laying of drains, 158.
Care to be taken when preparing branch-joints, 82.
Carpets in bath-rooms, 219.
Case of drains not laid to proper fall, 133.
Casing for pipes, 95.
Casting lead ears for rain-water pipe, 107.
Cast-iron drains, 144.
Cast-iron troughs for urinals, 249.
Casting frame, 22.
Casting lead astragals, 105.
Cast-lead acted on by sun's rays, 24.
Cast-lead traps, 283.
Cast sheet-lead, 21.
Cast v. milled lead, 24.
"Cat's-paw," 67.
Cause of sweaty joints, 70.
Cause of dirty manholes, 137.
Cause of drains leaking, 130, 144.
Cause of drain smells, 113.
Cause of hot water ceasing to run at taps on upper floors, 336.
Cause of joints leaking, 87.
Cause of lead breaking in sinks, 118.
Cause of lead pipe not being of equal thickness, 32.
Cause of noise in boilers, 313.
Cause of smells from a wash-hand basin, 230.
Cause of smells from vent pipes, 134.
Cause of stoppage in drains, 132.
Cement fillets to scullery sinks, 191.
Cement for bedding marble tops of wash-hand basins, 224.
Cement inside joints of drain pipes, 132.
Ceruse, 20.
Cerussite, 17.
Cesspool and trapping of house drain, 171.
Cesspool at foot of stairs, 182.
Cesspool trap, 181, 184.
Chalk line to mark edge of soiling, 115.
Change of direction in drains, 134.
Channel bends, 137.
Channel branches, 137.
Channel for hot-water pipes, 335.
Charcoal boxes, 123.
Charcoal boxes an obstruction to ventilation, 123.
Charcoal useless unless dry, 123.
Chase lined with cement, 101.
Chases for pipes, 99.
Chases for pipes too small, 100.
Chemical combinations of lead, 21.
Chemical properties of lead, 20.
Child's bath, 223.
Chipping holes in boilers, 312.
Chokage of dipstone trap, 139.

Cistern bottom, care to be taken when wiping, 116.
Cistern broken by frost, 119.
Cistern, how to line with one piece of lead, 108.
Cistern lined with lead, 26.
Cistern overflow connected with soil pipe, 263, 265.
Cistern-waste into water-closet trap, 273.
Clay in drain pipe joints, 133.
Claying iron drain joints, 147.
Clean servants fill trap gratings, 185.
Clean water passing through waste pipe of bath, 209.
Clear water-way through pipes, 40.
Close kitchen boiler, 307.
Coating lead pipes with tin, 32.
Cocks for baths, 209, 210.
Cocks for spray baths, 222.
Cock to empty cylinder, 324.
Cold countries and safety-valves, 318.
Cold plug fixed by putting into a hot washer, 120.
Cold supply to boiler, 313.
Cold water drawn from hot-water circulation pipes, 330
Cold-water supply-cistern to hot-water boiler, 336.
Collar for iron drains, 157.
Combined waste and overflow bath apparatus, 212, 213.
Combined soil and rain-water pipes, 261, 268.
Comfort of a bath, 209.
Common drain syphon-trap, 140.
Common way of filling up a bath, 209.
Compasses for fixing joints, 87.
Compasses with shave-hook, 80.
Compressed air in sewers, 123.
Compressed air passes through water-closet traps, 278 to 280.
Concrete in drain trenches, 130.
Concrete under drains, 157.
Coned joint to allow for expansion, 204.
Connection between boiler and cylinder flow pipes, 324.
Connection of bath and slop-sink-waste pipes, 216.
Connection of pipes with boiler, 313.
Contagious diseases and wash-hand basins, 243.
Conical swab, 36.
Contents of drain-syphon, 140.
Contracted waterway in pipes, 40.
Contrivances for keeping smells out of house drains, 125.
Cook grumbles at plumber, 101.
Cooling, shrinkage of lead when, 148.
Copper baths, 208, 210.
Copper-bit joints trimmed up, 93.
Copper-bit joint overcast, 92.

INDEX.

Copper-bit plumbers, 91.
Copper-bit seams, 38.
Copper-bit v. wiped joints, 91.
Copper-covered draining-boards, 192.
Copper ferrules on ends of soil pipes, 163.
Copper flaps, 215.
Copper pipe nails, 99.
Copper strainers in sinks, 121.
Cores of sand and salt, 78.
Corrosion of boilers, 315.
Cost of joint moulds in a large firm, 94.
Country plumbers' overcast joints, 77.
Country builders and brick drains.
Country mansion, scullery-sinks in a, 192.
Coupling unions for lead-encased tin pipes, 325, 326.
Coverings for hot-water cylinders, 322.
Covers to grease-traps, 187.
Cover-plates for iron drain openings, 154.
Cowls on vent pipes, 169.
Cracked sockets in iron drains, 148.
Cradled baths, 208.
Creed's patent for pipes, 29.
Cross-bars in sink washers, 120.
Crowding pipes in chases, 100.
Crushing of drain pipes, 130.
Curbs round gulley-traps, 196.
Currents of air in drains, 166.
Curtains for shower baths, 220.
Cylinder manholes, 323.
Cylinder system for hot water, 319.

D

Dampers to boiler flues, 309.
Danger of using wood shavings to catch solder, 72.
Danger of water entering an empty boiler when hot, 312.
D-traps, 282.
D-trap fixed for a bath-waste, 215.
D-traps for water-closets, 271.
Dead plate under boiler, 309.
Dead weight safety valve, 318.
Defective drains and rats, 127.
Defective joint between drain and soil pipe, 161.
Defective joint in soil pipe, 263.
Defective soil pipes inside a house, 182.
Details of a good grease-trap, 186.
Diarrhœa from smell of drains, 124.
Dipstone trap, 139.
Different kinds of baths, 208 to 223.
Different ways of ventilating drains, 175.
Difficulty in tracing source of smells from water-closets, 278, 279.
Difficulty of cleaning out boilers, 314.
Difficulties of drawing seams, 36.
Digging point of shave hook into lead, 108.
Dirty manholes, 137.

Dirty water in bath safes, 216.
Discharge cock to empty boiler, 324.
Disconnecting traps, necessity of, 138, 141.
Disease germs and traps, 138.
Disguising drain vents, 172.
Dishing wood for taft joint, 80.
Disinfectant distributor, 127.
Disinfectants should be applied to grease tanks, 190.
Disreputable tradesmen, 129.
Distance apart of tacks, 97.
Diversity of opinions on drain ventilation, 166.
Dr. Angus Smith's solution for iron pipes, 145.
Dr. Lyon Playfair on increased length of life, 125.
Doctors on sewer air, 124.
Dr. Richardson on sewermen, 124.
Double bends, 56.
Double boot boiler, 309.
Double elbow, 62.
Double elbows of square section, 63.
Double jacket grease interceptor, 189.
Double socket for iron drains, 157.
Draining boards, 192.
Drains air-bound, 127.
Drains as rubbish shoots, 154.
Drains at a banker's house, 130.
Drains broken by traffic in streets, 131.
Drains in adjoining house defective, 165.
Drains in country parsonage, 128.
Drains join sewer near the bottom, 127.
Drains laid in rocky ground, 130.
Drains laid in sandy or loose soil, 130.
Drains leaking outside the house, 164.
Drains leak through split pipe sockets, 133.
Drain machine, 136.
Drain manholes ventilated, 173.
Drains not laid to proper fall, 133.
Drains of random rubble, 128.
Drain pipes, 128.
Drain pipes as soil pipes, 260.
Drain pipe built in wall, 261.
Drain pipes, how to lay them, 132.
Drain pipes, how to select, 131.
Drain pipe joints, 132.
Drain pipe junctions, 135.
Drain plan of a house, 179, 180.
Drain-rods for clearing iron drains, 155.
Drains should be well flushed, 166, 176.
Drain smells laid on to the house, 263.
Drain smells pass through skylight, 268.
Drains tested by asphyxiator, 127.
Drain stopped with paper, 136.
Drains under new streets, 131.
Drains ventilated at intervals, 172.
Drains ventilated near the centre, 183.
Drain-syphon fixed unlevel, 140.

INDEX. 343

Drain traps, 184.
Drain trenches, 129.
Drain ventilation, 166 to 168.
Drain vents disguised, 172.
Drain vents and soil pipes, 166.
Drain ventilated through soil pipe, 181.
Draughts in bath-rooms, 218.
Drawing pipe seams, 36.
Drawing soapy water at sink, 217.
Drawn-lead traps, 284.
Drilling holes in boilers, 312.
Driving chisels into bench, 87.
Dry area round house connected with cesspool, 172.
Dry hair-felt for covering hot-water pipes, &c., 322.
" Duck's-foot " bend, 157.
Ductility of lead, 19.
Dummies with iron heads, 47.
Dummy and dresser best for making bends, 56.
Dummy-made bends, 47.
Dummy socket for rain-water pipes, 107.

E

EACH drain to have a trap, 138.
Ears strongly soldered to lead pipe, 107.
Earthenware slop sinks, 200, 201.
Effect of ice in a cistern, 119.
Elbows, 39.
Elbows in hot-water pipes, 331.
Elbows soldered in two heats, 59.
Elbows to fit over plinths, 64.
Elbows made out of sheet-lead, 61.
Elbows v. bends, 57.
Elbows v. bends in hot-water pipes, 331.
Elbows were made before bending introduced, 57.
Elbows with soldered angles inside, 67.
Empty pipes in frosty weather, 95.
Enamelled iron slop-sinks, 201.
Enamelled slate for urinals, 254.
Enclosed urinals, 256.
Enclosures for public urinals, 249.
Enclosures for wash-hand basins, 236, 237.
Enclosure for hot-water cylinder, 322.
End of a waste-pipe vent, 198.
End of drain always covered in rainy season, 127.
Engraved face-plates for cocks, 207.
Evaporation of water in gulley-traps, 122.
Evil of cap and screw improperly placed, 84.
Evil of stop cocks in hot-water circulating pipes, 330.
Evils of bobbins, 45.
Evils of buckles in pipes, 40.
Evils of a common drain-syphon, 140.

Evils of dipstone trap, 139.
Evils of pipe-hooks, 96.
Evils of pipes instead of bends, 134.
Evils of fixing boilers unlevel, 313.
Evils of trough water-closets, 302.
Evils of treadle-action flushing apparatus, 253.
Evils to avoid when drain-laying, 133.
Erratic circulation of hot water, 334.
Example of hot-water arrangements in a mansion, 326, 327, 336.
Examples of alterations made to hot-water pipes, 307, 329, 333, 334.
Examples of sanitary improvements, 179 to 182.
Excessive heat on boilers, results of, 312.
Expansion of hot-water pipes, 331.
Expansion of pipes with hot water, 96.
Expansion pipes, 318.
Expansion joint, 202, 204.
Expansion joints for hot-water pipes, 331.
Expansion socket for hot-water pipes, 331.
Expensive traps not always best, 188.
Experiment with tube on air-currents, 167.
Experiments with water on hot iron, 312.
Extracting silver from lead, 19.
Face-tacks, 96.
Factory chimneys as sewer-ventilators, 123.
Fancy wash-hand basins, 225.
Fat in traps, 185.
Faulty construction of water-closet and sink pipes, 277, 278.
Feed-cistern, 317.
Feed-cistern to kitchen boiler, 306.
Feed-pipe to boiler, 317.
Felt on pipes, 102.
Ferrules, how to fix, 89.
" Field's " flushing-tank used, 180.
Fifty gallons of water fail to remove piece of paper out of trap, 140.
Fire-bars under boiler, 309.
Fire in bath-room, 218.
Fire Insurance Societies and hot-water pipes, 335.
First patent for making pipes, 29.
Fixing astragals to lead pipes, 106.
Fixing boilers, 311.
Fixing brass unions, 89.
Fixing joints for wiping, 87.
Fixing lead pipes, 95.
Fixing pipes in shafts, 290.
Fixing pipes with screws, 101.
Fixings for hot-water cylinders, 321.
Fixings for hot-water pipes, 325, 332.
Fixing water-closet traps by rule of thumb, 286.
Flange at bottom end of soil pipe, 162.

Flanged ends to pipes, 28.
Flanged connections of pipes to boilers, 314.
Flapper v. dresser, 108.
Flap-valve in hot-water pipes, 330.
Flap-valve traps, 240.
Flasks for casting lead astragals, 106.
Floated seam of tacks to pipe, 98.
Floors covered with lead, 26.
Floors of bath-rooms, 219.
Floors under water-closets, 300.
Floors of urinals, 246.
Flues to boilers, 309, 311.
Flushing-cisterns for urinals, 258, 259.
Flushing of drains, 166.
Flushing-rims to water-closets, 293.
Flushing-rim wash-hand basins, 226.
Flushing urinal basins, 252.
Flush soldering at the angles on top edge of cistern, 110.
Fluted draining-boards, 192.
Folding urinals, 257.
Foot to trap, 140.
"Fossil meal" as cylinder cover, 322.
Four-inch joints on half-inch pipe, 75.
Four-inch trap bent out of pipe, 50.
Frame for sheet-lead casting, 22.
French lavatory, 243.
Friction between lead and lead, 113.
Frozen pipes and boiler, 318.
Fungoids growing in manholes, 173.
Fur in hot-water pipes, 323
Furniture in a bath-room, 219.
Fusibility of lead, 20.
Fusible plugs in boilers, 318.
Fustian for cloths, 76.

G

Galena, 17.
Galvanic action of lead and iron, 20.
Galvanic action on boiler, 315.
Galvanized-iron bends for hot water, 331
Galvanized-iron capping for sinks, 120.
Galvanized-iron pipe-nails, 99.
Galvanized iron pipes for drains, 145.
Garbage and rats, 127.
Garbage passes into traps, 185.
Gas-burner in vent pipe, 170.
Gases of decomposition and water-traps, 138.
Gases generated in drains, 169.
Gasfitter and pipes in chases, 100.
Gasfitters use copper-bit, 92.
Gas-pipe testing, 153.
Gauge-hook, 35.
Geometry an aid to plumbers, 60.
Geometry not dry work, 64.
Gland joint between soil pipe and iron drain, 164.

Gland joint for hot-water pipes, 331.
Glass back to urinal, 256.
Glazed socket pipes for drains, 128.
Glazed tiles round scullery sinks, 191.
"Glory hole" under scullery-sink, 190.
Good description of drain-syphons, 141.
Good mate necessary, 36.
Good tradesmen and bobbins, 44.
Gratings in sink washers, 120.
Gratings in slop-sinks, 200.
Gratings in washers for wash-hand basins, 232.
Gratings over air-inlets in fields, 171.
Gratings over manholes, 173.
Gratings reduce water-way of waste pipes, 121.
Gratings should be sunk below bottom of sinks, 120.
Gratings under urinals, 249.
Grated connections for wash-hand basins, 235.
Grease and kitchenmaids, 189.
Grease interceptor, 186.
Grease interceptor with outer jacket filled with water, 189.
Grease lifter in trap, 187.
Grease-tanks should be disinfected, 190.
Grease-trap, 8 feet long, 188.
Grease-trap fixed to receive discharges from wash-hand basin, 188.
Grease-trap fixed where no grease to intercept, 188.
Grease-traps necessary evils, 188.
Grease-traps patented, 187.
Grease-trap ventilation, 188.
Groove in iron pipe socket, 149.
Ground saturated with waste water, 196.
Gulley-trap with a valve trap, 123.
Gun-metal pipe-nails, 99.
Gun-metal unions for lead hot-water pipes, 325.

H

Half pipes for manholes, 137.
Hammer used for jointing iron pipes, 159.
Hammer and bolt bends, 42.
Hand-made pipes, 34.
Hand-made traps, 282, 283.
Hard bottom under drains, evils of, 130.
Hard lead, causes of, 18.
Hardware-sinks destructive to glass and crockeryware, 117.
Hard waters form fur in boilers, 314.
Health Exhibition, urinal at, 256.
Heat as an aid to create air-currents 167.
Heat of lead for joints in iron pipes, 148.
Heat of pipe when bending, 46.
Heavy lead bends, 28.
Heel-side of bends thin, 51.

INDEX.

Height of urinal basins, 254.
High pressure boiler, 306, 308.
Hole in lip of ladle, 36.
Holes in boiler, position of, 312.
Holes in boiler should be drilled, 312.
Holes in boilers, tapping, 314.
Hollow astragals, 106.
Hollow fillet in sink angles, 119.
Hollow piers for air-inlets to drains, 171, 172.
Hollow plugs for sinks, 120.
Hopper heads and waste pipes, 231.
Horizontal air-inlet drains, 173.
Horizontal cylinders for hot water, 338.
Horizontal pipes for hot-water circulation, 324.
Horizontal soil pipes, 271, 275.
Hospital spray, &c., bath, 222.
Hot closet in kitchen, 328.
Hot pipes warm water in cold pipes, 101.
" Hot plate " boiler, 310.
Hot-water boilers, 305.
Hot-water boilers supplied from street main, 322.
Hot water breaks lead soil pipes, 241.
Hot-water chamber, 319.
Hot-water circulation, 322.
Hot water circulating in branch pipes, 335.
Hot water circulating backwards, 334.
Hot-water circulate through cold-water cistern, 318.
Hot-water cisterns, 319.
Hot-water cistern connected with bath, 217.
Hot-water coil in bath-room, 218.
Hot-water cylinder, 320, 321.
Hot water drawn direct from cistern, 317.
Hot-water draw-off pipes, 329.
Hot-water fitting, 316.
Hot-water men, 332.
Hot water passes into cold-water cistern, 317.
Hot-water pipes air-bound, 329.
Hot-water pipes always in motion, 326.
Hot-water pipes assist drain ventilation, 181.
Hot-water pipe bends, 331.
Hot-water pipes beneath flooring, 335.
Hot-water pipes fixed on rollers, 336.
Hot-water pipes fixed in metal-lined channels, 335.
Hot-water pipes improperly fixed, examples of, 307, 329, 333, 334.
Hot-water pipes in cold countries prevent freezing, 101.
Hot-water pipes in chases near lead pipes, 100.
Hot-water pipes near woodwork, 335.
Hot-water pipes near drinking water, 101.

Hot-water unions want periodical examination, 325.
Hose for shower apparatus, 244.
Houses connected directly with sewer, 166.
House drains, 128.
House drains act as a gas retort, 166.
House drains not ventilated, 124.
House drains not trapped from sewers, 124.
House in a sewage bog, 165.
House owners and sanitary inspectors, 269, 270.
How a good plumber wipes a cistern, 112.
How cloths get torn, 91.
How lead pipes are made, 30.
How to line a cistern with ends soldered in, 113.
How lead-encased tin pipe is made, 33.
How lead pipes used to be cast, 27.
How milled-lead is made, 21.
How pipe seams are drawn, 36.
How to arrange channel bends in a manhole, 138.
How to arrange " weeping pipe," 272.
How to calculate thickness of sheet-lead, 26.
How to cast sheet-lead, 22.
How to calculate weight of sheet-lead, 26.
How to cast lead flanges on pipes, 28.
How to connect branch circulating pipes for hot water, 335.
How to connect flushing pipe to water-closet, 300.
How to connect pipes to boilers, 314.
How to connect pipes to cylinder, 323.
How to cut out sheet-lead for an elbow, 60.
How to cut a double elbow out of sheet-lead, 63.
How to fix a branch joint on a soil pipe, 288.
How to fix channel bends, 137.
How to fix underhand joint for wiping, 87.
How to get bruises out of pipes, 41.
How to heat pipes, 46.
How to join drain pipes, 133.
How to lash a sand bend, 53.
How to lay branches for drains, 135.
How to line cisterns with one piece of lead, 108.
How to line sinks, 118.
How to line a large cistern, 113.
How to make a pressure gauge, 152.
How to make an opening in an iron drain, 154.
How to make bends with two pieces of lead, 55.
How to make branch connections to hot-water pipes, 326.

346   INDEX.

How to make dummies, 48.
How to make overcast copper-bit joints, 93.
How to make pipe out of sheet-lead, 35.
How to make sand bends, 53.
How to make shave-hook for fitting to compasses, 81.
How to open hole for branch joint, 82.
How to prepare joints for wiping, 69.
How to prevent noise from boiler cover, 306.
How to prepare trenches for drains, 129.
How to prevent water in feed-cisterns getting hot, 306.
How to repair the lead in a sink, 118.
How to "set out" a water-closet trap, 286 to 290.
How to set out a bend or elbow, 57.
How to select a rasp, 71.
How to select drain pipes, 131.
How to shave a taft-joint, 80.
How to soil and shave a joint, 70.
How to start bending pipes, 42, 48.
How to test the flush of a water-closet basin, 301.
How to test gas pipes, 153.
How to tin pipe-ends, 72.
How to trim a joint, 72.
How white lead is made, 20.

I

If drains clean there are no smells, 172.
Ill-used pipes, 40.
Illustrations of bad design in hot-water work, 333.
Impetus of water through a water-closet trap, 283.
Importance of concrete round drains, 130.
Importance of geometrical knowledge, 60.
Improper connection of drain with sewer, 129.
Improper use of dresser, 109.
Improper use of manholes, 176.
Improper use of pantry-sinks, 197.
Improved drain-syphons, 140.
Improved public urinals, 248.
Improved water-closet for schools, 303.
Indiarubber cone, 204.
Indiarubber plug for bath-waste, 214.
Indiarubber plugs for wash-hand basins, 233.
Indiarubber ring, 203.
Indiarubber tubing for packing, 236.
Induced air-currents in drains, 169.
Inlet-pipes for a bath, 209.
Inspection-chambers, 136.
Inspection-pipes for iron frames, 155.
Interceptor trap, 195.

Inventors of water-closets, 291.
Invisible fixings for rain-water pipes, 104.
Iron bars under iron drains, 150.
Iron bends choked with zinc, 331.
Iron brackets to support lead pipe, 203.
Iron clamps, 148.
Iron drain branches, 156.
Iron drains and sanitarians, 143.
Iron drains built in walls, 150.
Iron drains fixed in walls, 150.
Iron drain jointing, 146.
Iron drains on brick piers, 149.
Iron drain pipes, thickness of, 145.
Iron drain pipes, weight of, 149.
Iron drains protected from rusting, 145.
Iron drains, tools used for laying, 158, 159.
Iron D-trap, 194.
Iron gratings in urinal floors, 254.
Iron pipes Bower-Barff'd, 145.
Iron pipes broken by settlement of ground, 150.
Iron pipes cut uneven, 147.
Iron pipes, difficult to paint, 103.
Iron pipes, galvanized, 145.
Iron pipe joints "set up," 146.
Iron pipes rust, 103.
Iron pipes should be examined inside before fixing, 158.
Iron pipes tested with hammer, 148.
Iron punch, 110.
Iron rain-water pipes used for drains, 145.
Iron rain-water pipes with lead bends, 107.
Iron rods for dummies, 47.
Iron soil pipes, 145.
Iron to make joints, 76.
Iron v. lead rain-water pipes, 31, 103.
Iron v. steel bolts for bending, 43.
Iron v. stoneware drains, 144.
Irrigation substituted for cesspool, 172.

J

Jack-of-all-trades, 269.
Jerry-built houses, hot-water work in, 316.
Joints and grease, 77.
Joints at unequal intervals, 84.
Joints cast in moulds, 94.
Joints crack when wiping them, 87.
Joints for lead-encased-tin pipes, 325.
Joints in hot-water pipes leak by expansion, 331.
Joints made with blow-pipe, 91.
Joints made with copper-bit, 91.
Joints made with lamps, 92.
Joints of lead to iron pipes, 146.
Joints on to brasswork, 74.

INDEX.   347

Joints rolled, 89.
Joints, solder falls off, 90.
Joints, table of length of, 75.
Joint between soil pipe and drain, 161 to 164.
Jointing drains, 132.
Jointing iron drains, 146.
Joint-making, 69.
Joint moulds and plumber's assistant, 94.
Joint of soil pipe to iron drain, 157.
Joint of trap to safe, 80.
Joint of water-closet to soil pipe, 296, 297.
Joint of 4-in. to 6-in. pipe, 88.
Joint shaving, 71.
Joint shown at Health Exhibition, 163.
Joint to allow for expansion, 202.
Joints to lavatory valves, 93.
Joints to waste unions, 93.
Joints trimmed at ends, 72.
Joint wiped with thin cloth, 73.
Journeyman plumber and bad work, 269.
Junction of brasswork to pipe not in centre of joint, 75.
Junction of house drain with sewer, 127.
Junction for drains, 135.

K

KEEP pipe-ends clean after shaving, 72.
Kitchenmaids and grease, 189.
Kitchenmaids and sore throats, 190.
Kitchen-range boilers, 305.
Knife to cut ends of soldering, 98.
Knot bends, 54.
Knuckle bends, 43.

L

LADIES' cloak-rooms, 258.
Lady complains of plumbers, 101.
Laminated lead, 25.
Lamp-holes in sewers, 122.
Lamp-posts as sewer-ventilators, 123.
Lamps for making joints, 92.
Land springs under houses, 150.
Lap-welded steam pipe, 316.
Large cesspool at nobleman's mansion, 172.
Large drains and small traps, 142.
Large drain-traps, 139.
Large D-traps, 271.
Large firms and joint moulds, 94.
Large grease trap, 188.
Large-headed nails for tacks, 99.
Large overflows to baths, 211.
Large v. small flushing-tanks, 181.
Lavatory valves, joints to, 93.
Lead alloys, 21.

Lead apparatus for bath-waste, 213.
Lead as a conductor of heat or electricity, 20.
Lead basis for some paints, 20.
Lead brittle when heated, 19.
Lead carbonate, 20.
Lead cements, 21.
Lead, chemical properties of, 20.
Lead collar for block joint, 79.
Lead collar to catch solder, 73.
Lead cut with dresser, 109.
Lead, ductility of, 19.
Lead-encased-tin hot-water pipes, 324.
Lead-encased-tin pipe, 32.
Lead flappers, 51.
Lead flasks and bends in small pipes, 52.
Lead for cisterns, 26.
Lead for lining vitriol chambers, 26.
Lead, fusibility of, 20.
Lead glazed pipes, 128.
Lead-headed nails 99.
Lead hot-water pipes, 324.
Lead in combination with chemicals, 21.
Lead in iron drain joints, 147.
Lead joints, 78.
Lead joints on iron pipes, 21.
Lead, Latin name for, 17.
Lead-lined baths, 208.
Lead, lustre of, 20.
Lead, malleability of, 19.
Lead, market forms of, 21.
Lead melting point, 19.
Lead on floors, 26.
Lead ores, 17.
Lead ore preparation, 17.
Lead ore smelting and roasting, 17.
Lead oxides, 20.
Lead, physical properties of, 19.
Lead, pigs, sheets, and pipes, 21.
Lead pipes, 27.
Lead pipes and painting, 103.
Lead pipes and baggy parts, 95.
Lead pipes easy to repair, 103.
Lead pipes fixed on iron brackets, 203.
Lead pipes formed into knot, 54.
Lead pipes injured by iron hot-water pipes, 100.
Lead pipes of unequal thickness, 32.
Lead plates, 21.
Lead, separating other metals from the ores of, 18.
Lead shavings in pipes, 71.
Lead shrinks when cooling, 28.
Lead socket and ears for square pipe, 107.
Lead, softness of, 20.
Lead, specific gravity of, 19.
Lead "squirting," 30.
Lead tacks on large pipes, 96.
Lead, tenacity of, 19.
Lead turned over edge of sink, 119.
Lead, uses of, 26.

348   INDEX.

Lead used on roofs, 26.
Lead v. iron rain-water pipes, 31, 103.
Lead, where found, 17.
Leakage of pipe in chase, 100.
Leakage of water out of a trap, 130.
Leaking drain joints make a large cavern, 130.
Length of cast-lead sheets, 25.
Length of drain pipes, 128.
Length of joints, 75.
Length of iron pipes, 146.
Light iron drain pipes, 145.
Light iron soil pipes, 261, 265.
Light v. heavy iron drain pipes, 149.
Lime-white inside iron pipes, 145.
Limited water supply to urinals, 258.
Lines for bends and elbows, 57.
Linen room warmed by waste heat, 328.
Lining cisterns, 108, 113.
Lining sinks, 117.
Litharge for leading glazing earthenware, 20.
Little finger burnt by solder when wiping, 73.
Locked air-tight covers, 176.
London sewers take rainfall, 123.
Long hopper water-closet, 301.
Long v. short bolts for bending, 43.
Looking-glass behind wash-hand basin, 244.
Loops on dummy handles, 47.
Louvred cap on air-inlet, 174.
Low-pressure boiler, 306.
Lump boilers, 309.
Lustre of lead, 20.

M

Machine for casting lead pipes, 27.
Machine-made pipes, 31.
Machine-made traps, 284.
Mahogany lagging round hot-water cylinders, 322.
Makers of water-closets, 294.
Making joints without irons, 76.
Malleability of lead, 19.
Mandrel for opening lead pipe, 203.
Mandrel inside branch joint, 88.
Manholes, 136.
Manholes built of common bricks, 138.
Manholes built of white glazed bricks, 138.
Manholes cleaned when dirty, 138.
Manhole covers should be air-tight, 175.
Manhole, how to arrange branch drain, 138.
Manholes in cylinders, 323.
Manholes lime-whited, 138.
Manholes rendered with cement, 138.
Manholes to boilers, 315.

Manholes to hot-water chambers, 320.
Manufacturers' v. shop-made bends for hot water, 331.
Marble tops for wash-hand basins, 224.
Marble urinal stalls, 254.
Market forms of lead, 21.
Massicot, 20.
Master in a rage with plumber, 101.
Mr. Shirley Murphy on sanitary improvements, 125.
Materials for sinks, 117.
Mates and dummies, 48.
Mechanical means for ventilating drains, 170.
Melted lead to heat bends, 47.
Melting point of lead, 19.
Men-servants and traps in cellars, 197.
Metals that get mixed with lead, 47.
Mica air-inlet valves, 176.
Milled-lead, 21.
Milled v. cast lead, 24.
Minium, 20.
Modern valve water-closets, 292.
Modern way of fixing scullery sink, 190.
Momentum of water through water-closet traps, 284.
Moths in felt coverings to hot-water pipes, 322.
Motion of hot-water pipes, 326, 331.
Moulds for casting joints, 94.
Mouldy smells in bathroom, 218.
Mushrooms growing inside bath enclosure, 219.

N

Nails with large head and flat shank, 99.
Nailing cistern angles, 115.
Narrow baths objectionable, 208.
Navvies v. tradesmen for laying drains, 146.
Necessity of disconnecting traps, 138.
Necessity of traps, 285.
Needle baths, 221.
New beginner at cistern soldering, 112.
New streets, drains under, 131.
Nickel-plated bath, 223.
Nickel-plated copper coil in a bath-room, 218, 337.
Nipple v. wiped joint, 94.
Noise in boilers, cause of, 313.
Noise of water in waste pipe heard in rooms, 204.
Noise of water in water-closet heard in rooms, 266.
Noise of boiler-cover, 306.
Number of sheets of lead cast in one day, 23.
Nursery scullery-sink, 197.
Novices and house examinations, 269.

O

Oak capping on sinks, 119.
Objections to lead pipes for hot water, 325.
Objections to long hopper water-closet, 301.
Objections to " wash-out " water-closets, 298.
Obstructions in iron drains, 155.
Offensiveness of pan water-closet, 298.
Offensive smells in urinals, 247.
Office urinals, 257.
Old and modern valve water-closets compared, 292.
Old-fashioned traps, 139.
Old hands at house examinations, 269.
Old lead partly pays for new in exchange, 118.
Old lead rain-water pipes, 103.
Old method of making large lead pipes, 27.
Old method of making pump barrels, 28.
Old specimens of wiped seam pipes, 34.
Old way of fixing water-closet traps, 271.
Old way of trapping scullery-sink, 184.
On air drains, 176.
On ball-traps, 184.
On bent drain pipes, how to lay them, 131.
On digging trenches for drains, 129.
On distorting water-closet traps, 285.
On dirty corners round scullery-sinks, 191.
On levelling drains, 133.
On emptying boilers, 324.
On enamelled iron sinks, 117.
On errors of judgment, 281.
On fitting ends of pipe for joining 69.
On fixing water-closet traps, 285 to 290.
On galvanized-iron sinks, 117.
On lead-lined sinks, 117.
On lids to sinks, 198.
On makers of bath-cocks. 211.
On mechanical traps, 184.
On patents for destroying evil effects of sewage gases, 123.
On porcelain sinks. 198.
On setting out water-closet traps, 286 to 290.
On shaving joints, 70.
On sink enclosures, 198.
On slate sinks, 117.
On stoneware sinks, 117.
On swilling floors, 186.
On the strength of an arch or cylinder, 131.
On tightly fixing hot-water pipes, 332.
On waste pipes, 198.
On waste pipes discharging over gratings, 195.

On wooden sinks, 171.
On various traps, 184.
Open channels for manholes, 137.
Open channels smell offensive, 197.
Open channels with side inlets, 137.
Opening for branch joint to be large, 83.
Opening for clearing stoppages in iron drains, 154.
Open manhole bends, 137.
Open range, 305.
Ornamental machine-made pipes, 31.
Ornamental swabbing, 38.
Ornamental tacks, 97.
Overcast copper-bit joint, 92.
Overcasting joints, 77.
Overflow from cistern into soil pipe, 263, 265.
Overflow pipes to baths, 211.
Overflow to sinks, 121.
Overflow to urinal basins, 252.
Overflow to wash-hand basins, 228.
Overflow to water-closets, 295.
Oxidation inside iron boilers, 314.
Oxidation of iron pipes, 103.
Oxide of lead, 20.
Oxygen, action of on lead, 20.

P

PAINTING iron pipes, 103.
Paint under-side of water-closet seats, 304.
Pan water-closet, 298.
Paper hangings in bath-rooms, 218.
Paper test for water-closet basins, 301.
Pasted paper on soldered seams, 98.
Patent bath-waste-valve, 214.
Patent burning, 38.
Patent drain syphon traps, 141.
Patent grease-traps, 187.
Patent slop-sinks, 201.
Patent " tip-up " wash-hand basins, 236.
Patents for pipe making, 29, 30.
Pattern for astragals, 106.
Pattinson's system of extracting silver from lead, 19.
Perforated cornice in bath-room, 218.
Pewter, 21.
Piece of soiled lead to catch solder, 72.
Piecework plumbers, 264.
Pig lead, 21.
Pipe astragals, 104.
Pipe bending, 39.
Pipe bent into a knot, 54.
Pipe brittle if made too hot, 46.
Pipe casing, 96.
Pipe ends dummied out, 69.
Pipe ends torn off when bending, 42.
Pipe enlarged at branch joint, 85.
Pipe filled with water before bending, 54

Pipe fixing, 95.
Pipe-hooks, 96.
Pipe-making machine, 30.
Pipe nails, 99.
Pipe nicked with pocket-knife, 72.
Pipe not made of pure lead, 47.
Pipes accessible for repairs, 101.
Pipe air-bound, 95.
Pipe bruised when moved about, 41.
Pipe bent cold, 41.
Pipe coated with Dr. Angus Smith's preparation, 145.
Pipe connections to hot-water cylinders, 323.
Pipe cut with wall-hooks, 95.
Pipes ill-used, 40.
Pipes in chases, 99, 100.
Pipes laid in concrete, 129.
Pipes lime-whited inside, 145.
Pipes painted distinctive colors, 204.
Pipes protected from freezing, 102.
Pipes rough inside, 28.
Pipe socket with current, 83.
Pipes suspended on blocks, 78.
Pipes to drain empty, 95.
Pipes used for hot water, 316.
Pipes with flanged ends, 28.
Plane and tools for casting sheet-lead, 22.
Plan of drains in a house, 178 to 180.
Plaster-moulds for astragals, 105.
Plug and washer in sink, 120.
Plug-waste for bath, 214.
Plug and washer should be below bottom of sink, 120.
Plug-wastes for wash-hand basins, 232, 233.
Plumbers and knowledge of drawings, 289.
Plumber's assistant and joint moulds, 94.
Plumber's difficulties, 100.
Plumber's designs, astragals, 105.
Plumbers' Exhibition at Kensington, 50.
Plumbers should start at connection of drains with sewer, 125.
Plumbers should have a knowledge of geometry, 60.
Plumbers and burning, 38.
Plumbers make dummies, 48.
Plumbers' work prepared in England for fixing in China, 289.
Plunger water-closets, 297.
Pocket-knife and trimming joints, 72.
Pocket-knife for trimming edges of soldering, 111.
Pointed compasses, 80.
Points to be attended to when soldering cisterns, 112.
Poisoning rats, 127.
Pools of sewage in drains, 143.

Polytechnic students gain prizes, 24, 25.
Porcelain gratings for bath-wastes, 214.
Portland cement bursts drain pipe sockets, 133.
Portland cement for jointing drain pipes, 133.
Position for cylinder manholes, 323.
Position for hot-water cylinder, 323.
Position for slop-sinks, 204.
Position for safety-valves, 318.
Position of bath-cocks, 209.
Position of branches in manholes, 138.
Position of cocks in a sink, 199.
Position of grease-traps, 187.
Position of hot-water cylinders, 321.
Position of holes in boilers, 312.
Position of lavatories, 236.
Position of return pipe to cylinder, 327.
Position of seams on elbow, 59, 66.
Position of sink-trap vent, 191.
Position of steam pipe from kitchen boiler, 306.
Position of valves for wash-hand basins, 229.
Position of vent pipes on hot-water circulating pipes, 330.
Position of water-closet, bad, 266.
Position of waste pipe vents, 198.
Pot for lead-casting, 22.
Practice required to make joints, 86, 94.
Precautions to be taken when laying drains, 132.
Precautions when cistern soldering, 111.
Preparing block joint, 78.
Preparing joints on iron pipes, 147.
Preparing joints and young plumbers, 69.
Pressure-gauge, 152.
Pressure of gas in a main, 153.
Pressure required to make lead pipes, 32.
Public authorities aggravate an evil, 123.
Public school water-closets, 301.
Public sewers, 122.
Public to blame for bad work, 269, 270.
Public urinals, 246 to 249.
P-traps, 283.
Pumps and overcast joints, 77.
Pyramid of paper in a drain, 136.

Q

QUACK doctors hand-bills, 246.
Quack plumbers, 91.

R

RADIATORS heated from hot-water pipes to sinks, &c., 328.

INDEX. 351

Rain-water pipes, 103.
Rain-water pipes act as vents, 124.
Rain-water pipes as soil pipes, 261, 268.
Rain-water pipes of iron used for drains, 145.
Rain-water troughs connected with soil pipe, 267, 268.
Rainfall runs into London sewers, 123.
Railway station soil pipes at, 276, 277.
Railway station urinals, 251.
Range water-closets, 303.
Rasping pipe-ends, 71.
Rats and defective drains, 127.
Rats and tide-flaps, 126.
Rats as scavengers, 127.
Rats cannot get out of iron or stone-ware drains, 127.
Rats pass into drains when low in sewer, 127.
Rats propagate under floors, 127.
Rats starved or poisoned, 127.
Reasons for not mentioning a number of traps, 184.
Reasons for ventilating sink-traps, 205.
Red lead as driers, 20.
Red-lead joints to soil pipes, 261, 262.
Reducing iron pipes, 158.
Registration of plumbers, 264.
Repairing lead sinks, 118.
Result of leaking drains, 130.
Reverberatory furnace, 18.
Ribbon joints, 94.
Right-angled junctions in manholes, 137.
"Ringing" of iron pipes, 148.
Risk with hydrogen gas, 38.
Rivetted boilers, 310.
Rivetted cylinders, 320.
Roasting lead ores, 18.
Rolled joints, 89.
Rollers under hot-water pipes, 336.
Rotten leaves in manhole, 173.
Round pipes, how made, 31.
Rubbish shoots of drains, 154.
Rule of thumb, 57.
Rule of thumb when fixing water-closet traps, 285, 286.
Rules *v.* guesswork, 64.
Rusting of pipe nails, 99.

S

Saddle boilers, 309.
Safe-traps and weeping pipes, 270, 271.
Safes under baths, 215.
Safety-plug in boilers, 318.
Safety-valves on boilers, 318.
Safe waste into soil pipe, 272.
Safe waste pipes, 272, 273.
Salt as a core for lead joints, 78.

Salt-glazed pipes, 128.
Sand bends, 53.
Sand bends made to a sharp radius, 53.
Sand bends thin outside, 54.
Sand core for lead joints, 78.
Sand cracks in cast-lead, 23.
Sand interceptors, 186.
Sand in traps, 185.
Sandy soil and drains, 130.
Sanitary engineers and manholes, 136.
Sanitary engineers and water-closets, 294.
Sanitarians and drain pipes, 138.
Sanitarians and stoneware drains, 143.
Sanitarian's trouble commences when laying the drains, 128.
Sanitary improvements, examples of, 179 to 182.
Scotch joints, 91.
Scouring bends with sand, 47.
Screw caps in iron drains, 155.
Screws for fixing pipes, 101.
Scrubbing brush to clean sinks, 119.
Scullery-sinks and dirty corners, 191.
Scullery-sink badly arranged, 185.
Scullery-sinks should be fixed on cantilevers, 190.
Second-hand water-closets, 264.
Sediment in boilers choke pipes, 312.
Self-cleansing syphon-traps, 141.
Self-closing valves for wash-hand basins, 227.
Separating other metals from lead, 18.
Separate vents to traps, 206.
Servants improperly use pantry sinks, 197.
Service pipes to boilers "air-bound," 329.
Setting up iron drain joints, 148.
Settlement of earth under drains, 130.
"Sets" of wash-hand basins, 241, 242.
Several dummies wanted on large works, 48.
Several pipes joined to water-closet trap, 274.
Sewer air, 124.
Sewage-bog round a house, 165.
Sewage-conduits should be water-tight, 165.
Sewage gases pass through water-traps, 124.
Sewage gases kept back by tide-valves, 126.
Sewage passes through a party-wall, 165.
Sewage soaking through house walls, 165.
Sewerage and sewage, 122.
Sewer-men and sewage gases, 124.
Sewer-men and drain connections, 129.
Sewer-men have rheumatism, 124.
Sewer vented in a party-wall, 123.
Sewer ventilated into factory chimneys 123.

352   INDEX.

Sewer ventilated through drain vent pipe, 170.
Sewers syphon water out of traps, 124.
Sewer ventilation, 122.
Sewn v. pinned cloths, 76.
Shaft for fixing pipes in, 290.
Shaft used as a soil pipe, 260.
Shampooing apparatus, 245.
Shape of channel bends, 137.
Shape of hoods for baths, 221.
Shape of hot-water cylinders, 320.
Shape of wash-hand basins, 225.
Shave-hook compasses, 80.
Shave-hook for cisterns, 115.
Shaving branch joints, 85.
Shaving round the pipes, 71.
Sheet-lead, 21.
Sheet-metal covering for wood-work near hot-water pipes, 335.
Ship plumbers, 56.
Shop-made v. manufacturers' bends for hot-water, 331.
Shower apparatus over wash-hand basin, 244.
Shower baths, 219.
Shower-bath enclosures, 220.
Side gulleys in streets, 122.
Silt box, 122.
Silver extracted from lead, 19.
Sink in butler's pantry, 195.
Sinks in hotels and clubs, 120.
Sinks in nursery and scullery, 198.
Sinks in upper stories should have overflows, 121.
Sinks for washing vegetables, 121.
Sinks lined with lead, 117.
Sinks of enamelled-iron, 117.
Sinks of galvanized-iron, 117.
Sinks of slate, 117.
Sinks of stoneware, 117.
Sinks, position of cocks, 199.
Sink waste pipe into water-closet trap, 278.
Sinks of wood, 117
Sir William Jenner on poisons contained in public sewers, 125.
Size of bath-safes, 215.
Size of branch hot-water pipes, 330.
Size of feed pipe to boiler, 317.
Size of return pipe to cylinder, 327.
Size of slop-sinks, 201.
Size of tacks for 4-inch pipe, 96.
Size of trap in bath-waste, 216.
Sizes of cast-iron and stoneware drains, 144.
Sizes of cylinder circulation pipes, 324.
Sizes of pipes between boiler and cylinder, 323.
Sizes of wash-hand basins, 225.
Sizes of waste pipe vents, 198.
Sitz bath, 221, 223.

Skilled labour required for iron drains, 146.
" Slag wool " covering for hot-water pipes, 322.
" Slag wool " on pipes, 102.
Slate aprons to urinals, 248.
Slate floors to water-closets, 300.
Slop and washup sink, 206.
Slop-sinks, 200.
Slop-sinks fixed over each other, 204.
Slip joint in soil pipe buried in wall 263.
Small ball-cock to feed-cistern, evils of, 317.
Small traps in large drains, 142.
Small vent pipes to soil pipes, 271.
Small v. large flushing-tanks, 181.
Smell of drains cause sickness, 124.
Smells from bath utensils, 217.
Smells from cesspool escape through bedroom floor, 268.
Smells from cesspool pass into house, 172.
Smells from grease-tanks pass through sink waste pipe, 190.
Smells from side gulleys in streets, 122.
Smells from soil pipes pass into bedrooms, 265, 267.
Smells from urinals, 254, 255.
Smells from vent pipes, cause of, 134.
Smells from wash-hand basin, 230.
Smells from water-closets, 295.
Smells pass through safe-trap, 270.
Smelting lead ores, 17.
Smoke passes through valve water-closets, 296.
Smoke test and trapless water-closets, 296.
Soap-trap, 238.
Soapy matters clog traps, 240.
Socket to allow for expansion in hot-water pipes, 331.
Softening lead, 18.
Softness of lead, 20.
Soft v. hardwood dressers, 47.
Soil scratched with iron, 37.
Soil inside pipe ends, 72.
Soil makes joints look dirty, 72.
Soil pipes, 260 to 290.
Soil pipe as drain vent, 166, 181.
Soil pipe broken by hot water, 241.
Soil pipes fixed inside the house, 164.
Soil pipe in angle of drawing-room, 265, 266.
Soil pipe insufficiently supported, cause of smells, 285.
Soil pipe jointed to drain, 161 to 164.
Soil pipes and drains tested with air, 152.
Soil pipes near windows, 263.
Soil should be touched, 111.
Solder an alloy of lead and tin, 21.

INDEX.   353

Soldered angle weak at edges, 109.
Soldered angles crack by expansion of lead, 111.
Soldered pipe should not be used for acids, 38.
Soldered seam bends, 55.
Solder getting into valves, 69.
Soldering angles of cistern, 110.
Soldering elbows, 58.
Soldering over a welted joint, 81.
Soldering safe to water-closet trap, 80.
Soldering tacks to pipes, 97.
Soldering running into pipe, 69.
Solder runs through joint, 92.
Solder should cover shaved parts on lead pipe, 74.
Solder under finger-nail, 111.
Space round iron drains, 149.
Sparge pipes in urinals, 247, 250.
Speaking tubes near lead pipes, 100.
Special-made bends for rain-water pipes, 107.
Special-made tools for jointing iron pipes, 159.
Special size of shave-hook for cisterns, 115.
Specific gravity of lead, 19.
Specimen of work, 24, 25.
Splash-board to sinks, 207.
Splash-sticks, 76.
"Splints" for joints, 89.
Split sockets in iron drains, 148.
Spoons falling into pipes, 120.
"Spoon-hook" for cisterns, 116.
Sponge swab, 36.
Sponge to clean joints, 76.
Spray apparatus for wash-hand basin, 245.
Spray baths, 221, 222.
Square hot-water cylinder, 319.
Square junctions for drains, 135.
Square junctions on iron drains, 154, 156.
Square lead pipe, how to fix, 107.
Square pipe and double elbow, 62.
Square pipes, how made, 31.
Square section bends, 67.
"Staffing and burning machine," 27.
Stagnation of water in traps, 195.
Stall urinals, 246 to 254.
Standing on water-closets, how to prevent, 303.
Starting point for sanitary plumber, 125.
Starving rats, 127.
Steam in bath-rooms, 210.
Steam in water-boiler, 307.
Steam pipe to kitchen boiler, 306, 307.
Steam pipes to hot-water cisterns, 318.
Steam-tight covers to hot-water cisterns, 317.
Stench from traps, 185.

Steps to water-closets, 275.
Steps to urinals, 256.
Stone blocks for joints, 78.
Stone drains, 128.
Stone sinks worn by pails and tubs, 120.
Stone sinks smell offensive, 191.
Stone troughs for urinals, 247.
Stoneware drains costly, 144.
Stoneware grease traps, 187.
Stoneware troughs for urinals, 249.
Stop-cocks in hot-water circulations, 330.
Stop-cock in supply-pipe to boiler, 324.
Stoppage in branch drain, 136.
Stoppage in drains, cause of, 132.
Stoppage in bends, 40.
Storm water passes from sewer into house, 129.
Strainers in sinks, 121.
Stranger uses 28lbs. of lead to 4-inch joint, 146.
Street driftings choke air-inlets, 177.
Strength of joints, 74.
"Strike," the, 22.
Strong fixing for square pipe, 107.
Study of traps refer to advertisements in sanitary newspapers, 185.
Stupid arrangement of waste pipes, 275.
Sulphuretted hydrogen in drains, 169.
Supply-valves to wash-hand basins, 226, 227.
Support for hot-water cylinder, 321.
Surface water-traps, 184.
Swan-neck bends, 56.
Sweaty joints, 70.
Syphon trap improperly fixed, 141.

T

Table of length of joints, 75.
Table of weights for iron drains, 149.
"Tacking" elbows, 58.
"Tacking" seams of lead pipes, 35.
Tacks and wall-hooks, 96.
Tacks fixed singly, 97.
Tacks fixed in pairs, 97.
Tacks, how to solder to pipe, 97.
Tacks soldered on face, 99.
Taft joints, 80.
Tan-pin, 69.
Taper drain pipes, 143.
Tapering bends, 56.
Tapering channel, 143.
Tarred yarn for drain joints, 133.
Tarred yarn for iron pipe joints 147.
Tearing cloths, 91.
Technical knowledge necessary, 281.
Templates for bends and elbows, 57.
Tenacity of lead, 19.

AA

INDEX.

Testing iron drain pipes, 148.
Testing iron drains with air, 152.
Testing iron drains with water, 152.
The plumber a scientist, 281.
Thermometer for shower-bath, 220.
The "syphon" applied to drain ventilation, 168.
Thin cloths for branch joints, 86.
Thin v. thick cloths, 74.
Thick cloths necessary for cisterns, 111.
Thickness of copper for hot-water cylinders, 320.
Thickness of iron drain pipes, 145.
Thickness of iron plate for hot-water cylinders, 320.
Thickness of pipe reduced by shaving, 73.
Thickness of sheet-lead, how to calculate, 26.
Throated traps, 282.
To avoid waste of solder, 73.
To bed cover over opening in iron drain, 154.
To empty cylinder, 324.
To get a bruise out of a pipe with copper wire, 55.
To heat bends by melted lead, 47.
To kill grease on pipe ends, 70.
To make moulds for casting astragals, 105.
Too large exposed surface in urinals, 248.
Tools for laying iron drains, 158, 159.
Tools for bending pipes, 51.
Tool-marks on bends, 47.
Tool-marks on lead, 110.
Tool-marks on pipes, 41.
To test heat of pipe, 46.
Touching cistern angles, 111.
Touching pipe ends, 72.
"Touch" or tallow, 22.
Trap at bottom of soil pipe, cause of smells, 280.
Trap like a cesspool, 181.
Trap made out of 1-inch pipe, 52.
Trapping of house drain, 171.
Trapping slop-sinks, 202.
Traps and disease germs, 138.
Traps for wash-hand basins, 237, 238.
Traps in areas, 193.
Traps in aristocratic mansions, 185.
Traps made by machinery, 284.
Traps made in two halves, 283.
Traps, necessity of, 285.
Traps of cast-lead, 283.
Traps should be same size as waste pipe, 52.
Traps should not be fixed in cellars or inside a house, 197.
Traps too large in the body, 142.
Traps useless without water in them, 138, 185, 281.

Traps under valve water-closets, 295.
Trapless water-closets, 296.
"Tide-flaps," 122, 125
"Tide-flaps" and rats, 126.
"Tide-flaps" **useless for keeping back smells**, 125.
Tide-valves, 126.
Tie-rods in hot-water chamber, 320.
Tiled floors under water-closets, 300.
Time to empty bath, 214.
Tinned-copper pipes, 316.
Tinned-copper sinks, 207.
Tinning lead pipes, 32.
Tinning pipe ends, 72.
"Tip-up" wash-hand basins, 235, 243.
Trapless closets and smoke test, 296.
Treadle-action flushing apparatus for urinals, 253.
"Trigger" waste-valve for wash-hand basins, 234.
Trimming soldered angles of cisterns, 111.
Trimming soldered seams, 37.
Troughs for tramps, 244.
Trough urinals, 248, 249.
Trough water-closets, 301.
Tube-boiler, 309.
Tubs jam plug in a sink, 120.
Tunnels under houses for drains to pass through, 150.
T-unions for branch pipes, 326.
Turning lead over edge of cistern, 110.
Type-metal, 18.

U

UNDER-side of drain joints left open, 165.
Unions on bath pipes, 211.
Unnecessary number of traps, 195.
Unprincipled builders, 264.
Unsuitable positions for water-closets, 289, 290.
Untrapped waste pipes, 196.
Unventilated bath-rooms, 217.
Upholsterers' hangings in bath-rooms, 218.
Upright v. underhand joints, 76.
Upright joints, how to **catch solder**, 73.
Urinals, 246 to 259.
Urinal basins, 251 to 253.
Urinals enclosed, 256, 257.
Urinal flushing-cisterns, 258, 259.
Urinals in a club-house, 255.
Urinals in a public institution, 254.
Urinal stalls, 254.
Urinals to fold up, 257.
Urinal under a wash-hand basin, 257.
Urinals want attention, 257, 258.
Urinettes, 258.
Usage of a "Field's" flushing-tank, 180.

INDEX. 355

Usage of upper water-closets affects others below, 278 to 280.
Use of cardwire, 70.
Use of drain-traps, 138.
Use of laminated lead, 26.
Use of geometry, 60.
Uses of lead, 26.

V

VALVE and regulator-supply to water-closets, 293.
Valve-closet with trap above floor, 296.
Valves for baths, 209, 210.
Valves for wash-hand basins, 227.
Valveless water-closets, 298.
Varieties of materials for sinks, 117.
Various kinds of baths, 208.
Various ways for ventilating slop-sink traps, 205.
Vegetable washers, 121.
Ventilating drain in a terrace house, 175.
Ventilating waste pipes, 198.
Ventilation of bath-rooms, 217.
Ventilation of drains, 166 to 168.
Ventilation of grease-traps, 188.
Ventilation of sewers affect house plumbing, 123.
Ventilation of sewers not plumbers' work, 123.
Ventilation of urinals, 246.
Ventilation of water-closets, 304.
Vent-arm to valve-box of valve water-closet, 294.
Vent pipes in party-walls, 123.
Vent pipe to prevent "air-binding," 191.
Vent pipe with gas burning inside, 170.
Vent for slop-sink traps, 205.
Vent-shafts in water-closets, 182.
Vent to hot-water circulating pipes, 330.
Vitriol chambers lined with lead, 26.
V-pieces cut out to form bend, 56.

W

Wall-hooks for pipes, 95.
Walls of urinals, 246.
Warm nails used for cistern angles, 115.
Want of practice in joint wiping, 86.
"Wash-down" water-closet, 299.
Wash-hand basins, 224 to 241.
Wash-hand basins fixed on brackets, 242.
Wash-hand basins in public schools, 243.
Wash-hand basins should not be in a bedroom, 236.
Wash-hand basin traps, 237, 238, 240.
"Wash-out" water-closets, 298.
Waste apparatus to set of urinal basins, 252.

Wash-up and slop-sink, 206.
Waste from safe into trap of water-closet, 272.
Waste heat from hot-water pipes used for warming purposes, 328.
Waste-holes of wash-hand basins, 229.
Waste-holes in urinal basins, 251.
Waste of solder, 73.
Waste pipes air-bound, 275.
Waste pipes discharging into an open channel, 196.
Waste pipes discharging into hopper heads, 231.
Waste pipes from water-closet safes, 273.
Waste pipes from wash-hand basins, 241, 242.
Waste pipes should discharge into interceptor-traps, 197.
Waste unions, joints to, 93.
Waste valves for baths, 212, 213.
Waste valves for wash-hand basins, 234.
Water-backs, 305.
Water bends, 54.
Water-closets, 291.
Water-closets and sanitary engineers, 294.
Water-closets and slops, 200.
Water-closets and their smells, 295.
Water-closet arrangements and architects, 289.
Water-closets difficult to ventilate, 182.
Water-closet doors, back spring to, 304.
Water-closet enclosures, 304.
Water-closet floors, 300.
Water-closet for public schools and institutions, 301.
Water-closet, how to test the flush, 301.
Water-closets in an hotel, 290.
Water-closets in bedrooms, 304.
Water-closets in one piece of earthenware, 298, 299.
Water-closet, long hopper, 301.
Water-closet makers, 294.
Water-closet not an ejector, 304.
Water-closet overflow pipe, 295.
Water-closet, "Pan," 298.
Water-closet ranges, 303.
Water-closet safe waste pipes, 272, 273.
Water-closet seats screwed down, 304.
Water-closet seat fixed to prevent standing upon, 303.
Water closets, second-hand, 264.
Water-closets should be ventilated, 304.
Water-closet traps, 282.
Water-closets, trough, 301.
Water-closet useless without water to flush it, 281, 304.
Water-closet, "washdown," 299.
Water-closet, "washout," 298.
Water-closets without enclosures, 299.
Water-closets without traps, 296.

AA 2

Water-closets valveless, 298.
Water coloured with iron rust, 314.
Water companies and basin-cocks, 229.
Water companies and bath-valves, 215.
Water discharges through drain-syphons, 140.
Water fall into traps, 142.
Water-flushed drains, 166.
Water from bath should flush drains, 209.
Water impregnated with gases becomes sewage, 124.
Water in areas passes through floor vent gratings, 194.
Water in basin of water-closet, 297.
Water in soil allowed to pass into sewage drains, 165.
Water-pressure test for iron pipes, 149.
Water-testing of iron drains, 152.
Water-traps and disease germs, 138.
Waste-bath, 221.
Way pipes should socket, 69.
"Weathering" of sink capping, 119.
Weight of cast sheet-lead, 24.
Weight of iron drain pipes, 149.
Weight of lead, how to calculate, 26.
Weight of lead used in various positions, 26.
Weight of lead used for cisterns, 108.
Weight of sheet-lead, 21.
Weight on bottom of wash-hand basin, 93.
Welded boilers, 310.
Welted branch joints, 85.
Welted joint, 94.
Welted joints on soil pipe, 81.
What to avoid when shaving a joint, 71.
Where lead is found, 17.
Where to attach cold supply to boiler, 323.

Which way to socket an elbow-joint, 59.
White lead, 20.
White lead, how made, 20.
White lead-making injurious to workers, 21.
White yarn for water mains, 147.
Why joints are overcast, 77.
Why pipes split, 32.
Wide recess for pipes, 101.
Width of seams for hand-made pipes, 35.
Wilkinson's patent for making lead pipes. 29.
Winch and bobbins strung on a rope, 46.
Wine bottles in manhole, 176.
Wiped tacks on pipe, 98.
Wiped v. copper-bit joints, 91.
Wiped seams to water-closet traps, 283.
Wiping branch joints, 85.
Wiping cloths, 76.
Wiping seams on pipes, 37.
Wiping upright joints, 90.
Wire crossbars in sink-washers, 120.
Wire gratings on air-inlets, 173.
Wood blocks for joints, 78.
Wood blocks inside branch joint, 88.
Wood fillets for fixing pipes on, 325.
Wood pattern for astragals, 106.
Wood shavings and bends, 46.
Wood shavings to catch solder, 72.
Woodwork near hot-water pipes, 335.
Wooden carriage for wash-hand basins, 224.
Wooden cases for sinks should have sloping sides, 118.
Wooden enclosure for shower-bath, 221.
Wooden fillet for fixing pipes, 95.
Wooden lath for taking dimensions of cistern, 113.
Wood v. iron splash-sticks, 76.

"THE PECKHAM"
(No. 1.)

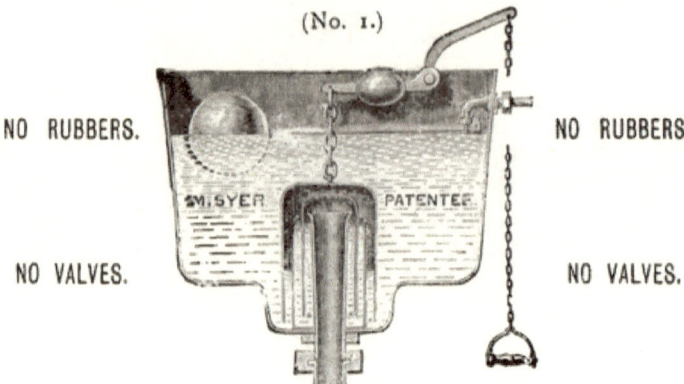

NO RUBBERS. NO RUBBERS.

NO VALVES. NO VALVES.

SYPHON CISTERN.

The above Cistern is extremely simple in its action, and has great flushing power. It flushes well at a low head of water. The flush is certain whether lever be held or not. There are NO VALVES, excepting ball valve, to get out of order, THUS DISPENSING with RUBBERS, WASHERS, LEATHERS, &c. Water cannot be made to run by holding down the lever. Flushing power increased towards the last. Specially adapted for Flush-out Closets: gives good two-gallon flush. Silent in action.

PRICES.

Cast Iron Cistern - - 16s.	Wrought Galvanized Iron - 13s.
Galvanized Cast Iron - - 21s.	

NOTICE.—IN THE SUPREME COURT OF JUDICATURE COURT OF APPEAL. HUMPHERSON v. SYER.—LORD JUSTICE COTTON, in giving judgment allowing the appeal of the Defendant from the judgment of Mr. JUSTICE KEKEWICH, said:—"That not only is the Defendant the Grantee of a previous Patent for the invention of the Double Cap, but that that was openly shown in his shop previous to the time when the Plaintiff took out his patent. Therefore, what the Plaintiff has done is this, he has taken the machine which was protected by the prior 'Letters Patent' of the Defendant, and which was exhibited publicly by the Defendant before the date of the Plaintiff's 'Letters Patent.' In my opinion, therefore, the Plaintiff's Patent fails—IT IS A BAD ONE, and this action ought to be dismissed." LORDS JUSTICES BOWEN and FRY concurred, and the appeal of the Defendant was accordingly allowed, with costs of the appeal, and in the Court below.

IMPORTANT NOTICE.—In consequence of the above decision, legal proceedings will be taken against anyone infringing Defendant's Patent or dealing with the same; and the public are warned against purchasing Syphon Cisterns of this pattern unless they bear the name of Milton Syer, Patentee, 36, Rye Lane, Peckham, London, S.E., and his Trade Mark, "The Peckham."

J. H. MOGGRIDGE, *Solicitor for the Defendant,*
4, Furnival's Inn, E.C.

Sole Maker—**MILTON SYER, 36, Rye Lane, Peckham, S.E.**

The Engineering and Building Record

AND

THE SANITARY ENGINEER.

ESTABLISHED 1877.

Conducted by HENRY C. MEYER.

DEVOTED TO

ENGINEERING & ARCHITECTURE.

OF SPECIAL INTEREST TO

Engineers, Architects, Builders, Contractors, Mechanics, and Municipal Officers.

The Treatment of Municipal Problems a Prominent Feature.

"It has been of incalculable value to the general public, whose interest it has always served."—*Cincinnati Commercial.*

"It may be regarded as the representative paper devoted to Architecture and Engineering."—*Boston Herald.*

"It stands as a fine example of clean and able journalism."—*Railroad Gazette.*

"A paper whose excellence and independence merit continued prosperity."—*Railroad and Engineering Journal.*

"Congratulate it upon the enviable position it has attained."—*American Machinist.*

Under date of January 9th, 1888, General M. C. Meigs, formerly Quartermaster-General U. S. Army, and recently Architect of the new Pension Building at Washington, wrote as follows:

1239, VERMONT AVENUE, WASHINGTON, D.C.,

January 9th, 1888.

THE ENGINEERING AND BUILDING RECORD.

DEAR SIRS: I enclose cheque for $5.00, for which please send me "Steam-Heating Problems" and "Plumbing and House-Drainage Problems." I will be obliged, also, for a copy of your No. 6, Volume XVII., January 7th, 1888, which is a capital number, just read and sent to a Western engineer, a friend, containing much in his line of work.

I have looked at the Index of Volume XVI. It is a marvellous list of knowledge made accessible to the profession at small cost to each subscriber.

I congratulate you upon producing for the Building trade one of the most copious and valuable instructors in safe and sanitary building science in all branches ever published.

Faithfully yours, M. C. MEIGS.

Pub'ished Saturdays at {82-84, Fulton Street, New York. {92 & 93, Fleet Street, London.} 20s. per year, 6d. per copy.

N.B.—An attractive feature is its series of critically selected Architectural Illustrations, artistically rendered and handsomely reproduced.

WINN'S PATENT
ACME SYPHON CISTERN.

Patented in
GREAT BRITAIN.

No. 1163.
PRICE—as drawn, 20s.
Galvanized, 28s.

Patented in
AMERICA.

No. 1163.
PRICE—as drawn, 20s.
Galvanized, 28s.

Approved and authorized for use by the Birmingham Water Works authorities, and by all the London and leading Provincial Companies. Also now largely used in Barracks, Workhouses, and Her Majesty's Prisons throughout the country, and adopted by the leading Railway Companies, while more than 29,000 have been sold to the trade.

This flushing-box is suitable for any flushing-rim basin, but especially for Winn's Improved "Free-flushing" Basin and Trap, which was designed to meet the want now generally felt for a good clean flushing pan and trap, constructed upon modern principles, at a moderate cost, the price being 11s. for cane and white, and 15s. for cane colour outside and ornamental flower ivy inside: a better quality, in superior porcelain, being 16s. white, and 20s. ornamental.

THE ARTISANS', LABOURERS', AND GENERAL DWELLINGS COMPANY, LIMITED,
34, Great George Street, London, S.W., *April 15th, 1889.*

Messrs. CHAS. WINN & CO., Birmingham.

DEAR SIRS,—We have about one thousand of your PATENT "ACME" CISTERNS in use at Noel Park. They work most satisfactorily, and, although somewhat exposed in several water-closets, have sustained no injury from the severe weather of the past winter.

Yours truly, (Signed) R. E. FARRANT,
Deputy-Chairman & Managing Director.

CHARLES WINN & CO., BIRMINGHAM,
Manufacturers of Sanitary Appliances.

LONDON OFFICE: 41, HOLBORN VIADUCT, E.C.

GLASGOW: 26, RENFIELD STREET.

THE "BOSTEL" SANITARY APPLIANCES
Are so Registered as a TRADE MARK.

They have stood the sure test of time, and are established as reliable.

The Brighton Closet is well known for its *simplicity* and extraordinary cleanliness.

None of the *Appliances—Closets, Cisterns, Urinal,* &c., are made for *Cheapness,* but for Quality of *Material* and *Workmanship,* and *Reliability*—these considered, they are amongst *the Cheapest.*

Can be obtained of any good Plumbers' Material Dealer, or Plumber, or of the Inventor and Proprietor—

D. T. BOSTEL, REGD. PLUMBER,

24, Charing Cross, London, S.W.

THE PRINCIPLES
OF
VENTILATION AND HEATING
AND
THEIR PRACTICAL APPLICATION

BY

JOHN S. BILLINGS, M.D., LL.D. (Edinb.),

Surgeon U.S. Army.

PROFUSELY ILLUSTRATED.

This interesting and valuable series of papers, originally published in THE SANITARY ENGINEER, have been re-arranged and re-written, with the addition of new matter.

The volume is published in response to the general demand that these important papers should be issued in a more convenient and permanent form, and also because almost all the reliable literature on this subject has been furnished by English Authors, and written with reference to the climate of England, which is more uniform and has a higher proportion of moisture. The need of a book based upon the conditions of the American climate is therefore apparent.

The following will indicate the character of the subject-matter:

Expense of Ventilation—Difference Between "Perfect" and Ordinary Ventilation — Relations of Carbonic Acid to the Subject—Methods of Testing Ventilation.

Heat, and some of the Laws which govern its Production and Communication—Movements of Heated Air—Movements of Air in Flues—Shapes and Sizes of Flues and Chimneys.

Amount of Air-Supply Required—Cubic Space.

Methods of Heating: Stoves, Furnaces, Fire-Places, Steam and Hot-Water.

Scheduling for Ventilation Plans—Position of Flues and Registers—

Means of Removing Dust—Moisture, and Plans for Supplying It.

Patent Systems for Ventilation and Heating—The Ruttan System—Fire-Places—Stoves.

Chimney-Caps—Ventilators — Cowls—Syphons—Forms of Inlets.

Ventilation of Halls of Audience—Fifth Avenue Presbyterian Church—The Houses of Parliament—The Hall of the House of Representatives.

Theatres—The Grand Opera-House at Vienna — The Opera-House at Frankfort-on-the-Main — The Metropolitan Opera-House, New York—The Madison Square Theatre, New York—The Criterion Theatre, London—The Academy of Music, Baltimore.

Schools.

Ventilation of Hospitals—St. Petersburgh Hospital—Hospitals for Contagious Diseases—The Barnes Hospital —The New York Hospital—The Johns Hopkins Hospital.

Forced Ventilation — Aspirating Shafts — Gas-Jets — Steam Heat for Aspiration—Prof. Trowbridge's Formulæ — Application in the Library Building of Columbia College—Ventilating-Fans—Mixing-Valves.

The book is free from unnecessary technicalities, and is not burdened with scientific formulæ.

It is invaluable to Architects, Physicians, Builders, Plumbers, and those who contemplate building or re-modelling their houses.

Large 8vo. Handsomely Bound in Cloth. Price 15s. Postage Paid.

Address: BOOK DEPARTMENT,

THE ENGINEERING AND BUILDING RECORD,

92 and 93, Fleet Street, London.

A. W. REID & CO.,

MAKERS OF

BATHS, LAVATORIES, WATER-CLOSETS, URINALS, PLUMBERS'
BRASS WORK, SANITARY EARTHENWARE, FIRECLAY SINKS,
PATENT FOLDING LAVATORIES FOR HOTELS, RAILWAY CARS,
STEAMSHIPS, &c.,

69, ST. MARY AXE, LONDON, E.C.

SOLE MANUFACTURERS OF

PEARSON'S TWIN-BASIN WATER-CLOSET

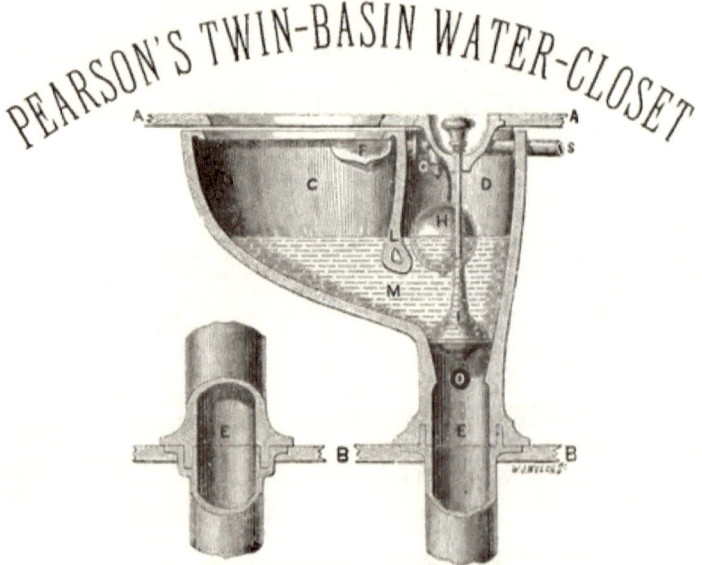

WE respectfully invite attention to **Pearson's Patent "Twin-Basin" Water-Closet**. It is simple in construction, is entirely without that complicated mechanism by which most other closets are worked, and consequently less liable to get out of order, the very **important features**, however, being **cleanliness** and **protection against** foul air from the **Sewer**.

THE closets have been thoroughly tested by many years' use in Hospitals, Schools, Railway Stations, Hotels, Factories, Warehouses, Mansions, and Cottages, with the very best results. They have been tried and approved by the highest Sanitary Authorities, by Architects and by the Medical Profession.

Below is an extract from one of the numerous Testimonials received.

"SURVEYOR'S OFFICE, GUY'S HOSPITAL, LONDON, 10th February, 1877.

"Taken altogether, I have seen nothing to equal the **Pearson's Closet**, or I should not have introduced them for use in the Hospital.

(Signed) "ARTHUR BILLING, *Surveyor to the Hospital.*"

For Diagrams and List of Prices, apply to the Manufacturers as above.

GOLD MEDAL AWARDED AT THE INTERNATIONAL HEALTH EXHIBITION, LONDON 1884.

SECOND EDITION.

Plumbing & House-Drainage Problems;
OR,
Questions, Answers, and Descriptions from THE SANITARY ENGINEER.

With 142 Illustrations.

[FROM THE PREFACE.]

"A feature of THE SANITARY ENGINEER is its replies to questions on topics that come within its scope, included in which are Water-Supply, Sewage Disposal, Ventilation, Heating, Lighting, House-Drainage, and Plumbing. Repeated inquiries concerning matters often explained in its columns suggested the desirability of putting in a convenient form for reference a selection from its pages of questions and comments on various problems met with in house-drainage and plumbing, improper work being illustrated and explained as well as correct methods. It is therefore hoped that this book will be useful to those interested in this branch of sanitary engineering."

TABLE OF CONTENTS:

DANGEROUS BLUNDERS IN PLUMBING.

Running Vent Pipe in Improper Places—Connecting Soil Pipes with Chimney-Flues—By-Passes in Trap-Ventilation, &c. *Illustrated.*
A Case of Reckless Botching. *Illustrated.*
A Stupid Multiplication of Traps. *Illustrated.*
Plumbing Blunders in a Gentleman's Country House. *Illustrated.*
A Trap Made Useless by Improper Adjustment of Inlet and Outlet Pipes. *Illustrated.*
Unreliability of Heated Flue as a Substitute for Proper Trapping. *Illustrated.*
Need of Plans in Doing Plumbing-Work.

HOUSE-DRAINAGE.

City and Country House-Drainage—Removal of Ground-Water from Houses—Trap-Ventilation—Fre h-Air Inlets—Drain-Ventilation by Heated Flues—Laying of Stoneware Drains.
Requirements for the Drainage of Every House.
Drainage of a Saratoga House. *Illustrated.*
Ground-Water Drainage of a Country House. *Illustrated.*
Ground-Water Drainage of a City House. *Illustrated.*
Fresh-Air Inlets.
The Location of Fresh-Air Inlets in Cities. *Illustrated.*
Fresh-Air Inlets. *Illustrated.*
Air-Inlets on Drains.
The Proper Way to Lay Stoneware Drains.
Risks Attending the Omission of Traps and Relying on Drain-Ventilation by Flues. *Illustrated.*
The Tightness of Tile-Drains.
Danger of Soil Pipe Terminals Freezing unless Ends are without Hoods or Cowls.
Objection to Connecting Bath-Waste with Water-Closet Trap.
How to Adjust the Inlets and Outlets of Traps. *Illustrated.*
How to Protect Trap when Soil Pipe is used as a Leader.
Size of Ventilating Pipes for Traps.
How to Prevent Condensation Filling Vent Pipes.
Ventilating Soil Pipes.
How to Prevent Accidental Discharge into Trap Vent Pipe.
Why Traps should be Vented.

MISCELLANEOUS.

Syphoning Water through a Bath-Supply. *Illustrated.*
Emptying a Trap by Capillary Attraction. *Illustrated.*
As to Safety of Stop-Cocks on Hot Water Pipes.
How to Burnish Wiped Joints.
Admission to the New York Trade Schools.
Irregular Water Supply. *Illustrated.*
Hot Water from the Cold Faucet, and how to Prevent it. *Illustrated.*
Disposal of Bath and Basin Waste Water.
To Prevent Corrosion of Tank Lining.
Number of Water-Closets Required in a Factory.
Size of Basin Wastes and Outlets.
Tar-Coated Water Pipe Affect Taste of Water.
How to Deal with Pollution of Cellar Floors.
How to Heat a Bathing Pool.
Objections to Galvanized Sheet-Iron Soil Pipe.
To Prevent Rust in a Suction Pipe.
Automatic Shut-Off for Gas Pumping Engines when Tank is Full. *Illustrated.*
Paint to Protect Tank Linings.
Vacuum Valves not always Reliable.
Size of Water Pipes in a House.
How to Make Rust Joints.
Covering for Water Pipes.
Size of Soil Pipe for an ordinary City House.
How to construct a Sunken Reservoir to Hold Two Thousand Gallons.
Where to Place Burners to Ventilate Flues by Gas Jets. *Illustrated.*
How to Prevent Water Hammer.
Why a Hydraulic Ram does not Work.
Air in Water Pipes.
Proper Size of Water-Closet Outlets.
Is a Cement Floor Impervious to Air?
Two Traps to a Water-Closet Objectionable.
Connecting Bath Wastes to Water-Closet Traps. *Illustrated.*
Objections to Leaching Cesspool and need of Fresh-Air Inlet.
The Theory of the Action of Field's Syphon.
How to Disinfect a Cesspool.
Drainage into Cesspools.
Slabs for Pantry Sinks—Wood v. Marble.
Test for Well Pollution.
Cesspool for Privy Vault.
Corrosion of Lead Lining.

Size of Flush Tank to deal with Sewage of a Small Hospital.
Details of the Construction of a House Tank. *Illustrated*
The Construction of a Cistern under a House.
To protect Lead Lining of a Tank, and Cause of Sweating.
Stains on Marble.
Lightning Strikes Soil Pipes.
Will the Contents of a Cesspool Freeze?
Bad Tasting Water from a Coil. *Illustrated*.
How to fit Sheet-Lead in a Large Tank.
Why Water is "Milky" When First Drawn.
Material for Water Service Pipes.
Carving Tables. *Illustrated*.
Is Galvanized Pipe Dangerous for Soft Spring Water.
How to Arrange Hush Pipes in Cisterns to Prevent Syphoning Water Through Ball Cock.
Depth of Foundations to Prevent Dampness of Site.
Where to Place a Tank to get Good Discharge at Faucet.
Self-Acting Water-Closets. *Illustrated*.
Wind Disturbing Seal of Trap.
How to Draw Water from a Deep Well.
Cause of Smell of Well Water.
Absorption of Light by Gas Globes.
Defective Drainage. *Illustrated*.
Fitting Basins to Marble Slabs. *Illustrated*.
Intermediate Tanks for the Water Supply of High Buildings. *Illustrated*.
How to Construct a Filtering Cistern. *Illustrated*.
Objections to Running Ventilating Pipe into Chimney-Flue.
Size of Water Supply Pipe for Dwelling-House.
Faulty Plan of a Cesspool. *Illustrated*.
Connecting Refrigerator Wastes with Drains. *Illustrated*.
Disposing of Refrigerator Wastes. *Illustrated*.
Pumping Air from Water-Closet into Tea Kettle as Result of Direct Supply to Water-Closets. *Illustrated*.
Danger in Connecting Tank Overflows with Soil Pipes.
Arrangement of Safe Wastes. *Illustrated*.
The Kind of Men Who do not Like the Sanitary Engineer.
What is Reasonable Plumbers' Profit.

HOT-WATER CIRCULATION IN BUILDINGS.

Bath Boilers. *Illustrated*.
Setting Horizontal Boilers. *Illustrated*.
How to Secure Circulation Between Boilers in Different Houses. *Illustrated*.
Connecting One Boiler with Two Ranges. *Illustrated*.
Taking Return Below Boiler. *Illustrated*.
Trouble with Boiler.
An Ignorant Way of Dealing with a Kitchen Boiler. *Illustrated*.
Returning into Hot Water Supply Pipe. *Illustrated*.
Where should Sediment Pipe from Boiler be connected with Waste Pipe?
Several Flow Pipes and one Circulation Pipe. *Illustrated*.
How to run Pipes from Water Back to Boiler. *Illustrated*.
Hot-Water Circulation when Pipes from Boiler pass under the Floor. *Illustrated*.
Heating a Room from Water Back.
The Operation of Vacuum and Safety Valves. *Illustrated*.
Preventing Collapse of Boilers.
Collapse of a Boiler. *Illustrated*.
Explosion of Water Backs.
A Proposed Precaution against Water Back Explosions. *Illustrated*.
The Bursting of Kitchen Boilers and Connecting Pipes. *Illustrated*.
Giving out of Lead Vent Pipes from Boilers in an Apartment House. *Illustrated*.
Connecting a Kitchen Boiler with One or more Water Backs. *Illustrated*.
New Method of Heating Two Boilers by One Water Back. *Illustrated*.
Plan of Horizontal Hot Water Boiler. *Illustrated*.

HOT WATER SUPPLY IN VARIOUS BUILDINGS.

Kitchen and Hot-Water Supply in the Residence of Mr. W. K. Vanderbilt, New York. *Illustrated*.
Kitchen and Hot-Water Supply in the Residence of Mr. Cornelius Vanderbilt, New York. *Illustrated*.
Kitchen and Hot-Water Supply in the Residence of Mr. Henry G. Marquand, New York. *Illustrated*.
Kitchen and Hot-Water Supply in the Residence of Mr. A. J. White. *Illustrated*.
Hot-Water Supply in an Office Building. *Illustrated*.
Kitchen and Hot-Water Supply in the Residence of Mr. Sidney Webster. *Illustrated*.
Plumbing and Water Supply in the Residence of Mr. H. H. Cook. *Illustrated*.

Large 8vo. Cloth, 10s. post paid.

Address, BOOK DEPARTMENT,

THE ENGINEERING AND BUILDING RECORD,

92 and 93, Fleet Street, London.

WATER-WASTE PREVENTION:

Its Importance and the Evils Due to Its Neglect.

With an Account of the Methods Adopted in Various Cities in Great Britain and the United States.

By HENRY C. MEYER, Editor of THE ENGINEERING AND BUILDING RECORD.

With an Appendix.

EXTRACT FROM PREFACE.

During the summer of 1882 the Editor of THE ENGINEERING AND BUILDING RECORD carefully investigated the methods employed in various cities in Great Britain for curtailing the waste of water without subjecting the respective communities to either inconvenience or a limited allowance. The results of this investigation appeared in a series of articles entitled "New York's Water-Supply," the purpose being to present to the readers of THE ENGINEERING AND BUILDING RECORD such facts as would stimulate public sentiment in support of the enforcement of measures tending to prevent the excessive waste of water so prevalent in American cities, and especially the City of New York, which was then suffering from a short supply. Numerous requests for information, together with the recent popular agitation in connection with a proposition to increase the powers of the Water Department of New York City with a view to enabling it to restrict the waste of water, have suggested the desirability of reprinting these articles in a more convenient and accessible form, with data giving the results of efforts in this direction in American cities since the articles first appeared, so far as they have come to the author's notice.

TABLE OF CONTENTS:

CHAPTER I.—CONDITION OF NEW YORK'S WATER-SUPPLY.—Mr. Thomas Hawksley on Advantages of Waste-Prevention—Condition of Water-Supply in England Thirty Years Ago—Means Adopted to Prevent Waste in Great Britain—Norwich the First City in England to Adopt Measures of Prevention—London: the Practice There.

CHAPTER II.—GLASGOW.—District Meters Tried as an Experiment—Results of Experiments—Prevalence of Defective Fittings—Testing and Stamping of Fittings—Rules Governing Plumbers' Work.

CHAPTER III.—MANCHESTER.—History of Waste-Prevention Measures—Methods of House-to-House Inspection—Duties of Inspectors—Methods of Testing and Stamping Fittings.

CHAPTER IV.—LIVERPOOL.—Change from Intermittent to Constant Supply—Method of Ascertaining Locality of Waste by Use of District Meters—Method of House Inspection—Method of Testing Fittings.

CHAPTER V.—PROVIDENCE AND CINCINNATI.—Review of Measures to Prevent Water-Waste in the Unites States prior to 1882—Providence, R. I.: Results following the General Use of Meters—Cincinnati: Methods of House Inspection with the Aid of the Waterphone—Results Attained.

CHAPTER VI.—NEW YORK.—Measures Adopted by the Department of Public Works prior to 1882.

CHAPTER VII.—GENERAL CONCLUSIONS.—Points to be Considered in Adopting Measures for Large Cities.

APPENDIX.—POINTS SUGGESTED IN THE CONSIDERATION OF VARIOUS METHODS.—Water-Waste Prevention in Boston in 1883 and 1884—Results Attained—Waste-Prevention in New York City—Liverpool Corporation Water-Works Regulations—Glasgow Corporation Water-Works Regulations—Description of Standard Fittings—Penalties for Violations—Cisterns v. Valve Supply to Water-Closets in New York City—New York Board of Health Regulations Concerning Water-Supply to Water-Closets—Letters from Water-Works Authorities Sustaining the Action of the New York Board of Health in Requiring Cistern-Supply to Water-Closets—Extracts from Report of Boston City Engineer on Wasteful Water-Closets—Proposed Water-Rates on Water-Closets in New York—Resolutions of the New York Board of Health Endorsing the Proposed Water-Rates for Water-Closets—Excerpts from Articles Explaining Methods of Arranging Water-Supply to Water-Closets to Secure the Minimum Water-Rate in New York (with Illustrations.)

Large 8vo. Bound in Cloth, 5s.

Sent (post paid) on receipt of price. Address,

BOOK DEPARTMENT,

THE ENGINEERING AND BUILDING RECORD,

92 and 93, Fleet Street, London, E.C.

JOHN SMEATON, SON & Co.
PLUMBERS,
SANITARY, HYDRAULIC, VENTILATING, AND HOT-WATER ENGINEERS,

56, Great Queen Street, Lincoln's Inn Fields,

CATALOGUE FREE.   LONDON, W.C.   CATALOGUE FREE.

DRAINAGE, WATER SUPPLY & SANITARY PLUMBING
ON THE
MOST IMPROVED SYSTEMS,
RECEIVE SPECIAL ATTENTION.

The "Nautilus" Closet.

These Closets are supplied with handsomely decorated brass cast or polished wood patent syphon risterns, lead down service pipes, and hinged seats to enable them to be used as urinals and slop sinks. They can be had in colours at a slight extra cost. The "Nautilus" is strongly recommended as thus the ordinary unsightly wood casing is entirely dispensed with; it is specially adapted for W.C. rooms of small dimensions, where space is an object. Of the two types of the "Nautilus" we strongly recommend the "Rapid Flush" as the most sanitary.

PRICES.
From   £3 3s.
To .. ..   £6 6s.

JUST OUT

SOME DETAILS
OF
WATER-WORKS CONSTRUCTION.
By W. R. BILLINGS,
Superintendent of Water-works, at Taunton, Mass.

WITH

Illustrations from Sketches by the Author.

INTRODUCTORY NOTE.

Some questions addressed to the Editor of THE ENGINEERING AND BUILDING RECORD AND THE SANITARY ENGINEER by persons in the employ of new water-works indicated that a short series of practical articles on the Details of Constructing a Water-Works Plant would be of value; and, at the suggestion of the Editor, the preparation of these papers was undertaken for the columns of that journal. The task has been an easy and agreeable one, and now, in a more convenient form than is afforded by the columns of the paper, these notes of actual experience are offered to the water-works fraternity, with the belief that they may be of assistance to beginners and of some interest to all.

TABLE OF CONTENTS.

CHAPTER I.—MAIN PIPES.—Materials—Cast-Iron—Cement-Lined Wrought Iron—Salt-Glazed Clay—Thickness of Sheet-Metal—Methods of Lining—List of Tools—Tool-Box—Derrick—Calking Tools—Furnace—Transportation—Handling Pipe—Cost of Carting—Distributing Pipe.

CHAPTER II.—FIELD WORK.—Engineering or None—Pipe Plans—Special Pipe—Laying out a Line—Width and Depth of Trench—Time-Keeping Book—Disposition of Dirt—Tunnelling—Street Piling.

CHAPTER III.—TRENCHING AND PIPE-LAYING—Caving—Tunnelling—Bell-Holes—Stony Trenches—Feathers and Wedges—Blasting—Rocks and Water—Laying Cast-Iron Pipe—Derrick Gang—Handling the Derrick—Skids—Obstructions Left in Pipes—Laying Pipe in Quicksand—Cutting Pipe.

CHAPTER IV.—PIPE-LAYING AND JOINT-MAKING.—Laying Cement-Lined Pipe—"Mud" Bell and Spigot—Yarn—Lead—Jointers—Roll-Calking—Strength of Joints—Quantity of Lead.

CHAPTER V.—HYDRANTS, GATES, AND SPECIALS.

CHAPTER VI.—SERVICE PIPES.—Definition—Materials—Lead v. Wrought Iron—Tapping Mains for Services—Different Joints—Compression Union—Cup.

CHAPTER VII.—SERVICE PIPES AND METERS—Wiped Joints and Cup-Joints—The Lawrence Air-Pump—Wire-drawn Solder—Weight of Lead Service Pipe—Tapping Wrought-Iron Mains — Service Boxes—Meters.

Handsomely Bound in Cloth.

Sent (post paid) on receipt of 10s.   Address,

BOOK DEPARTMENT,

THE ENGINEERING AND BUILDING RECORD,

92 and 93, Fleet Street, London, E.C.

A WARM BATH IN 10 MINUTES.
Cost of Gas, 1d.

The New Patent "Calda."
Gas no contact with Water.

GAS BATHS from £4 12s. 6d.
PATENT "RELIANCE" Gas Conservatory Boilers.
FIXED INSIDE WITH PERFECT SAFETY.

G. SHREWSBURY,
122, Newgate Street, London, E.C.

BASEMENTS, AREAS, COAL CELLARS, AND ALL UNDERGROUND PASSAGES AND WORKS, *Lighted and Ventilated most effectively.*

WALSH AND SON'S
NEW PAVEMENT LIGHT (REGD.)
CHEAPEST IN THE MARKET.
*May be seen at our New Premises—*13, *PENNY STREET, BLACKBURN.*

PARTICULARS AND PRICES ON APPLICATION.

NOTICE.

This space is reserved for the announcement of

H. P. SKIDMORE,

OF

ATLAS TUBE WORKS, NETHERTON, NEAR **DUDLEY,**

Manufacturer of

WROUGHT-IRON TUBES for GAS, STEAM, WATER, &c.

PRICE, 15s. POSTAGE PAID.

Steam-Heating Problems;

OR,

Questions, Answers, and Descriptions

RELATING TO

STEAM-HEATING AND STEAM-FITTING,

FROM

THE SANITARY ENGINEER.

With One Hundred and Nine Illustrations.

PREFACE.

THE SANITARY ENGINEER, while devoted to Engineering, Architecture, Construction, and Sanitation, has always made a special feature of its departments of Steam and Hot-Water Heating, in which a great variety of questions have been answered and descriptions of the work in various buildings have been given. The favour with which a recent publication from this office, entitled " Plumbing and House-Drainage Problems," has been received suggested the publication of " STEAM HEATING PROBLEMS," which, though dealing with another branch of industry, is similar in character. It consists of a selection from the pages of THE SANITARY ENGINEER of questions and answers, besides comments on various problems met with in the designing and construction of Steam-Heating apparatus, and descriptions of Steam-Heating work in notable buildings.

It is hoped that this book will prove useful to those who design, construct, and have the charge of Steam-Heating apparatus.

CONTENTS:

BOILERS.

On Blowing off and Filling Boilers.
Where a Test-Gauge should be Applied to a Boiler.
Domes on Boilers; whether they are necessary or not.
Expansion of Water in Boilers.
Cast *v.* Wrought Iron for Nozzles and Magazines of House-Heating Boilers.
Pipe-Connections to Boilers.
Passing Boiler-Pipes through Walls; how to Prevent Breakage by Settlement.
Suffocation of Workmen in Boilers.
Heating Boilers. (A Problem.)
A Detachable Boiler-Lug.
Isolating-Valve for Steam-Main of Boilers.
On the Effect of Oil in Boilers.
Iron Rivets and Steel Boiler-Plates.
Proportions for Rivets for Boiler-Plates.
Is there any Danger in using Water Continuously in Boilers?
Accident with Connected Boilers.
A Supposed Case of Charring Wood by Steam-Pipes.
Domestic Boilers Warmed by Steam.

VALUE OF HEATING-SURFACES.

Computing the Amount of Radiator-Surface for Warming Buildings by Hot Water.
Calculating the Radiating-Surface for Heating Buildings—The Saving of Double-Glazed Windows.
Amount of Heating-Surface Required in Hot-Water Apparatus Boilers and in Steam-Apparatus Boilers.
Calculating the Amount of Radiating-Surface for a Given Room.
How much Heating-Surface will a Steam-Pipe of Given Size Supply?
Coils *v.* Radiators and Size of Boiler to Heat a Given Building.
Calculating the Amount of Heating-Surface.
Computing the Cost of Steam for Warming.

RADIATORS AND HEATERS.

A Woman's Method of Regulating a Radiator (Covering it with a Cosey).
Improper Position of Radiator-Valves.
Hot-Water Radiator for Private Houses.
Remedying Air-Binding of Box-Coils.
How to Use a Stove as a Hot-Water Heater.
" Plane " *v.* " Plain " as a Term as Applied to Outside Surface of Radiators.
Relative Value of Pipe on Cast-Iron Heating Surface.
Relative Value of Pipe on Steam-Coils.
Warming Churches (Plan of placing a Coil in each Pew.)
Warming Churches.

PIPE AND FITTING.

Steam-Heating Work—Good and Indifferent.
Piping Adjacent Buildings: Pumps *v.* Steam-Traps.
True Diameters and Weights of **Standard Pipes.**
Expansion of Pipes of Various **Metals.**
Expansion of Steam Pipes.
Advantages Claimed for Overhead Piping.
Position **of Valves on** Steam-Riser Connection.
Cause of Noise in Steam Pipes.
One-Pipe **System** of Steam-Heating.
How to Heat Several Adjacent Buildings with a Single Apparatus.
Patents on Mill's System of Steam-Heating.
Air-Binding in Return Steam Pipes.
Air-Binding in Return Steam Pipes and Methods to overcome it.

VENTILATION.

Size of Registers to Heat Certain Rooms.
Determining the Size of Hot-Hair Flues.
Window Ventilation.
Removing Vapour from Dye-House.
Ventilation of Cunard Steamer "Umbria."
Calculating Sizes of Flues and Registers.
On Methods of Removing **Air** from Between Ceiling and **Roof of a** Church.

STEAM.

Economy **of using Exhaust Steam for** Heating.
Heat of Steam **for** Different Conditions.
Superheating Steam by the use of Coils.
Effect of Using a Small Pipe for Exhaust Steam-Heating.
Explosion of a Steam-Table.

CUTTING NIPPLES AND BENDING PIPES.

Cutting Large Nipples—Large **in Diameter and** Short in Length.
Cutting Crooked Threads.
Cutting a Close Nipple out of **a Coupling after** a Thread is Cut.
Bending Pipe.
Cutting Large Nipples.
Cutting Various Sizes of Thread **with a Solid** Die.

RAISING WATER AUTOMATICALLY.

Contrivance for Raising Water in High Buildings.
Criticism of the Foregoing and Description of Another Device for a Similar Purpose.

MOISTURE ON WALLS, &c.

Cause and Prevention of Moisture on Walls.
Effect of Moisture on Sensible Temperature.

MISCELLANEOUS.

Heating Water in Large Tanks.
Heating **Water for Large Institutions and** High City **Buildings.**
Questions Relating **to Water-Tanks.**
Faulty Elevator-Pump **Connections.**
On Heating Several Buildings **from one Source.**
Coal Tar Coating for Water Pipe.

Filters **for Feeding House-Boilers. Other** Means of **Clarifying Water.**
Testing **Gas Pipes for Leaks and** Making Pipe-Joints.
Will Boiling Drinking-Water Purify it?
Differential Rams for Testing Fittings **and** Valves.
Percentage of Ashes in Coal.
Automatic Pump-Governor.
Cast-Iron Safe for Steam-Radiators.
Methods **of** Graduating Radiator **Service** According to the Weather.
Preventing Fall of Spray from **Steam-Exhaust** Pipes.
Exhaust-Condenser for Preventing Fall **of** Spray from Steam-Exhaust Pipes.
Steam-Heating Apparatus and Plenum (Ventilation), System in Kalamazoo Insane Asylum.
Heating and Ventilation of a Prison.
Amount of Heat Due to **Condensation** of Water.
Expansion-Joints.
Re-setting of House-Heating **Boilers—a Possible** Saving of Fuel.
How to Find the Water**-Line of Boilers and** Position of Try-Cocks.
Low-Pressure Hot-Water **System for Heating** Buildings in England **(Comments by** *The Sanitary Engineer*).
Steam-Heating Apparatus in Manhattan Company's and Merchants' Bank Building, New York.
Boilers in Manhattan Company's and Merchants' Bank Building, with Extracts from Specifications.
Steam-Heating Apparatus in Mutual **Life** Insurance Building on Broadway.
The Setting of Boilers in Tribune Building, New York.
Warming and Ventilation of West Presbyterian Church, New York City.
Principles of Heating-Apparatus, **Fine Arts** Exhibition Building, Copenhagen.
Warming and Ventilation of Opera House at Ogdensburg, N. Y.
Systems of Heating Houses **in Germany and** Austria.
Steam Pipes under New York Streets—Difference Between Two Systems Adopted.
Some Details of Steam and Ventilating Apparatus used on the Continent of Europe.

MISCELLANEOUS QUESTIONS.

Applying Traps to Gravity Steam-Apparatus.
Expansion of Brass and Iron Pipe.
Connecting Steam and Return Risers at their Tops.
Power Used in Running Hydraulic Elevators.
On Melting Snow in the Streets by Steam.
Action of Ashes Street Fillings on Iron Pipes.
Arrangement of Steam-Coils for Heating Oil-Stills.
Converting **a** Steam-Apparatus into a Hot-Water Apparatus and Back Again.
Condensation Per Foot of Steam-Main when laid Under Ground.
Oil in Boilers from Exhaust Steam, and **Methods** of Prevention.

Address, BOOK DEPARTMENT,

THE ENGINEERING AND BUILDING RECORD.

92 and 93, Fleet Street, London.

THE FIFTEENTH VOLUME

OF

The Engineering and Building Record

AND

THE SANITARY ENGINEER.

(December 4th, 1886—May 28th, 1887.)

Aside from the weekly record of events of special interest to Engineers, Architects, Municipal Officers, Mechanics, and Contractors, the following of the numerous special articles are mentioned as of permanent interest to Municipal Engineers and Water-Works Superintendents.

ENGINEERING.

The Series on Builders' and Contractors' Engineering and Plant, which are illustrated Articles in Detail, of the Construction of the Equitable Building and St. Patrick's Cathedral in New York City; of the Raising of the old U. S. Court House in Boston; of the Dredging Scows and Machinery used on several Government Works; of the Hoisting and other Machinery used on the Elevated Railroad in Brooklyn of the Machinery used in the Construction of the Suburban Elevated Railroad, of New York.

Building Construction and Details, describing the Practice in the Eastern and Western Parts of the United States and of Europe; of interest both to the Engineer and Architect.

The Engineering at the Lawrenceville School. Including description in detail, with Illustrations of the Drainage, Sewerage, Water-Supply, Heating and Ventilation and Plumbing of the Work.

Recent Water-Works Construction in the United States. A Series of Illustrated Articles Descriptive of Works now Building or just Completed.

The New Croton Aqueduct for New York City is Described as it Progresses in Articles of Great Value to the Engineer

There are many Descriptions, Discussions, and Notes of Interesting Water-Works, Undertakings in America and Europe. These are generally Illustrated, and, with the Reviews of Reports of Water-Works Officers, make up a Valuable History of Current Undertakings.

Modern Sewer Construction and Sewage Disposal. A Series of Papers by Edward S. Philbrick, Mem. Am. Soc. C. E., on the Modern Theory and Practice of Sewer Work.

Recent Sewer Construction Contains a Number of Illustrated Articles descriptive of the most important Sewer Work, such as that at Newark, N.J., now in progress.

Pavements and Street Railroads is a series of Papers on the Construction and Maintenance of Roadways.

In the Natural Gas-Supply of Pittsburg and Vicinity is given a very fully Illustrated Account of the Mechanical Means used in Applying Natural-Gas to Manufacturing and Domestic Purposes.

In Addition to the Serials, there are many Articles on General Engineering, Water-Works, Sewerage, Pavements, and other Topics of Interest to Engineers, Contractors, and Builders.

DOMESTIC ENGINEERING.

This Department relates more particularly to topics connected with the welfare of the individual. Under it there are, in this volume Descriptions of the Heating by Steam and Hot Water of notable buildings in the United States and Canada, such as the Buildings for the State, War, and Navy Departments at Washington, the High School at Honesdale, Pa., the Post-Office at Woodstock, N.B., and others.

Descriptions of Plumbing.

A Series of Articles on the Theory and Practice of Hot-Water Heating, by "Thermus."

A Discussion of the Practicability of Heating Railway Cars by means which will not incur the risk of burning passengers in case of collision, with descriptions of several new Systems.

Editorials and Notes on the Preservation of Health by Purity of Water-Supply, Proper Sewerage, and Similar Means.

Reviews of Reports of Boards of Health, and Books on Sanitary Topics.

CONTRACTING INTELLIGENCE.

This Department is a very complete Record, week by week, of projected Works in Water-Supply, Sewerage, Gas, Railroad Construction, &c., of great value to the Engineer, Contractor, Builder, and Merchant. Returns of projected Buildings are also made by Special Correspondents from all parts of the United States for each issue.

The Proposals give the earliest information of projected works from the Government Departments, Municipal Bureaus, and Private Undertakings.

Bound in Cloth with Index, 15s. Postage Paid. THE ENGINEERING AND BUILDING RECORD, 92 and 93, Fleet Street, London.

THE SIXTEENTH VOLUME
OF
The Engineering and Building Record
AND
THE SANITARY ENGINEER.

(June 4, 1887—November 26, 1887.)

Aside from the weekly record of events of special interest to Engineers, Municipal Officers, Mechanics, and Contractors, the following of the numerous special articles are mentioned as of permanent interest to Municipal Engineers and Water-Works Superintendents:

ENGINEERING.

Location of Plant at Shafts on New Croton Aqueduct. (*Two Illustrations*.)
Recent Water-Works Construction—East Orange and Bloomfield, N.J., Water Companies. (*Three Illustrations*.) Water-Works at Ware, Mass. (*Four Illustrations*.) Water-Works at Calais, Me. (*Three Illustrations*.)
Pavements and Street Railroads—Continuation of this series, in which the question of wood pavements in London is fully discussed.
New Croton Aqueduct. No. XIII. Disc. for Measuring Cross-section in Tunnel. (*Nine Illustrations*.)
Tipple for Dumping Cars on the New Croton Aqueduct. (*Six Illustrations*.)
Modern Sewage Disposal and Engineering. By E. S. Philbrick, M. Am. Soc. C. E. (*Two Illustrations*.)
Sweetwater Dam and Irrigation Experience in Southern California. (*One Illustration*.)
Repair and Maintenance of Roads. By W. H. Wheeler, C.E.
Report of the Disposal of Sewage in the City of Worcester, Mass.
Receiving and Catch Basins at Waterbury, Conn. (*Four Illustrations*.)
Testing of Portland Cement for the Harbour Works at Calais and Boulogne. By F. Guillain.
Carrying Water-Mains Across the River at Ekhart, Ind. (*Two Illustrations*.), and at Grand Rapids, Mich. (*Three Illustrations*.)
Filtration or Subsidence. By J. D. Cook, C.E.
Special Report of the Chicago Drainage and Water-Supply Commission.
Driven-Well System as a Source of or Means of Obtaining a Water-Supply.
Recent Sewer Construction. Chiswick Sewage Works. (*Three Illustrations*.)
Burial of Sewage and Refuse. (Criticism on an Address by Dr. G. V. Poore, of London.
The Molteno Reservoir at Cape Town, Africa. Some Details of Water-Works Construction. By William R. Billings, C.E. (*Four Articles of this Series, with illustrations, have appeared*.)
Accident on the New Croton Aqueduct—Collapse of Bulkhead. (*Four Illustrations*.)
New Water-Works Tunnel, Chicago—Abstract of Specifications. (*One Illustration*.)
Description of Water-Tower at Franklin, Mass. (*Four Illustrations*.)
Wreck of Seneca Falls Stand-Pipe. (Description and *Four Illustrations*.)
Bursting of Little Falls Reservoir. (Description.)
Remarkable Meeting of Headings on the New Croton Aqueduct.

Six Years' Experience with Memphis Sewers. Special Report to THE ENGINEERING AND BUILDING RECORD, by Rudolph Hering, with Editorial Comment.

DOMESTIC ENGINEERING.

(This Department is of Special Interest to Water-Works Superintendents and Plumbers.)
Hot-Water Heating and Fitting. By Thermus (This Series Continued.)
Description of Plumbing—Kitchen Boiler Arrangement—Residence of H. C. Fahnestock, Esq. (*Two Illustrations*.)
Kitchen Boiler in Diocesan House, New York.
Equitable Building, New York. Description of Plumbing. (*Four Illustrations*.)
Bath in the Residence of Mr. E. H. Wales. (*One Illustration*.)
Comparative value of Steam and Hot Water for Transmitting Heat and Power. By Chas. E. Emery.
Domestic Engineering—Army Mess Hall at Davids Island. (*Four Illustrations*.)
Novel Pipe-Joints or Couplings for Natural Gas.
Plumbing — Hot-Water Circulation from Kitchen to Top Floor of Building.
Foot-Vents, their Location and Termination. Giles Smith.
Trade Schools and Technical Education in their Relation to the Plumber of the Future.
House Drainage Regulations of Haverhill, Mass.
Specimens of Bad Plumbing Discovered by the New York Board of Health. (A Series, Illustrated.)
Conanicut Park Fever Outbreak.
Is a Trap on a Main Drain of a Building a Necessity? Fresh-Air Inlets, their Location and Termination. (Paper by Richard Murphy and James A. Gibson.)
Rules for Figuring Steam-Heating Surfaces. Plumbing Violations. (*Several Illustrations*.)
Revised Plumbing Regulations, New York Board of Health.
The Fitting Up of Hot-Water Boilers in English Plumbing Practice. (Three Articles, Illustrated.)
Equitable Building Plan, Showing Domestic Engineering Plants, Including Boilers, Engines, Hydraulic Pumps and Elevators, Dynamos, Pneumatic Service, Heating Mains, &c. (Seven Articles *with Illustrations*.)
Plumbing in the Residence of Mr. Francis Lyne Stetson. (*Three Illustrations*.)
Washington, D.C., Plumbing Regulations. (Controversy Over Them.)
Plumbing in Residence of Mr. W. F. Weld, Brookline, Mass. (*Six Illustrations*.)

Bound in Cloth, with Index, 15s.   Postage Paid.   THE ENGINEERING AND BUILDING RECORD, 92 and 93, Fleet Street, London.

HEATING APPARATUS.
J. & W. WOOD,

MEDALS AWARDED. Birmingham Street Foundry
AND
Hot-water Engineering Works,
STOURBRIDGE.

IMPROVED EXPANSION JOINT.

MANUFACTURERS OF THE

IMPROVED EXPANSION JOINT HOT-WATER PIPES, COILS, BOILERS, &c.

OUR HOT-WATER JOINTS HAVE GAINED UNIVERSAL APPROBATION.

Estimates for Heating Public Buildings, Greenhouses, &c., free.

WRITE FOR CATALOGUE.

HOT WATER

HEATING APPARATUS

BATH & RANGE BOILERS,
HOT-WATER BOILERS,
PIPES, JOINTS,
VALVES, BATHS,
PUMPS,
&c., &c.

Inventor, Patentee, Manufacturer,

J. ATTWOOD.
LATE CO-PARTNER OF JONES & ATTWOOD

Engineer, Ironfounder,
STOURBRIDGE.
THE ONLY ATTWOOD IN THE TRADE.

Illustrated Catalogue Post Free.

ATTWOOD'S IMPROVED EXPANSION JOINT.

COMPACT.
NEAT.
DURABLE.
CHEAP.

HUNDREDS OF THOUSANDS SOLD AND IN USE.

CHURCHES, SCHOOLS, MANSIONS, GREENHOUSES, CONSERVATORIES, &c., HEATED COMPLETE. ESTIMATES FREE.

www.ingramcontent.com/pod-product-compliance
Lightning Source LLC
Chambersburg PA
CBHW030405230426
43664CB00007BB/755